Process Analysis and Simulation in Chemical Engineering

Iván Darío Gil Chaves
Javier Ricardo Guevara López
José Luis García Zapata
Alexander Leguizamón Robayo
Gerardo Rodríguez Niño

Process Analysis and Simulation in Chemical Engineering

 Springer

Iván Darío Gil Chaves
Chemical and Environmental Engineering
Universidad Nacional de Colombia
Bogotá, Colombia

José Luis García Zapata
Processing Technologies
Alberta Innovates - Technology Futures
Edmonton, AB, Canada

Gerardo Rodríguez Niño
Chemical and Environmental Engineering
National University of Colombia
Bogotá, Colombia

Javier Ricardo Guevara López
Y&V - Bohórquez Ingeniería
Bogotá, Colombia

Alexander Leguizamón Robayo
Chemical Engineering
Norwegian University of Science
   and Technology
Trondheim, Norway

ISBN 978-3-319-79201-9   ISBN 978-3-319-14812-0   (eBook)
DOI 10.1007/978-3-319-14812-0

Springer Cham Heidelberg New York Dordrecht London
© Springer International Publishing Switzerland 2016
Softcover re-print of the Hardcover 1st edition 2016

Printed on acid-free paper

Springer International Publishing AG Switzerland is part of Springer Science+Business Media
(www.springer.com)

Preface

This book has been developed as an alternative to illustrate the use of process simulation as a tool for the analysis of chemical processes and for the conceptual design of unit operations. In the last years, chemical process simulation has become of significant importance due to the evolution of computing tools, which have opened a wider spectrum of possibilities in the use of applications for process integration, dynamic analysis, costs evaluation, and conceptual design of reaction and separation operations. All above added to the need of performing calculations in a fast way in order to focus in the analysis of the obtained information and on other relevant aspects such as safety, green engineering, economic profitability, and many other factors that make the solutions of engineering more competitive.

Process simulation is a discipline transversal to all the areas of chemical engineering. The development of many engineering projects demands simulation studies since the preliminary feasibility analysis, conceptual design, detailed design, until the process operation. For that reason, the generation of new process supported in the simulation requires the integration of concepts of chemical engineering and the breeding of innovation abilities. All that integration redounds in controllability studies and dynamic analysis, energy integration, and optimization, which aim to achieve the goals of environmental protection, process safety, and product quality.

The last academic reform of the chemical engineering program of the National University of Colombia has punted to a more integral formation, addressed to the reinforcement of the new areas that have become a part of the modern chemical engineering. Modeling and simulation of chemical processes is one subject that appears as a natural response to the always more imperative exigency of the working market in the process design. This book brackets the fundamental concepts of process simulation and brings together in a concise way some of the most important principles of the chemical engineering to apply them in the use of process simulators. To achieve it, the book has been sectioned in ten chapters. The first one attempts to introduce the fundamentals of process simulation in steady state. In the second chapter, thermodynamics and basic criteria for the selection of property packages are addressed. The third and fourth chapters cover to fluid dynamics and heat exchangers, respectively. There the fundamentals that support the calculation of pressure and temperature change operations are illustrated. In Chap. 4, the last part is dedicated to heat integration and the tools available to design and evaluate heat exchanger networks. The fifth chapter deals with chemical reactors making emphasis in the importance of this kind of equipment in the conceptual design of the process. In consequence, with the objective of giving a systematic order from the learning point of view, the sixth chapter regards the operations of vapor–liquid separation, presenting the theory behind the shortcut and rigorous calculations. The last four chapters involve transversal or special topics, beginning with process optimization in Chap. 7, dynamic analysis of processes in Chap. 8, solids operations in Chap. 9 and, finally, some cases of study related to integrated problems with specific applications are addressed in the tenth chapter.

According to the goals of the instructor, a typical 16-week semester is enough to cover the entire text. The topics presented in the chapters are organized in an inductive way, starting from the more simple simulation up to some additional and advanced complex problems. As the reader can note, the first four chapters are focused on the fundamentals with special emphasis on thermodynamics and simulation convergence. After that, the core of a chemical process, i.e., chemical reactors, is studied. Column operations and advance distillation technologies are presented after reactors in order to show the importance of solving the separation problems that naturally appear at the outputs of chemical reaction operations. Finally, specific topics are developed to illustrate the application of process simulation in a global way. At National University of Colombia, we teach the one-semester, three-credit course Modeling and Simulation of Chemical Processes. Students attend two 2-h sessions each week. During the sessions, the first part is dedicated to present and discuss some theoretical aspects related to specific problem simulations, then in the last part of the sessions students are encouraged to solve computer simulation exercises, working in pairs. So, examples from each chapter can be used to be solved in exercise session part. Additional problems are proposed at the end of the chapter to be discussed in additional sessions or as homework problems. In this way, some chapters can be covered in a week and others can be covered in 2 weeks.

This book is addressed to undergraduates in chemical and process engineering, as a support for the development of courses such as process simulation, process design, process engineering, plant design, and process control. Nonetheless, some sections can also be used in fundamental subjects like chemical reactions and mass transfer operations. The material presented here has been in part developed by the authors and, in part, compiled by Professor Iván Gil in the lecture of Modeling and Simulation of Chemical Processes, in the undergraduate program of chemical engineering at the National University of Colombia. Also, his experience as instructor of the lecture of Process Plant Design at Andes University, as well as the experience resulting of the development of some research projects and the interaction with real problems of the industry, has influence in the final contents of the book. The solved examples aim to ensure the understanding of all the presented topics and invite the reader to look deeper with more complex and elaborated examples. It is expected that the guidelines and examples will allow the effective usage of commercial process simulators and become a consulting guide for all engineers involved in the development of process simulations and computer-assisted process design.

The authors state their most sincere thanks to the Department of Chemical and Environmental Engineering of the National University of Colombia, campus Bogotá, for the lessons learned and the support of all the professors in the department, generating a pleasant environment of work and cooperation. It is also important to mention the support of teaching assistants from Computing Laboratory for Process Analysis and Design, as well as the students from Modeling and Simulation of Chemical Processes, and Process Control lectures. Particularly, we would like to thank some of our graduate students at National University of Colombia: Paola Bastidas, Nelson Borda, Francisco Malagón, Andrés Ramirez, Edward Sierra, Santiago Vargas, and Karen Piñeros, who helped us in proofreading and the development of some tests for simulation examples.

Bogotá, Colombia   Iván Darío Gil Chaves
Bogotá, Colombia   Javier Ricardo Guevara López
Edmonton, AB, Canada   José Luis García Zapata
Trondheim, Norway   Alexander Leguizamón Robayo
Bogotá, Colombia   Gerardo Rodríguez Niño

Contents

1   Process Simulation in Chemical Engineering   1
   1.1   Introduction   1
   1.2   Chemical Process Simulators   2
   1.3   Types of Process Simulators   3
         1.3.1   Sequential Modular Simulators   4
         1.3.2   Simultaneous or Equation Oriented Simulators   5
         1.3.3   Hybrid Simulators   6
         1.3.4   Aspen Plus® and Aspen Hysys®   7
   1.4   Applications of Process Simulation   8
         1.4.1   Computer-Aided Design   8
         1.4.2   Process Optimization   9
         1.4.3   Solution of Operating Problems   9
         1.4.4   Other Applications   9
   1.5   Convergence Analysis 10
         1.5.1   Convergence Methods 10
         1.5.2   Problems with Simple Recycles 16
         1.5.3   Partitioning and Topological Analysis 17
         1.5.4   Nested Recycles 18
   1.6   Introductory Example 20
         1.6.1   Problem Description 20
         1.6.2   Simulation Using Aspen HYSYS® 20
         1.6.3   Simulation Using Aspen Plus® 25
   1.7   Sensitivity Analysis 38
         1.7.1   Sensitivity Analysis in Aspen Plus® 38
         1.7.2   Sensitivity Analysis in Aspen HYSYS® 43
   1.8   Design Specifications 45
   1.9   Summary 47
   1.10 Problems 47
   References 50

2   Thermodynamic and Property Models   53
   2.1   Introduction   53
   2.2   Ideal Model   54
   2.3   Equations of State   54
         2.3.1   Redlich–Kwong   54
         2.3.2   Soave–Redlich–Kwong   55
         2.3.3   Peng–Robinson   56
   2.4   Activity Coefficient Models   57
         2.4.1   Van Laar Model   57
         2.4.2   Wilson Model   58
         2.4.3   NRTL (Nonrandom Two Liquids)   59
         2.4.4   UNIQUAC   59
         2.4.5   UNIFAC   61
   2.5   Special Models   62
         2.5.1   Polymeric Systems   62
         2.5.2   Electrolytic System   63
   2.6   Integration of the Activity Models with Equations
         of the State   68
   2.7   Selection of Thermodynamic Model   68
         2.7.1   Selection of the Property Model   70
         2.7.2   Selection of the Properties Model   73
         2.7.3   Validate the Physical Properties   73
         2.7.4   Describe Additional Components
                to the Database   74
         2.7.5   Obtain and Use Experimental Data   75
   2.8   Example of Property Model Selection   76
   2.9   Example of Phase Diagram   81
         2.9.1   Simulation in Aspen HYSYS®   82
         2.9.2   Simulation in Aspen Plus®   86
         2.9.3   Results Comparison   87
   2.10   Example of Parameter Adjustment   88
         2.10.1 Example Using an Activity Coefficient Model   88
         2.10.2 Example Using an Equation of State   91
         2.10.3 Comparison and Results Analysis   93
   2.11   Hypothetical Components   93
         2.11.1 Usage in Aspen HYSYS®   93
         2.11.2 Usage in Aspen Plus®   97
   2.12   Summary   98
   2.13   Problems   100
   References   101

3   Fluid Handling Equipment   103
   3.1   Introduction   103
   3.2   General Aspects   103
         3.2.1   Background   104

	3.2.2	Piping	106
	3.2.3	Pumps	110
	3.2.4	Compressors and Expanders	111
3.3	Modules Available in Aspen Plus®		112
3.4	Modules Available in Aspen HYSYS®		113
3.5	Gas Handling Introductory Example		114
	3.5.1	Problem Description	114
	3.5.2	Simulation in Aspen HYSYS®	115
	3.5.3	Results Analysis	127
3.6	Liquid Handling Introductory Example		130
	3.6.1	Problem Description	130
	3.6.2	Process Simulation	130
	3.6.3	Results Analysis	135
3.7	Summary		136
3.8	Problems		136
References			137

4 Heat Exchange Equipment and Heat Integration 139
4.1	Introduction		139
4.2	Types of Programs Available		139
4.3	General Aspects		140
	4.3.1	Shortcut Calculation	141
	4.3.2	Rigorous Calculation	142
	4.3.3	Calculation Models	149
4.4	Modules Available in Aspen Plus®		151
4.5	Modules Available in Aspen HYSYS®		151
	4.5.1	Thermodynamic Heat Exchangers	151
4.6	Introductory Example		155
	4.6.1	Problem Description	155
	4.6.2	Simulation in Aspen Plus®	156
	4.6.3	Simulation in Aspen HYSYS®	156
	4.6.4	Simulation in Aspen Exchanger Design and Rating®	164
	4.6.5	Results Analysis	171
4.7	Process Heat Integration		175
	4.7.1	Introduction	175
	4.7.2	Theoretical principles	176
	4.7.3	Aspen Energy Analyzer	181
4.8	Summary		188
4.9	Problems		189
References			192

5 Chemical Reactors 195
5.1	Introduction		195
5.2	General Aspects		195
	5.2.1	Chemical Reaction	196

	5.2.2	Stoichiometry	196
	5.2.3	Conversion	196
	5.2.4	Selectivity	197
	5.2.5	Reaction Kinetics	197
	5.2.6	Kinetic of Heterogeneous Reactions	198
5.3	Equations for Reactor Design		199
	5.3.1	Continuous Stirred Tank Reactor	199
	5.3.2	Plug Flow Reactor (PFR or PBR)	199
	5.3.3	Batch Reactor (Batch)	200
5.4	Modules Available in Aspen Plus®		200
5.5	Available Modules in ASPEN HYSYS®		201
5.6	Introductory Example of Reactors		201
	5.6.1	Problem Description	201
	5.6.2	Simulation in Aspen Hysys®	211
	5.6.3	Results Analysis	217
5.7	Propylene Glycol Reactor Example		220
	5.7.1	General Aspects	220
	5.7.2	Process Simulation in Aspen Plus®	221
	5.7.3	Results Analysis	224
5.8	Methanol Reforming Reactor		225
	5.8.1	Problem Description	225
	5.8.2	Simulation in Aspen Plus®	226
	5.8.3	Simulation in Aspen Hysys®	232
	5.8.4	Analysis and Results Comparison	236
5.9	Summary		236
5.10	Problems		236
References			240
6	**Gas–Liquid Separation Operations**		**241**
6.1	Introduction		241
6.2	Available Modules in Aspen Plus®		242
	6.2.1	Shortcut Methods	242
	6.2.2	Rigorous Methods	244
6.3	Modules Available in Aspen Hysys®		251
	6.3.1	Predefined Columns	252
	6.3.2	Shortcut Calculation Model	252
	6.3.3	Column Interface	253
6.4	Distillation Introductory Example		253
	6.4.1	Problem Description	253
	6.4.2	Simulation in Aspen Plus®	255
	6.4.3	Simulation in Aspen Hysys®	260
	6.4.4	Results Analysis and Comparison	268
6.5	Absorption Introductory Example		272
	6.5.1	Problem Description	272
	6.5.2	Process Simulation	272

6.6   Enhanced Distillation   275
        6.6.1   Residue Curves Map   277
        6.6.2   Extractive Distillation   291
6.7   Nonequilibrium Models   301
        6.7.1   Nonequilibrium Model Example   310
6.8   Columns Thermal and Hydraulic Analysis   325
        6.8.1   Thermal Analysis   326
        6.8.2   Hydraulic Analysis   328
        6.8.3   Application Exercise   328
6.9   Summary   337
6.10   Problems   338
References   340

7   Process Optimization in Chemical Engineering   343
7.1   Introduction   343
7.2   Formulation of Optimization Problem   344
        7.2.1   Degrees of Freedom   344
        7.2.2   Objective Function   345
        7.2.3   Classification of Optimization Problems   346
7.3   Optimization in Sequential Simulators   348
        7.3.1   General Aspects   349
7.4   Introductory Example   350
        7.4.1   Aspen Plus® Simulation   351
        7.4.2   Sensitivity Analysis   363
        7.4.3   Results   365
7.5   Summary   366
7.6   Problems   366
References   368

8   Dynamic Process Analysis   371
8.1   Introduction   371
8.2   General Aspects   371
        8.2.1   Process Control   373
        8.2.2   Controllers   375
8.3   Introductory Example   379
        8.3.1   Dynamic State Simulation   380
8.4   Gasoline Blending   385
        8.4.1   Steady State Simulation   385
        8.4.2   Dynamic State Simulation   395
        8.4.3   Disturbances   395
        8.4.4   Recommendations   398
8.5   Pressure Relief Valves   400
        8.5.1   General Aspects   400
        8.5.2   Application Example   400
        8.5.3   Dynamic State Simulation   402

8.6   Control of the Propylene Glycol Reactor   406
8.7   Control of Distillation Columns   414
        8.7.1   General Aspects   414
        8.7.2   Distillation Column Example   415
8.8   Summary   420
8.9   Problems   423
References   424

9   Solids Operations in Process Simulators   425
     9.1   Introduction   425
     9.2   General Aspects   425
             9.2.1   Separation or Classification   426
             9.2.2   Comminution   429
             9.2.3   Filtration   431
             9.2.4   Crystallization   432
             9.2.5   Particle Size Distribution Meshes   432
     9.3   Modules in Aspen Plus®   432
     9.4   Modules in Aspen HYSYS®   432
     9.5   Crusher Introductory Example   434
             9.5.1   General Aspects   434
             9.5.2   Simulation in Aspen Plus®   434
             9.5.3   Results Analysis   440
     9.6   Solids Handling Example   440
             9.6.1   General Aspects   440
             9.6.2   Simulation in Aspen Plus®   440
             9.6.3   Results Analysis   444
     9.7   Summary   445
     References   446

10   Case Studies   447
     10.1   Introduction   447
     10.2   Simulation of Nylon 6,6 Resin Reactor   447
             10.2.1   Problem Description   447
             10.2.2   Polymerization Reaction Kinetics   449
             10.2.3   Continuous Production   455
             10.2.4   Batch Production   457
             10.2.5   Results Comparison   460
     10.3   Azeotropic Distillation of Water–Ethanol Mixture
             Using Cyclohexane as Entrainer   462
             10.3.1   General Aspects   462
             10.3.2   Process Simulation   463
             10.3.3   Convergence Recommendations   475
     10.4   Ethylene Oxide Production   477
             10.4.1   Process Description   477
             10.4.2   Aspen HYSYS Simulation   479

10.5   Economic Evaluation Using Aspen Icarus®.   487
        10.5.1   General Aspects.   487
        10.5.2   Simplifications.   490
        10.5.3   Aspen Icarus® Simulation.   491
        10.5.4   Results Analysis.   509
     References.   512

Index.   515

About the Authors

Iván Darío Gil Chaves is a Professor of Chemical Engineering at the Department of Chemical and Environmental Engineering at National University of Colombia—Sede Bogotá. He received B.S. and M.Sc. degrees from National University of Colombia. He obtained his Ph.D. in Chemical Engineering at the University of Lorraine (France) and National University of Colombia (under joint supervision). Gil has participated in some industrial projects in the area of process design and control; mainly, he has collaborated with representatives of Aspen Technology in Colombia in advanced process control applications. He was also an instructor at Andes University in Colombia. Currently, he teaches university courses in modeling and simulation, process control, reaction engineering, and process design. In addition, he presents some short courses in advanced process control and process synthesis and optimization. Dr. Gil is coauthor of several publications in peer review journals on process design and control. His research interests include biofuels, with emphasis on fuel ethanol and the use of extractive distillation to dehydrate ethanol-water mixtures; modeling, simulation, and control of reaction and separation operations; nonlinear geometric control; and vapor liquid equilibrium.

Javier Ricardo Guevara López holds a B.Sc. in Chemical Engineering from National University of Colombia. He is a process engineer with experience developing conceptual, basic, and detailed engineering for the oil and gas industries. Javier Ricardo Guevara López works as process engineer in Y & V—Bohorquez Ingeniería SAS in Colombia. His interests include process design, process simulation, gas processing, relief systems, and upgrading of gas plants.

Gerardo Rodríguez Niño is a Professor of Chemical Engineering at the Department of Chemical and Environmental Engineering at National University of Colombia—Sede Bogotá since 1988. He teaches Mass Transfer Operations, Material Balances, and Chemical Engineering Laboratory courses in the area of Unit Operations. He obtained B.S., M.Sc., and Ph.D. degrees in Chemical Engineering from National University of Colombia. He was the chief of the chemical engineering laboratories for five years and in charge of coordinating the laboratory essays for industrial applications. He has participated in different projects for enterprises such as Preflex, Vaselinas de Colombia S.A., Coljap, Cyquim de Colombia, Carboquímica S.A., Epsa and Alcalis de Colombia. In the last few years, he has developed research projects in ethanol dehydration by extractive distillation with ethylene-glycol and glycerine for the production of fuel alcohol, and in the study of the usages of fusel oil generated in ethanol production. His research interests include distillation design, esterification, catalysis, and essential oils production.

Alexander Leguizamón Robayo is a graduate student at Norwegian University of Science and Technology (NTNU). He holds a B.Sc. in Chemical Engineering from National University of Colombia and is currently earning his Master's degree in Chemical Engineering at NTNU. Alexander has worked in distillation process design and simulation. He carried out a design project of a formaldehyde plant with Dynea in Lillestrøm, Norway. He is currently studying with the process systems engineering group at NTNU and working on a plantwide process control project for sweetening of syngas for the Swedish chemical company Perstorp.

José Luis García Zapata is a researcher with the Heavy Oil and Oilsands group at Alberta Innovates Technology Futures (AITF) in Edmonton, Canada. He holds B.Sc. and M.Sc. degrees in Chemical Engineering from the National University of Colombia (2010) and the University of Alberta (2013), respectively. García has worked as a Process Engineer for BRINSA S.A., a salt and chlor-alkali company in Colombia, and has been responsible for the design and commissioning of pilot scale reactors for heavy oil upgrading at the University of Alberta. His interests include process design, pilot plant, produced water treatment, and upgrading of heavy oil. He is registered as an Engineer-in-Training in Alberta.

Chapter 1
Process Simulation in Chemical Engineering

1.1 Introduction

Chemical process simulation aims to represent a process of chemical or physical transformation through a mathematic model that involves the calculation of mass and energy balances coupled with phase equilibrium and with transport and chemical kinetics equations. All this is made looking for the establishment (prediction) of the behavior of a process of known structure, in which some preliminary data of the equipment that constitute the process are known.

The mathematical model employed in process simulation contains linear, nonlinear, and differential algebraic equations, which represent equipment or process operations, physical–chemical properties, connections between the equipment and operations and their specifications. These connections are summarized in the process flow diagram.

Process flow diagrams are the language of chemical processes. Between them the state of art of an existing or hypothetical process are revealed. Thereby, the process simulators are employed for the interpretation and analysis of information contained in the process flow diagrams in order to foresee failures and evaluate the process performance. The analysis of the process is based on a mathematic model integrated by a group of equations that associate process variables such as temperature, pressure, flows, and compositions, with surface areas, geometrical configuration, set points of valves, etc.

In most of the simulators the solution of the equations system is made linearly, solving each unit separately and moving forward in the system once the variables required for the calculation of the next unit are known. However, that process is useless when there are stream recycles in the system since some of the variables to calculate are required for the process initialization.

An alternative solution for that type of problems consists in taking one stream as *tear stream*. That means assuming the initial values of that stream to start the calculations; later on, based on the assumed information, each of the following units

© Springer International Publishing Switzerland 2016

I.D.G. Chaves et al., *Process Analysis and Simulation in Chemical Engineering*,
DOI 10.1007/978-3-319-14812-0_1

is solved obtaining new values for the parameters of the tear stream. Subsequently, the new values help to repeat these calculations again and again, until the difference between the initial and the calculated values fulfill a given tolerance; that point is known as *convergence*.

1.2   Chemical Process Simulators

A process simulator is software used for the modeling of the behavior of a chemical process in steady state, by means of the determination of pressures, temperatures, and flows. Nowadays, the computer programs employed in process simulation have broad in the study of the dynamic behavior of processes, as well as to the control systems and their response to perturbations inherent to the operation. In the same way, software to perform equipment sizing, cost estimation, properties estimation and analysis, operability analysis and process optimization, are now available in the market; all those characteristics can be observed in the Aspen Engineering Suite, and some of them are presented in last chapters.

Process simulators allow:

- Predict the behavior of a process
- Analyze in a simultaneous way different cases, changing the values of main operating variables
- Optimize the operating conditions of new or existing plants
- Track a chemical plant during its whole useful life, in order to foresee extensions o process improvements

The appearance and development of digital informatics determined the advance of different areas of the human knowledge. Chemical engineering was a part of that development, particularly in the application to the process simulation. The first attempts of mathematical modeling refer back to the 1950s with the debut of FORTRAN language (FORmula TRANslating). Afterwards, in the 1970s, appears the first process simulator, known as FLOWTRAN, which would frame the begin of an ceaseless research work principally performed by the academy, in some cases financed by the industry, and routed to make the process operation more profitable and to access to the evaluation of different alternatives in a relative short time.

Process simulators are built in libraries of subroutines or models, generally developed in FORTRAN, C++, or Visual Basic, that conform algorithms for the solution of equations. The subroutines or models are known as "procedures," "models," or "blocks." To make an effective use of the simulators, process engineers must know the guidelines and assumptions of the models provided by each simulator. Those assumptions are described in the user manuals. Furthermore, it is always important to take into account the criteria used in the specification of phase equilibrium and the models used for that purpose, as the accuracy of the obtained results of a simulation is affected by them.

The commercial and academic process simulators more divulged actually are, among others: SPEED UP®, ASPEN PLUS®, DESIGN II®, HYSYM®, ASPEN HYSYS®, CHEMCAD®, and PRO II®.

1.3   Types of Process Simulators

Process simulators are classified according to the simulation strategy that they use to set the mathematical model that represents the process to simulate. The simulation strategy refers to the way in which the problem of the model solution is boarded. Generally the strategy depends on the complexity of the model and the calculation mode. The first one, understood as the different existing possibilities, since linear to sophisticated models with equations of mass, energy and *momentum* transfer rates. The second, referred to the information (input variables) that is necessary to specify to solve the model in terms of the remaining information (output variables).

The subroutines of a process simulator are computer programs supplied initially with vectors containing the information corresponding to the feed streams of the process and some of its parameters. The subroutine takes the vectors, interprets the information, and looks for the appropriate model to solve the problem. The results are, basically, the product streams of the process. Thus, the subroutines permit working with two calculation modes in a process simulator:

- Design mode: according to the required process conditions, a desired performance is used as starting point to find the process or equipment specifications that allow the accomplishment of those conditions.
- Rating mode: according to some design specifications provided to the simulator, the performance of the process or equipment is evaluated to meet some specific conditions of the process.

The two fundamental strategies used in the solution of simulation problems are the *sequential strategy* and the *simultaneous strategy*.

The concept of sequential simulation comes from the necessity of calculating different process units, which are part of a flow diagram, in a rating mode. In this mode, some of the values of the feed streams and the parameter specifications of each one of the units must be known. As it is impossible to specify all the streams entering to all the units in a simultaneous way, it becomes necessary to use the outcoming values obtained from the calculation in one unit as input information in the next unit. In that way a sequential order of calculation is established to solve one-by-one all the units of the process. The calculation order is fixed automatically by the simulator, ensuring that it would be consistent with the information flow. In almost all the cases the calculations begin in the equipment for which the feed streams values and equipment parameters are known. In Fig. 1.1 the typical calculation order in a flow diagram is shown with arrows. The calculation order normally matches with the direction of mass flow and is modified or interrupted

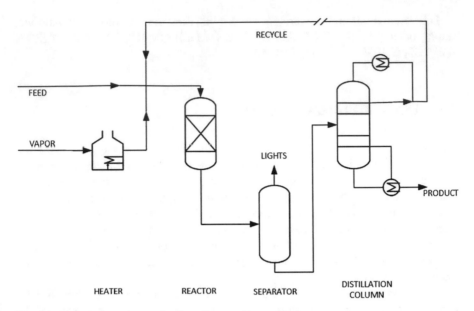

Fig. 1.1 Calculation scheme of a flow diagram. *Source*: Authors

with the appearance of recycles, that make mandatory the implementation of an iterative process.

A simulation becomes more complex as the number of recycles increases, thus it becomes more difficult to implement in a successful way. For that reason, a second option for the solution of the model consists on taking all the equations at a time and builds a unique simulation model by the simultaneous solution of all the equations. In this way it is not indispensable to evaluate all the equipment units from the input values and their parameters, allowing the manipulation of design or evaluation specifications without any distinction.

1.3.1 Sequential Modular Simulators

These are simulators in which each process unit is represented with a module in which the operation model and the numeric algorithm used for the calculation of the outgoing streams are included. The modules are totally independent from each other and the information flow for the calculation in the simulation coincides with the "physical flow" in the plant. The advantage of using modules is that each equations system is solved using its own methodology.

The module of each process unit must contain routines with the models and procedures for the solution from a set of predetermined variables. Additionally, when the process to simulate has several recycles caused by recirculation streams and counter-current operations, the executer module follows a methodology composed by three stages:

- Partitioning: recycle detection
- Tearing: selection of tear streams, over which the iterative process is made
- Ordering: establishment of a logical sequence for the operation units

The recycles owe their existence to all those processes in which reversible reactions and competitive phenomena take place. In the simulators there are specific subroutines for the calculation of recycles and the pursuit of convergence. Between them an initialization value for the tear stream is established, the calculation is executed and, finally, the values are compared and new initialization values are given until the convergence is achieved. The better-known methods of convergence in the simulators are Wegstein, successive substitutions (or direct iteration), Newton–Raphson and Broyden Quasi-Newton. The Wegstein method is employed in those situations in which the "successive substitutions" fail or a high number of iterations is required, as will be explained in Sect. 1.5.

The recycles constitute an extra unit in the simulation flow diagrams. It should be noticed that it must exist a convergence algorithm that allow the adjustment of the recycles and, therefore, their computing must be performed separately from the normal calculation units associated with unit operations. Generally, the recycles are not installed, that means, they must not be specified as calculation blocks in a simulation. Nonetheless, as will be seen further in Sect. 1.6.2, in the specific case of Aspen Hysys® some blocks that denote the recycle streams over which the iteration is executed, that is, the tear streams selected for the problem solution must be defined.

Summarizing, the principal features of a sequential modular simulator are following:

- Contains calculation libraries and routines
- The iteration variables are in the recycle streams
- The individual models are solved in an efficient way
- It is easily understandable for engineers that are not "simulation specialists" due to the existing correspondence between the mass flow and the calculation sequence
- It involves complex convergence methods (direct substitution, Wegstein, etc.)
- The information entered by the user (related with streams and equipment) is easily verifiable
- The design problems are easier to solve (parameter selection)

1.3.2   Simultaneous or Equation Oriented Simulators

In this type of simulators the mathematic model that represents the process is set building a large algebraic equations system. Here modules representing process units or subsystems may also exist, but these do not contain numerical methods for the calculation of output values; contrary, they possess the required information to provide the equations that represent the mathematic model.

The process model is made of the sum of the models of all the units that comprise the process or plant in the simulation. Due to the compilation and clustering philosophy of all the equations that build the process, this type of simulators are known as "Equation oriented" or "Equation based." The concept of modules is maintained here in order to facilitate the interaction with the user and to allow the specification of the required information for the problem.

The main problem associated to the concept of simultaneous or equation oriented solution is the convergence of the system and the consistency of the solutions found. Thus the highly nonideal systems that correspond to the chemical plants models could, for instance, produce multiple solutions. Additionally, the numeric solution of problems consisting of large equations systems requires proper initialization; this is close to a solution surrounding (status).

Summarizing, the principal features of the equation oriented simulators are:

- Each unit is represented by the equations that model it. The model is the integration of all the subsystems
- The distinction between process variable and operative parameters disappears. In consequence the design problems are simplified
- Simultaneous solution of the system of (nonlinear) algebraic equations
- Higher convergence velocity
- They need a better initialization (the higher the complexity of the problem to solve the better the initialization to be provided)
- The higher the complexity, the less the reliability in the results and the more the convergence problems (solutions without physical meaning)
- Easier to use for "nonspecialists"

1.3.3   Hybrid Simulators

This type of simulators uses a strategy mixture of the sequential and simultaneous ones. Each iteration consists of two steps: a first model solution employing a sequential strategy, and the upgrading of the lineal coefficients to find a solution with a simultaneous strategy.

The simulation begins with a sequential step in which the output variables are determined from input information and parameters in an initial scanning of the flow diagram. Then a simultaneous step starts for the solution of linearized models in a second scanning of the process.

Some simulators of high industrial application appeal to a hybrid strategy to ensure convergence, even in the worst cases.

1.3.4   *Aspen Plus® and Aspen Hysys®*

Aspen Plus® and Aspen Hysys® are process simulators in steady state used in the prediction of the behavior of a process or a set of unit operations, through the existing relationships between them. The relations and connections standing in the process determine, over the mass and energy balances, the phase and chemical equilibrium and the chemical transformation rates. In this way, it is possible to simulate the behavior of existing or projected plants, with the objective of improving the design specifications or increase the profitability and efficiency of an operation in process.

Between the main functions that can be found in this simulators are:

- Generation of plots and tables
- Performing of sensitivity analysis and cases of study
- Sizing and rating of equipment
- Experimental data adjustment
- Analysis of pure components and mixtures properties
- Study of residue curve maps
- Process optimization
- Estimation and regression of physicochemical properties
- Dynamic analysis of processes

Aspen Plus® is located between the group of simulators using the sequential *strategy*, in the same way as other simulators such as PRO II® and CHEMCAD®. Thus, it is composed by a group of simulation or program units (subroutines or models) represented through blocks and icons, to which the pertinent information must be provided to solve the mass and energy balances. However, it is important to mention that in the last versions of this simulator the possibility to work with the simultaneous or equation oriented strategy has been included, allowing the modeling of systems and processes much more complex, highly integrated and with a high number of recycles.

Aspen Hysys® is a process simulator widely used in an industrial level, especially in conceptual design and detailed engineering, control, optimization and process monitoring stages in a project. The most important applications of Aspen Hysys® correspond to the industries of oil and gas processing, refineries, and some industries of air separation. All these practices take advantage of this simulator architecture that permits the integration of the steady-state and dynamic models in an only unit. In this way, it is possible to bring together the stages of process design with the rigorous analysis of the dynamic behavior and the control of the same, to evaluate in a direct way the effects that the decisions in the detailed design step have over the dynamic and controllability of the process.

1.4   Applications of Process Simulation

Process simulation is a tool for process and chemical engineers that can be used in the execution of repetitive tasks or in activities of high complexity that must be solved in relatively short times.

The various applications that process simulation has found are result of the necessity of:

- Making a better use of the energy resources
- Minimizing the operating costs and the emission of waste streams that may be contaminant
- Increasing the yield and process efficiency
- Improving the process controllability
- Propelling the teaching of process design

Some of the principal applications of process simulation are discussed as follows.

1.4.1   Computer-Aided Design

The steady-state simulation of mass and energy balances constitutes itself the center piece of the computer-aided design. The principal reasons are: (a) the results of the calculations in the design stage are necessary for further stages; (b) during the design, in order to meet the economical and operation restrictions, the information changes dynamically, in such a way that it is necessary to adjust and actualize the result of the balances in a continuous way; (c) such a large quantity of information is generated that the only way to administrate it is if it is consolidated through the process simulations and the cases of study developed with the simulations (Yee Foo et al. 2005).

Process simulation allows making a study of different process alternatives (flow diagrams) in order to determine in a reasonable time which of them are not feasible. Besides, a flexible simulator develops different cases of study in the search of an optimal configuration of the process, making possible that the design moves to more advanced stages rapidly. In the same way, a more flexible simulator can be used to conclude about to rival technologies, to design or evaluate the more adequate operating configuration for a process or to plan in the most economical way the laboratory and plant experiments required for the design.

1.4.2   Process Optimization

The optimization of chemical processes has its origin in the linear programming at the beginning of the 1960s. This task has as fundamental goal the comparison of different alternatives to select the best according to some process response criteria. In an optimization process it is important to identify the independent characteristics that lead to different results (independent variables) and the variables that make possible the measurement of the relative excellence of a solution (dependent variable). The set of interactions between the dependent variables that conduce to a response is known as objective function, in which the parameters of the system relate with each other.

In a general way, the optimization of a process brings on the minimization of the operating costs, the energy consumption and the contaminant emissions, or to the maximization of the yields and operation productivity.

1.4.3   Solution of Operating Problems

A process plant never operates under the design operating conditions, either because the composition of the raw materials is different from the one considered initially, because the environmental conditions are not contemplated in the design or because the configuration of the plant or some equipment is modified as consequence of the materials availability or costs. It is often used an overdesign in engineering stages in order to have spare capacity for future plant modifications.

Process simulation permits to predict the effect of changes in the operating conditions over other process variables and, in that way, to establish control points more favorable, through dynamic simulation. Likewise, it simplifies the supervision of the changing conditions in large periods of time; for example, the deterioration of the random packing in a distillation column, the fouling in heat exchangers or the catalyst deactivation.

1.4.4   Other Applications

The commercialization of turnkey processes demands the demonstration—by the concerns—of the capabilities of the technologies that they are buying. Such demonstrations are made using process simulations, since it is impossible to find prototypes or pilot plants similar to the design to be commercialized.

The teaching of process design is an action that may be more enriching when a simulator is used, as it allows to evaluate different alternatives and to solve various cases of study without recurring to a large quantity of calculations that make the task tedious and impractical. Additionally, the integration of creativity and the application of concepts of engineering in the solution of each one of the cases of study is achieved.

Among other applications of simulation are the revamping studies, the validation of models through the adjustment of process data, the planning of plant operations, and the studies of flexibility of a process design.

1.5   Convergence Analysis

As previously mentioned, the solution strategy for a process simulation problem is defined by the way in which the calculation over the flow diagram of the process is boarded. In this task, one of the most important actions is the establishment of the simulation convergence, what implies that all the process units, the tear streams, and the global mass balance have reached convergence. This requires the selection of the best calculation sequence and the identification of the streams over which the iterative process should be performed to achieve convergence. Some aspects related with the available methodologies to reach convergence in process simulations and the way to determine the tear streams in a flow diagram are discussed briefly below (Babu 2004; Schad 1994; Towler and Sinnott 2008).

1.5.1   Convergence Methods (Babu 2004 ; Dimian 2003; Seider et al. 2004)

Flow sheets that are solved using process simulators are composed by tear streams, design specifications, and, in some cases, optimization calculations that must be solved with iterative methods. All this implies solving systems of nonlinear algebraic equations that in the modular simulators as well as in those equation oriented comprehends the solution of the equations that describe the unit operations, the physical properties, and the equations proper from the diagram topology (Babu 2004).

The convergence calculation of a flow sheet can be expressed mathematically as the minimization of the function $f(x)$ (Eqn. (1.1)) that represents the difference between the values estimated at the beginning of each iteration and the calculated values after making a complete track of the calculation sequence in the flow sheet. Said otherwise, the goal of the convergence is find such a vector that once the calculations are initiated with those values, the exact same vector is obtained as result. Formally, it is about finding the solution of $x = g(x)$, where x represents the vector of the initialized variables and $g(x)$ the calculation function over the diagram, that when is applied on x can lead to different values (case in which there is no convergence) or same values (meaning that the convergence was reached).

This can be expressed as follows:

$$f(x) = x - g(x) \tag{1.1}$$

where f, g, and x are vectors of dimension $n \times 1$.

The convergence methods are intended to achieve $f(x) = 0$. In that direction plenty of alternatives have been proposed and are available in process simulators. Each one of them will be described shortly.

The direct substitution method is the simplest one of the iterative methods and consists basically in substituting the estimated with the calculated values according to the equation

$$x_{k+1} = g(x_k) \tag{1.2}$$

Theoretically it is only possible to achieve convergence when the values of the Jacobian of the residual functions are less than 1. This condition is difficult to reach, reason why this method is slow, requires a large number of iterations and only conduces to convergence when the initialization is made with values near to the final solution (Dimian 2003; Seider et al. 2004).

The *Wegstein* method is applicable to the solution of equation systems that are by default in most of the simulators. There a linear extrapolation of the direct substitution is made through the equation

$$x_{(k+1)} = qx_k(1 - q)g(x_k) \tag{1.3}$$

where q is an acceleration parameter with value varying between -5 and 0, with the intention of giving stability to the iteration. When q is 0, the Eqn. (1.3) becomes the Eqn. (1.2) (direct substitution method). This method is applicable to the solution of multivariable problems with the assumption that there is no linkage between the variables, what is not completely true.

The *Secant* method uses a linear approximation of the Jacobian. In this case, compared with the Wegstein method, the number of iterations is reduced in a lot of problems. This method can be used for one variable, discontinuous or at-convergence problems.

Another of the methods employed to reach the convergence is the *Broyden* method that solves directly the Eqn. (1.1). In this case, the Jacobian is updated using algebraic calculations performed directly over the elements of a matrix and not by the inversion of the same. This method is useful for the convergence of multiple design specifications, tear streams, or a combination of both, especially when there is a high interdependence between tear streams and design specifications.

The method of *Newton* is a modification of the *Newton–Raphson* method for the simultaneous solution of nonlinear algebraic equations. Its implementation allows putting limitations to the variables and includes a search method that improves the stability of the iteration. This is a much more robust method than *Broyden*, and is used when the later does not produce good results. However, it is restricted to a limited number of variables.

1.5.1.1 Newton-Type Methods

Considering the general problem in the standard form $f(x) = 0$, where x is a vector of n real variables and $f()$ is a vector of n real functions. If a guess for the variables exist x', a Taylor series expansion can be developed about x' to extrapolate the solution point x^*. Each element of the vector $f()$ can be written as:

$$f_i(x^*) \equiv 0 - f_i\left(x'\right) + \frac{\partial f_i(x')}{\partial x^T}\left(x^* - x'\right) + \frac{1}{2}\left(x^* - x'\right)^T \frac{\partial^2 f_i(x_i)}{\partial x^2}\left(x^* - x'\right) + \quad i$$
$$= 1, \ldots n$$

or

$$f_i(x^*) \equiv 0 = f_i\left(x'\right) + \nabla f_i\left(x'\right)^T\left(x^* - x'\right) + \frac{1}{2}\left(x^* - x'\right)^T \nabla^2 f_i\left(x'\right)\left(x^* - x'\right)$$
$$+ \ldots i$$
$$= 1, \ldots n$$

Here $\nabla f_i(x)$ and $\nabla^2 f_i(x)$ are the gradient vector and Hessian matrix of the function $f_i(x)$, respectively. If the series is truncated at the second term, we have:

$$f\left(x' + p\right) \equiv 0 = f\left(x'\right) + J\left(x'\right)p$$

where $p = (x^* - x')$ is a search direction vector and the matrix J with elements:

$$\{J\}_{ij} = \frac{\partial f_i}{\partial x_j}$$

i and j are row and columns elements, respectively. This matrix is called *Jacobian*; if this matrix is nonsingular (has an inverse), p can be solved for directly and this is a *linear approximation* to the solution of nonlinear equations.

$$p = \left(J\left(x'\right)\right)^{-1} f\left(x'\right)$$

This relation allows for the use of a recursive algorithm to find the solution vector x^*, starting with an initial guess x^0 and using k as iteration counter, the solution is obtained by:

$$p^k = \left(J(x^k)\right)^{-1} f\left(x^k\right) \quad x^{k+1} = x^k + p^k$$

This method counts with some desirable convergence properties. Specifically, a fast rate of convergence close to the solution, it converges at a quadratic rate, for further details check the text by Biegler et al. (1999).

However, this fast rate of convergence only occurs if the method performs reliably. And this method can fail in high complexity simulation problems. Conditions required for the convergence of this method are (Biegler et al. 1999):

- The functions $f(x)$ and $J(x)$ exist and are bounded for all values of x
- The initial guess x^0 must be close to the solution.
- The matrix $J(x)$ must be nonsingular for all values of x

Some improvements for Newton's method aiming to overcome these shortcomings have been developed. These are discussed next.

Bounded Functions and Derivatives

By inspection, equations can be rewritten to avoid division by zero and undefined functions. Also, new variables can be included through additional equations.

Closeness to Solution

It is seldom easy to ensure a starting point "close" to the solution. As a result, starting from a poor guess one needs to control the step size of the Newton method to ensure that progress is made towards a solution. Therefore, the Newton step is modified to be only a fraction of the step predicted by the Newton iteration:

$$x^{k+1} = x^k + \alpha p^k$$

where α is a fraction between zero and one. A strategy is set to choose a step size automatically. This approach helps in providing a reliable convergence for the Newton's method. This method is known as the *Aramijo Line Search*. However this algorithm is not flawless, it can turn the matrix $J(x^k)$ into singular if a step size is not found after a few iterations, making the whole method fail. However, there are additional methods as the *Levenberg–Marquardt method* and the *Powell dogleg method* that are effective for the solution problems that are ill conditioned (their Jacobian matrix is singular or nearly singular).

Another solution alternative are the *Continuation Methods*, unlike the above methods, these do not attempt to solve equations by driving $f(x)$ to zero. Conversely, in these methods all the functions are evaluated at an initial guess $f(x_0)$ and then solve for a simpler problem, say: $f(x) - 0.9 f(x_0) = 0$. This problem should not be difficult to solve for the equation solver in use. If there is success in the solution of the problem with 0.9, then this *continuation parameter* is reduced to 0.8 and repeat, finally reducing it to zero, thus solving the original equation. These methods require an approach to select adequate *continuation parameters* and are computationally costly.

Methods That Do not Require Derivatives

The methods shown above based on the Newton method require the calculation of a Jacobian matrix at each iteration. This task can be very time consuming. A simpler alternative to an exact calculation of the derivatives is to use a finite difference approximation, given by:

$$\left(\frac{\partial f}{\partial x_j}\right)_{x^k} = \frac{f\left(x^k + he_j\right) - f\left(x^k\right)}{h}$$

where each element i of the vector e_j is given by: $(e_j)_i = 0$ if $i \neq 0$ or $= 1$ if $i = j$, and h is a scalar normally chosen from 10^{-6} to 10^{-3}. This approach requires additional n iterations.

Other alternative are the *Quasi Newton methods* where the Jacobian is approximated based on differences in x and $f(x)$, obtained from previous iterations. Here the motivation is to avoid evaluation and decomposition of the Jacobian matrix. The flagship of this type of methods is the *Broyden* method, it has been widely used in process simulation, especially in small equation systems. This approach is popular for flash calculations and recycle convergence in flow sheets. Though not as fast as Newton method, this method has a fast convergence rate.

1.5.1.2 First Order Methods

This family of methods is characterized by not evaluating or approximating the Jacobian matrix and their simplicity in structure. Their convergence is slow (at a linear rate). These methods are developed by Biegler et al. in a fixed point form $x = g(x)$, where x and $g(x)$ are vectors of n stream variables. These methods are most commonly used to converge recycle streams, there x represents a guessed tear stream and $g(x)$ is the calculated value after executing the units around the flow sheet.

Direct Substitution Methods

The simplest fixed point method is *direct substitution*. Here $x^{k+1} = g\left(x^k\right)$ with and initial guess x^0. The convergence of this properties method for the n dimensional case can be derived from the contraction mapping theorem. For the fixed point function, the Taylor series expansion is

$$g\left(x^k\right) = g\left(x^{k+1}\right) + \left(\frac{\partial g}{\partial x}\right)_{x^{k+1}}^T \left(x^k - x^{k+1}\right) + \dots$$

And if $\frac{\partial g}{\partial x}$ does not vanish, it is the dominant term near the solution, x^*, then:

$$x^{k+1} - x^k = g(x^k) - g(x^{k+1}) = \left(\frac{\partial g}{\partial x}\right)^T_{x^{k+1}} (x^k - x^{k+1})$$

And for

$$x^{k+1} - x^k = \Delta x^{k+1} = \Gamma \Delta x^k \text{ with } \Gamma = \left(\frac{\partial g}{\partial x}\right)^T$$

The normed expressions can be written as

$$||\Delta x^{k+1}|| \leq ||\Gamma|| \, ||\Delta x^k||$$

From this expression one can tell that the convergence is linear but the speed is related to $||\Gamma||$. By using the Euclidean norm, then $||\Gamma|| - |\lambda|^{max}$, which is the larger eigenvalue of Γ in magnitude. Now by recurring iterations we have the following relation.

$$||\Delta x^k|| \leq (|\lambda|^{max})^k ||\Delta x^0||$$

And a necessary and sufficient condition for convergence is that $|\lambda|^{max} < 1$. This relation is known as *contraction mapping*. The speed of convergence depends on how close $|\lambda|^{max}$ is to zero. When $|\lambda|^{max}$ approaches 1 the number of iterations rise substantially.

Relaxation Methods

For problems where $|\lambda|^{max}$ is close to one, *direct substitution* is limited and converges slowly. The fixed point function $g(x)$ can be altered so that it reduces $|\lambda|^{max}$, the general idea is to modify the fixed point function to:

$$x^{k+1} = h(x^k) \equiv \omega g(x^k) + (1 - \omega)x^k$$

ω is chosen adaptively depending on changes in x and $g(x)$. The two more commonly used fixed point methods or recycle convergence are the *dominant eigenvalue method (DEM)* and the *Wegstein iteration*. *Wegstein* method works well on flow sheets where components do not interact strongly (e.g., single recycles without reactors), interacting recycles and components may cause problems with this method.

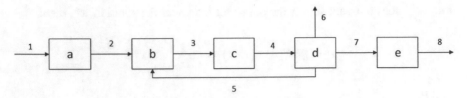

Fig. 1.2 Scheme of a simple system with recycle. *Source*: Adapted from Dimian (2003)

1.5.2   Problems with Simple Recycles

Suppose that there is a process composed by five interconnected units (a, b, c, d, and e) and with a recycle stream like shown in Fig. 1.2. In this case can be appreciated that it is necessary to use a tear stream, then in order calculate unit a the results of the stream 5 are required, values that can only be known after the calculation of units b, c, and d, which at the same time depend of the results of unit a, that means, of stream 5.

Intuitively it is possible to affirm that the tear stream corresponds to stream 5; when values for stream 5 are supposed, unit a can be calculated and then the units sequence b, c, and d, in such a way that new values for stream 5 are obtained. With these values units a, b, c, and d are recalculated. That process is repeated until the absolute value of the difference between the "last" and the "new" values for the stream 5 is less than the established tolerance, that means, until the convergence is reached. In this way, it can be said that the calculation corresponding to the modules representing each one of the units is summed in an only calculation block corresponding to the iterative method that permits reaching convergence over the tear stream (Fig. 1.3). In the latter case is evident that in order to interrupt the cycle not only the stream 5 can be taken as tear stream, but also streams 2, 3, and 4 break the cycle and could be taken also as tear streams. However, one of these streams cannot be chosen arbitrarily, since the first rule for that is that the stream should be preferably before the units that require a careful initialization. It can be observed that unit a must be initialized before any other, and due to the fact that the stream 1 would be defined, it would only be possible to take stream 5 as tear stream.

The complexity of the problem would increase if now a recycle connecting operation e with operation b is considered; in that case an additional recycle would be formed. For the solution of such a problem two tear streams can be used: 5 and 7; additionally it is possible to take the couple of streams 2 and 6, as well as 2 and 7. Nonetheless, all the presented solutions need simultaneous convergence of both streams to solve the problem. Another alternative to solve the problem consists on using only one tear stream. If the case is analyzed carefully it can be seen that streams 3 and 4 are common to both recycles, and when one of them is chosen as tear stream both cycles are interrupted in such a way that it is possible to calculate the whole system in a sequential way.

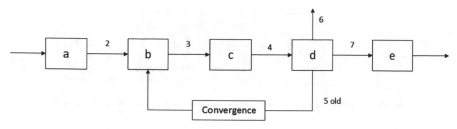

Fig. 1.3 Scheme of a system with one tear stream. *Source*: Adapted from Dimian (2003)

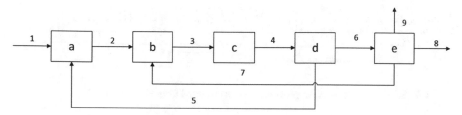

Fig. 1.4 Scheme of a system with two recycles. *Source*: Adapted from Dimian (2003)

In conclusion, problems with recycles can be solved sequentially using tear streams. However, these problems do not have a unique solution sequence, as it is possible to iterate over the cycles using different configurations (Fig. 1.4).

1.5.3   Partitioning and Topological Analysis

As previously approached, it is possible to solve a large variety of problems using tear streams. However, in the real life many systems do not have a few recycles, but, conversely, are complex systems consisting of many stages interconnected with each other. The process of finding the tear streams in the system may become troublesome and therefore the usage of more advanced methods is required.

As a first approximation to this type of systems it is possible to simplify the problem using partitions. This process consists in reducing the solution of a complex system to the solution of multiple subsystems having independent input and output streams according to their precedence. In this process the final subsystems cannot be further expressed as simpler subsystems.

To illustrate the partition process the system shown in Fig. 1.5 is considered. It involves 12 units and 5 recycle streams. Through the partition method, which establishes that each partition must have independent inputs and outputs and not contain simpler partitions, the system can be decomposed in six partitions, as can be observed in Fig. 1.6. In this way from a problem of 12 units and 5 recycles, one of six partitions, from which the more complex consists of four units and three

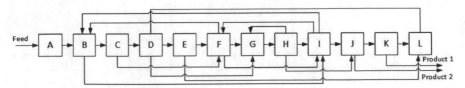

Fig. 1.5 Scheme of a system with multiple recycle. *Source*: Adapted from Dimian (2003)

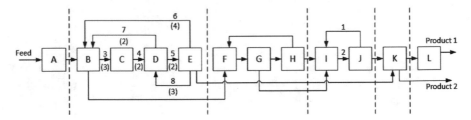

Fig. 1.6 Scheme of the partition process. *Source*: Adapted from Dimian (2003)

recycles, is obtained. As can be seen, with the partitions process a very useful simplification can be achieved in order to solve much more complex systems.

In many cases, even doing partitioning, the problem results very complicated and it becomes necessary to appeal to two tools to know: the topological analysis and linear programming.

With that objective function and some restrictions given for the system it is possible to optimize the solution in such a way that any of the following conditions is accomplished: using the less possible quantity of tear streams, having a minimal cost of the streams using evaluation factors, or minimizing the number of times that the recycles are calculated. However, it is necessary to evaluate another type of variables, such as the equipment having a sensible initialization, the difficulty in initializing values for the streams, and the sensitivity of the following units, among others.

In following sections an example of the mentioned topological analysis is presented.

1.5.4 Nested Recycles

In Fig. 1.7, a standard problem in process simulation, known as the *Cavett's* Problem (Dimian 2003), is presented. This consists of three nested loops of mixers and ash separators. In the same way, in Table 1.1 the cycle matrix is reported. In it the streams present in each recycle have been marked with the number "1." For example, the stream Z_1 is present in loops 1 and 2; thus, it has been signaled with "1" in them. The stream S_2 has been signaled in loop 2. The last column of the table

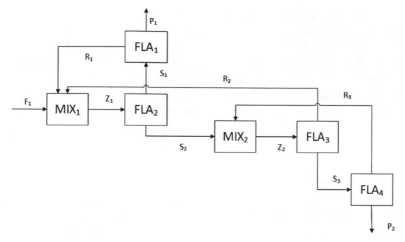

Fig. 1.7 Scheme of a system with nested recycles. *Source*: Adapted from Dimian (2003)

Table 1.1 Topological analysis for the *Cavett's* problem

	Loops where it is present			
Stream	1	2	3	Total loops
Z_1	1	1	0	2
S_2	0	1	0	1
S_1	1	0	0	1
Z_2	0	1	1	2
S_3	0	0	1	1
R_1	1	0	0	1
R_2	0	1	0	1
R_3	0	0	1	1

Source: Adapted from Dimian (2003)

shows the number of loops that has been "crossed" by a stream, in such a way that the information can be used to evaluate the tearing of the process.

An obvious option must consider three tear streams for the three recycles. For instance, streams R_1, R_2, and R_3, which cross one loop at a time, could become a first alternative. Another possibility would be considering streams R_1, S_2, and R_3, where S_2 would replace R_2. It is important to mention that the flow diagram can also be solved using only two tear streams, like Z_2 and Z_3, since each one of them "breaks" two loops. In the same form, R_3 or S3 could be used to replace Z_2. The most important fact to consider here is that all the streams selected as tear streams must be present in all the loops of the diagram. In that way can be assured that the iteration process would lead to the solution of the existent recycles. Additionally, as experience has shown, in general it is more appropriate to use as tear streams those that come in the process calculation units or modules of difficult initialization.

The above indicates that there are different arrangements or combinations of streams that can be feasible; nonetheless, each of them can behave in a different way in a simulation. It can be proved that when the direct substitution method is employed, that means, the combination of the streams R_1, R_2, and R_3, the problem reaches convergence faster than when the combination Z_1 and Z_2 is used, since despite this combination has less tear streams, the number of appearances of them in the loops (4) overcomes the number of loops in the problem (3).

1.6   Introductory Example

1.6.1   Problem Description

As a first approximation to steady-state process simulators, a simple problem consisting in modeling a phase separator with one recycle, as shown in Fig. 1.8, is proposed. The process comprises a stream of heated high pressure gases, typically coming from an exothermal reactor, that are cooled down using a recycled cold liquid stream. The liquid recycle is used to transfer heat by direct contact between the effluent of the reactor and the liquid, in such a way that the later would vaporize and its heat of vaporization is absorbed leading to the gases cooling.

In both simulators, the icons are installed in the process flow sheet and, through them; the mass and energy flows in the process are specified. The specification of the operating conditions is made in the sheets available for that purpose, which allows determining the degrees of freedom required to solve the simulation problem.

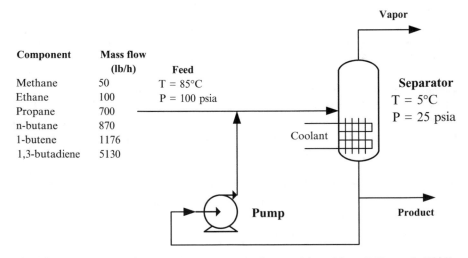

Fig. 1.8   Flow diagram of the introductory example. *Source*: Adapted from Seider et al. (2004)

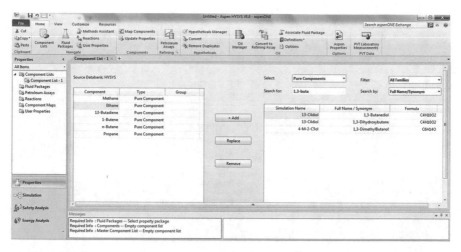

Fig. 1.9 Components selection window in Aspen HYSYS®. *Source*: Adapted from Aspen HYSYS®

1.6.2   Simulation Using Aspen HYSYS®

To start with the modeling of this process, as first step, after opening the simulator, is clicking on the *New* button. A window with the name *Properties* opens, in which the information regarding properties methods, components, and chemical reactions are specified. Initially the components to work with are defined. For this a components list is added with the button *Add*. Here a menu including a large quantity of components available in the databank of the simulator appears. There the components present in Fig. 1.8 are selected (Fig. 1.9).

Later, the component specification window is closed and the thermodynamic model, with which the calculation of the different streams and equipment will be carried on, is selected. Going to the tab *Fluid Packages*, a new properties package can be added using the *Add* button and a menu where the different thermodynamic models can be selected is displayed; for this case the SRK equation of state is employed (Fig. 1.10). Once the model is selected the window can be closed.

Then, it is possible to go to the simulation environment by clicking in the *Simulation* button. A picture of the simulation environment is shown in Fig. 1.11. The first thing to do here is adding a material stream from the equipment palette (blue stream), to then define the feed stream according to the conditions given in Fig. 1.8; the name of the stream is *Feed*. To define the composition it is necessary to go to the corresponding tab, click on the *Edit* button. In the displayed window the *Mass Flow* option is selected and the values reported in Fig. 1.12 are introduced.

After defining the composition and the conditions of the *Feed* stream (the name is modified in the cell *Stream name*), it becomes dark blue, indicating that the stream is completely defined. Next a mixer (named *Mixer*) is taken from the object palette, and the *Feed* stream is connected to the mixer input (to connect a stream the

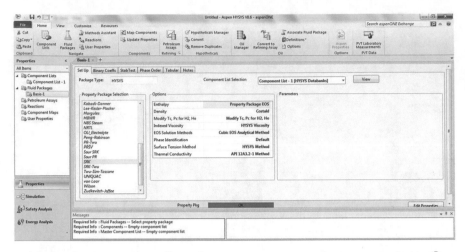

Fig. 1.10 Thermodynamic model selection window. *Source*: Adapted from Aspen HYSYS®

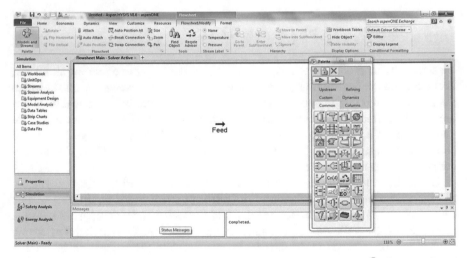

Fig. 1.11 Displayed window of the simulation environment in Aspen HYSYS®. *Source*: Adapted from Aspen HYSYS®

Ctrl key must be pressed and the output of the stream is taken to the mixer input or can be). Afterwards the stream coming out of the mixer is specified. This can be made in the flow sheet or in the specification window of the mixer in the *Outlet* box. This stream name is, by default, *1* (Figs. 1.13, 1.14, and 1.15).

In this way all the system is calculated, that means, the mixer with the corresponding input and output streams. Now stream *1* is connected to a phase separator (which in the object palette has the name *Separator*). An energy stream, *Qsep*, is connected to the phase separator (keeping pressed the Ctrl key, the purple

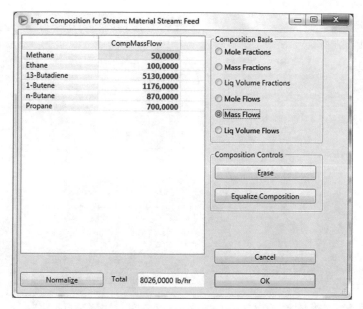

Fig. 1.12 Feed stream composition in Aspen HYSYS®. *Source*: Adapted from Aspen HYSYS®

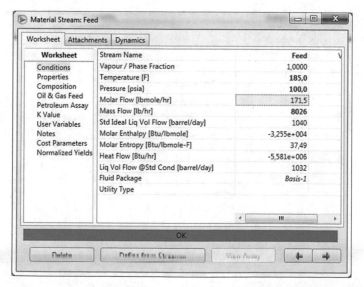

Fig. 1.13 Conditions of the feed stream. *Source*: Adapted from Aspen HYSYS®

square of the equipment can be taken). Two outcoming mass streams are also taken: one from the superior part of the equipment (gas output, *Vapor* stream) and the other one from the inferior part (liquid output, *Liquid* stream). The definition of the operating conditions of this separator is made in a specification window.

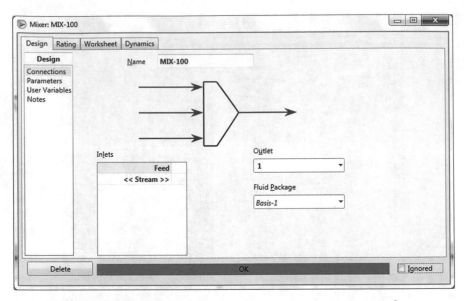

Fig. 1.14 MIX-100 Mixer module window. *Source*: Adapted from Aspen HYSYS®

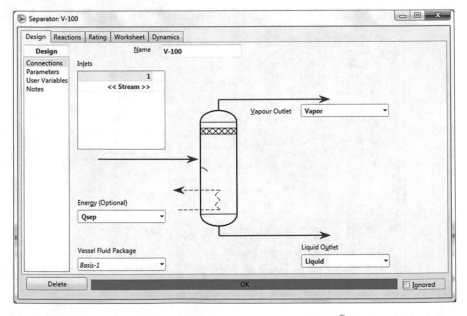

Fig. 1.15 Window of the phase separator module in Aspen HYSYS®. *Source*: Adapted from Aspen HYSYS®

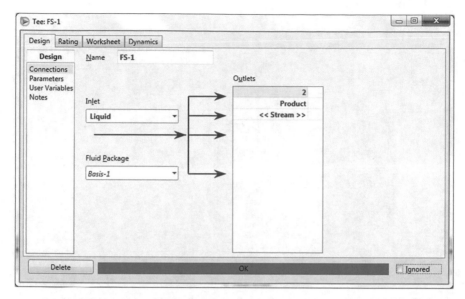

Fig. 1.16 Recycle fractions of the module *FS-1*. *Source*: Adapted from Aspen HYSYS®

For this, double click is made on the separator icon. Initially the temperature of one of the outcoming streams is specified; in this case 5 °C is specified for the *Vapor* stream. This value is inserted in the *Worksheet* tab. It is also necessary to specify a pressure drop at the entrance of 75 psi, what can be made in the *Parameters* option of the *Design* tab. In this way the operating conditions of the phase separator are 5 °C and 25 psi. Next it is necessary to add a unit to split the liquid stream; this unit is taken from the operations palette and has the name *Tee*. It is installed in such a way that the liquid output of the separator is the input stream of the operation and two output streams are given. One of these streams is the *Product* stream, while the other goes to the pump to be recycled (Stream 2). However, it is necessary to define in which proportion the stream must be splitted; for that, in the *Parameters* tab the operation is adjusted to a ratio of 0.5. Change the operation name to *FS-1* (Figs. 1.16 and 1.17).

Here a pump is included; to this operation an energy stream must also be added (*Duty*), and additionally a pressure increase of 75 psia must be defined. The specified energy stream is required for the simulator to calculate the needed energy to make the pressure change. The stream outcoming from the module is stream *R*.

By doing all the above, the system must be calculated until this point (Fig. 1.18).

In principle it could be thought that to close the recycle it would be enough to connect the stream *R* to the mixer. However, it cannot be made in that way since no calculation algorithm would be included to solve the recycle problem. As mentioned before, in Aspen HYSYS® a calculation block must be installed separately representing the convergence calculation over a stream that makes part of a recycle, and it is there where one of the iterative methods mentioned in Sect. 1.5.1 are executed. To achieve a proper specification of the flow diagram, an operation is

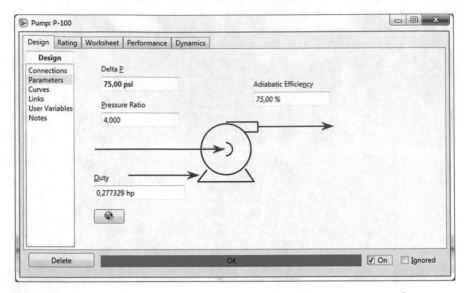

Fig. 1.17 Specifications for the pump *P-100*. *Source*: Adapted from Aspen HYSYS®

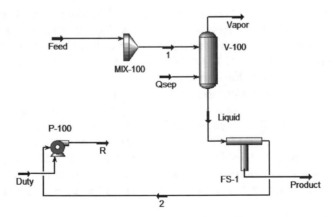

Fig. 1.18 Current Process Flow Diagram in Aspen HYSYS®. *Source*: Adapted from Aspen HYSYS®

taken from the objects palette. To it the stream *R* is connected and a stream *R** is plugged as output. This is illustrated in Fig. 1.19.

Finally the stream *R** is connected to the mixer *MIX-100* and, after some iterations, the whole system converges. The results of the simulation can be observed on each stream and equipment. The results of the convergence are reported in the inferior part of the screen in the white text box (Fig. 1.20).

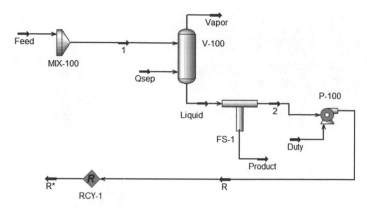

Fig. 1.19 Process Flow Diagram including the recycle operation. *Source*: Adapted from Aspen HYSYS®

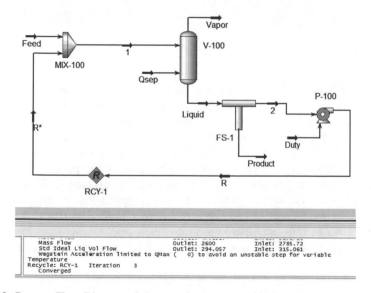

Fig. 1.20 Process Flow Diagram of the complete system with the recycle and convergence. *Source*: Adapted from Aspen HYSYS®

1.6.3   Simulation Using Aspen Plus®

Aspen Plus® is another simulation package that permits the calculation of different operation units. It has a large quantity of options and tools that make it one of the most potent programs for the solution and analysis of different chemical and biochemical processes.

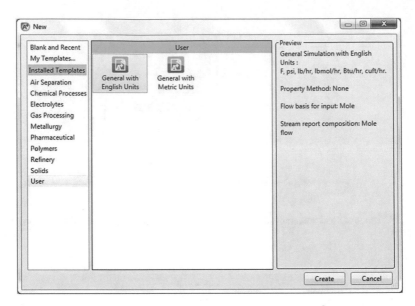

Fig. 1.21 First steps in Aspen Plus®. *Source*: Adapted from Aspen Plus®

For the solution of this problem, first the program must be opened (the route to follow is: *Start>Programs>AspenTech>Process Modeling>Aspen Plus>Aspen Plus User Interface*). After that, it can be directly accessed to the files previously worked on there, to a blank simulation or to a template. For this case the template is selected. With this option the simulation loads some default files containing groups of units and databanks of thermodynamic models that are appropriate for the type of application to model. The window that shows up posteriorly and that can be seen in Fig. 1.21 permits to select the units and the type of application to be made. For this example a simulation with English units (*General with English Units*). Note that in the upper right part, the simulator reports fundamental information of the simulation with respect to the units, property method, flow basis, and stream composition report.

After doing click on *Create* button a windows to specify components and property methods is opened. In the upper menu bar a section with the *Run Mode* is available. Three run modes can be selected:

- *Analysis*: property analysis of pure components and mixtures. It is used also to evaluate accuracy of the models.
- *Estimation*: estimation of properties of components or missing which are not available in the simulator databank by means of molecular structure.
- *Regression*: calculation of model parameters by regression of experimental data.

On the left lower section of the screen four options can be identified: Properties, Simulation, Safety Analysis, and Energy Analysis. Here we will start by clicking on *Simulation* in order to specify the process flow diagram.

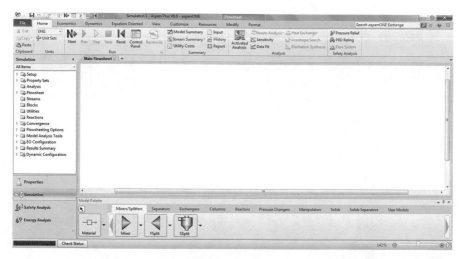

Fig. 1.22 Aspen Plus® interface. *Source*: Adapted from Aspen Plus®

Fig. 1.23 Different representation options for a same operation in Aspen Plus®. *Source*: Adapted from Aspen Plus®

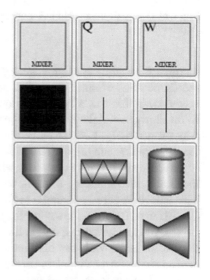

In this environment multiple options can be appreciated: in the lower part a bar with different units and type of streams that Aspen Plus® can calculate; in the upper parts, under the menu *Home* basic options (save, edit, view, etc.). Also sections for *Units* specification, *Run* the simulation, *Summary* of input and output information, process *Analysis*, and *Safety Analysis*, are available (Figs. 1.22 and 1.23).

Now, the different equipment that will be part of the simulation can be selected. These are: a mixing operation (Mixer), a phase separator (*Separators>Flash*), a flow splitter (*Mixers/Splitters>FSplit*), and a pump (*Pressure Changers>*Pump). It is necessary to clear up that in this equipment selection, Aspen Plus® offers a set of

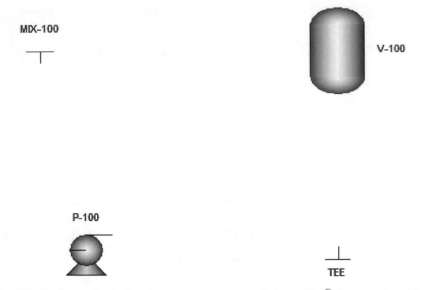

Fig. 1.24 Equipment for the introductory example process in Aspen Plus®. *Source*: Adapted from Aspen Plus®

pictures for a same apparatus, among which it is possible to choose. The different representations of same equipment do not necessarily symbolize other calculation methods. In this way all the operations of the initial problem can be inserted in the Main Flowsheet. After having all the apparatus, they must be accommodated in the way shown in Fig. 1.24. To rotate them it must be right-clicked on the object, and in the displayed menu the option *Rotate Icon* must be selected.

Afterwards, it is necessary to insert the streams involved in the process. For that the block corresponding to *Material Streams* is selected from the inferior panel. This selection permits seeing a set of red and blue arrows coming in and out from the different units, as shown in Fig. 1.25.

The red arrows correspond to those streams that are *mandatory* for the calculation of the unit, while the blue ones are optional streams. For the mixer, two red arrows are given, one input and one output, since they are the only required to do the mass and energy balances of the unit; that means, that after inserting these streams an additional feed stream must be added with a blue arrow. In these cases a mistake can occur frequently, and therefore it is recommended to keep the mouse pressed on the left click until the arrow changes of color. Finally it can be moved until the point in which the entrance to the equipment is desired. This procedure can also be made for the outgoing streams (Fig. 1.26).

After all the connectivity between the units is performed, the resulting diagram can be observed in Fig. 1.27.

It must be remembered that Aspen Plus® permits the modification of the names of streams and equipment. For that it must be right-clicked on the object and the

Fig. 1.25 Equipment connectivity for the introductory example in Aspen Plus®. *Source*: Adapted from Aspen Plus®

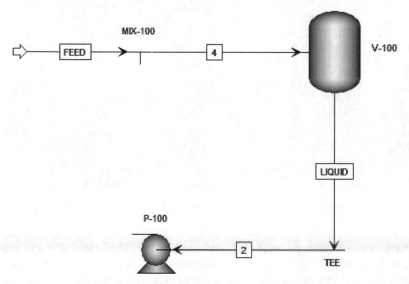

Fig. 1.26 First connection between streams in Aspen Plus®. *Source*: Adapted from Aspen Plus®

option *Rename Stream* or *Rename Block* must be selected, according to the case. It can also be made by selecting the stream or equipment and pressing Ctrl + M.

Afterwards, the button *Next* is clicked. The simulator asks if the user wants to proceed with the following step. When *Accept* is selected, the simulator displays a

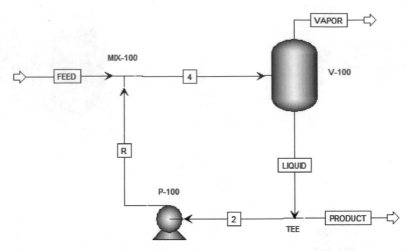

Fig. 1.27 Complete Process Flow Diagram in Aspen Plus®. *Source*: Adapted from Aspen Plus®

Fig. 1.28 Window for components search in Aspen Plus®. *Source*: Adapted from Aspen Plus®

window, where the components can be selected. In the inferior part of it appears the button *Find*, which allows using the component browser and add them to the simulation. In that way the components reported in Fig. 1.28 can be added.

Again, the button *Next* can be used, so the interface for the selection of the property package is displayed. There, different types of processes and property methods for its calculation can be selected. In this case the *SRK* method is chosen. The initial selection qualifies as the global for the selected process and will be used in the simulation of all the unit operations. However, in some situations, there are operations involving substances with particular behavior that cannot be modeled

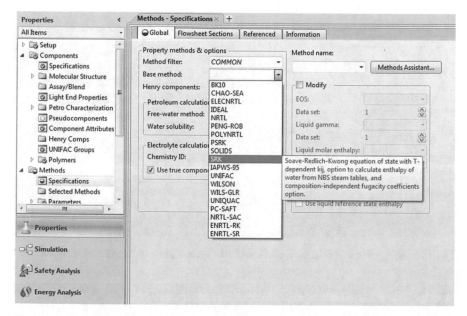

Fig. 1.29 Window for the selection of the thermodynamic model. *Source*: Adapted from Aspen Plus®

with the same properties method. In that case it is necessary to employ a specific model for that operation, generating and additional model. The additional models can be loaded in the *Referenced* Tab that can be found two tabs to the right of *Global*. Those models can be used posteriorly in a specific operation.

Later, when *Next* is clicked, a new window is displayed. There the binary parameters of the system for each one of the selected components are shown. It is always important to verify that the simulator has in its data bank the right values. If it is not the case, the results from the calculations would not be satisfactory (Figs. 1.29 and 1.30).

When the button *Next* is pressed again, the simulator asks if the user wants to move to the next required stage between different options. Here *Go to Simulation environment* option is selected. This option is accepted and, in that way, the simulator asks for the specifications of the stream *Feed*. In that screen the specifications must be introduced in the same way as shown in Fig. 1.31.

Again the *Next* button is selected and now the simulator requests the specification of the units. First the flow splitter, where a value of 0.5 must be given in the *Split fraction* cell for any of the streams. Verify that the specification is made in the stream *Product*, since later a sensitivity analysis will be made over this stream, and the split fraction must be given as an independent variable.

For the mixer no modifications are required, reason why click can be made directly on *Next*. There is no requirement of specification as the simulator can calculate the operating conditions for that equipment through the mass and energy

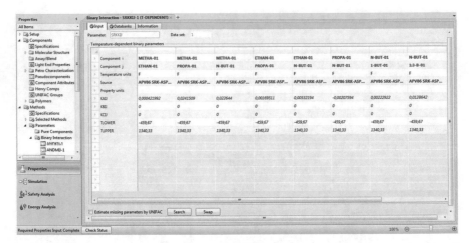

Fig. 1.30 Binary parameters of the model used in the system. *Source*: Adapted from Aspen Plus®

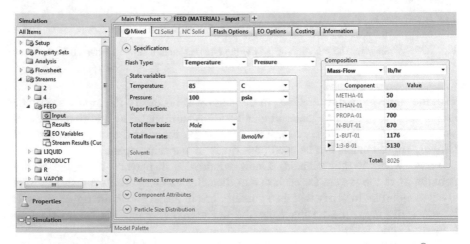

Fig. 1.31 Window of specifications for stream *Feed*. *Source*: Adapted from Aspen Plus®

balances at the conditions of the input and output streams. Note that initially no values of stream *R* are given, reason why an iterative process is given there when the calculation of this equipment is made (Fig. 1.32).

For the pump only a discharge pressure of 100 psia must be specified. Lastly, for the phase separator, a temperature of 5 °C and 25 psia is given. When *Next* is selected again, the simulator asks if the user desires the simulation to be executed. Do click in *Accept*. A Control panel can now be appreciated, where the calculation sequence used by the simulator, as well as the number of iterations, the errors and the warnings found in the calculation can be seen (Fig. 1.33).

Immediately after the calculation starts, the *Control panel* window appears, where the advance of the calculation algorithm, errors and warnings can be

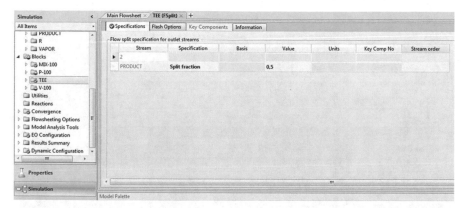

Fig. 1.32 Specification of the flow splitter in Aspen Plus®. *Source*: Adapted from Aspen Plus®

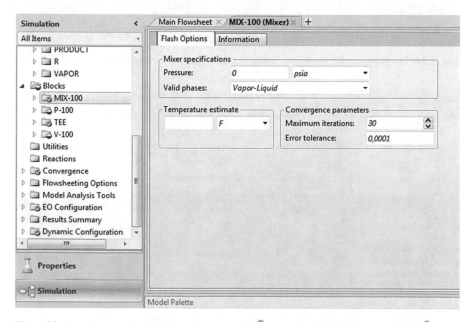

Fig. 1.33 Specification for the Mixer in Aspen Plus®. *Source*: Adapted from Aspen Plus®

followed step by step. When no errors were found, it will inform that the simulation was performed appropriately with the message: *No Errors or Warnings generated*. Observe that a determined number of iterations is required to achieve convergence and that, as mentioned previously, the simulator established a calculation sequence in one step defined as *Computation order* and that can be found under the Calculation Sequence in the left part of the Control Panel (Figs. 1.34, 1.35, 1.36, 1.37, and 1.38).

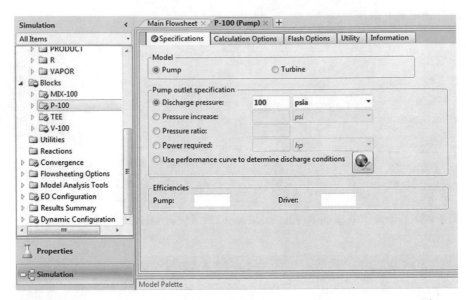

Fig. 1.34 Specifications for the pump in Aspen Plus®. *Source*: Adapted from Aspen Plus®

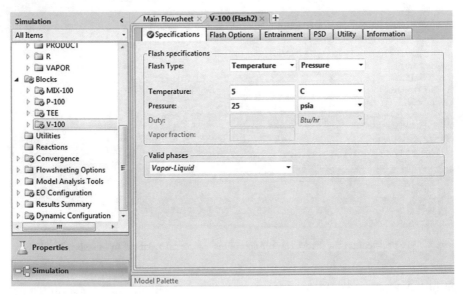

Fig. 1.35 Specifications for the phase separator in Aspen Plus®. *Source*: Adapted from Aspen Plus®

In that way the simulation is solved and it is possible to access to the unit and streams results with the option *Data Browser*, where the *Results summary* tab can be found and used to see the obtained results after the calculation. When the button to examine the results is pressed, a tab summarizing all the results of streams and

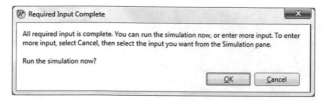

Fig. 1.36 Confirmation window to star the calculation in Aspen Plus®. *Source*: Adapted from Aspen Plus®

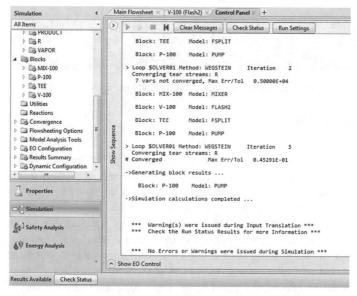

Fig. 1.37 Control Panel window in Aspen Plus®. *Source*: Adapted from Aspen Plus®

units, as well as the convergence information, appears. In each folder the values of the operating conditions and thermodynamic properties can be revised. It is recommended to the reader to verify the flows of vapor and liquid coming out of the phase separator *V-100*, as well as the quantity of heat required to be removed from that equipment according to the flow split fraction specified for the bottoms stream of the phase separator.

To observe the convergence results, follow the route: *Convergence>Convergence>$olver01* that is available in the left side of the window. If the user wants to change the tear streams, go to *Convergence>Tear*, where streams from the diagram, that the user finds convenient for the analysis, can be selected. If the streams do not accomplish the requirements, the simulator will complete the calculations with streams suggested by the system through the performance of a topological analysis.

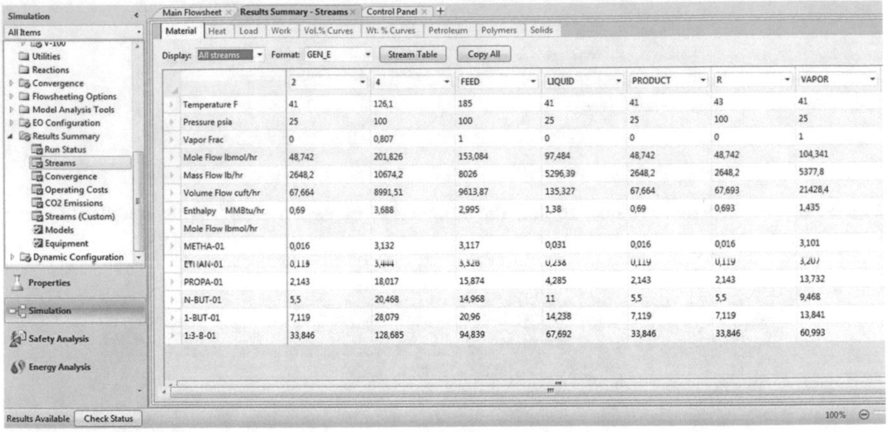

	2	4	FEED	LIQUID	PRODUCT	R	VAPOR
Temperature F	41	126,1	185	41	41	43	41
Pressure psia	25	100	100	25	25	100	25
Vapor Frac	0	0,807	1	0	0	0	1
Mole Flow lbmol/hr	48,742	201,826	153,084	97,484	48,742	48,742	104,341
Mass Flow lb/hr	2648,2	10674,2	8026	5296,39	2648,2	2648,2	5377,8
Volume Flow cuft/hr	67,664	8991,51	9613,87	135,327	67,664	67,693	21428,4
Enthalpy MMBtu/hr	0,69	3,688	2,995	1,38	0,69	0,693	1,435
Mole Flow lbmol/hr							
METHA-01	0,016	3,132	3,117	0,031	0,016	0,016	3,101
ETHAN-01	0,119	3,444	3,328	0,238	0,119	0,119	3,207
PROPA-01	2,143	18,017	15,874	4,285	2,143	2,143	13,732
N-BUT-01	5,5	20,468	14,968	11	5,5	5,5	9,468
1-BUT-01	7,119	28,079	20,96	14,238	7,119	7,119	13,841
1:3-B-01	33,846	128,685	94,839	67,692	33,846	33,846	60,993

Fig. 1.38 Simulation results in Aspen Plus®. *Source*: Adapted from Aspen Plus®

1.7 Sensitivity Analysis

In Aspen Plus® an algorithm for the selection of tear streams is incorporated, in such a way that they are selected automatically. Nonetheless, not all the times the selection is the best, and it is precise to revise and correct the selection of those streams, as the case may be. In the case of Aspen HYSYS®, the selection of the tear variables is made by the user and is determined by the location of the *Recycle* operations in the diagram. In the previous section, a cooling and separation process of a hot gas stream coming from a highly exothermic reaction was simulated. As mentioned, the goal of recycling part of the liquid that leaves the phase separator is cooling by direct contact the gases in the feed. Now a sensitivity analysis will be performed to establish the effect that the flow split fraction (recycle size) has over the temperature at the entrance of the phase separator and the heat to be removed in that equipment. This information may be important to determine the more efficient point of operation for the recycle. Later on a design specification that would make possible to establish a gas flow at the exit of the phase separator will be made (Fig. 1.39).

1.7.1 Sensitivity Analysis in Aspen Plus®

To perform a sensitivity analysis in Aspen Plus®, the folder *Data Browser* must be accessed. There the navigation tree with all the folders corresponding to the simulation will be displayed. If the simulation was already run, the results can be accessed and the corresponding folders will appear in blue. In the navigation tree the folder *Model Analysis Tools/Sensitivity* is selected, and there the information about the variable to be manipulated and the response variable can be entered.

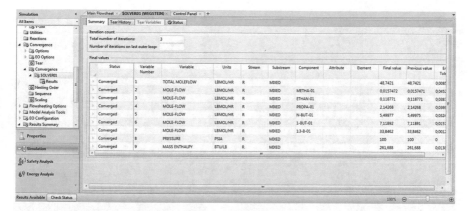

Fig. 1.39   Convergence results of the simulation in Aspen Plus®. *Source*: Adapted from Aspen Plus®

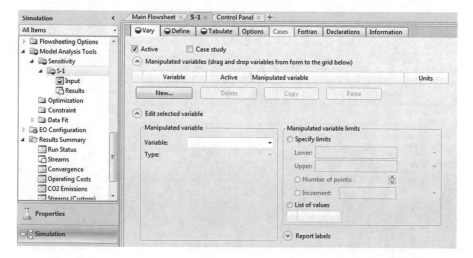

Fig. 1.40   Specification window for the sensitivity analysis. *Source*: Adapted from Aspen Plus®

By clicking in the button *New* a new sensitivity analysis is created and will be named by Aspen as *S-1* by default. Now three tabs named *Vary*, *Define*, and *Tabulate* will appear, as can be seen in Fig. 1.40.

In the tab *Vary*, the variables to manipulate and the variation interval can be modified. In this tab there are also filters for the variables definition, and, additionally, the number of calculations to be performed in the defined interval can be specified. In this exercise, the variable to modify is the recycling fraction from the bottom stream of the separator *V-100* and the variation interval is defined between 0.1 and 0.95 using increments of 0.05.

Afterwards, in the tab *Define* the response variables are declared. To do that the button *New* must be pressed, after that a window like the one presented in Fig. 1.41

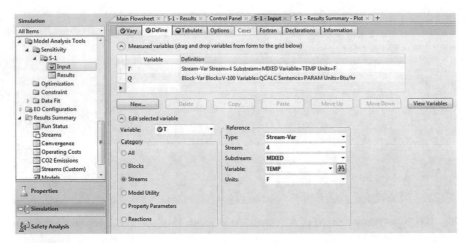

Fig. 1.41 Variable definition window for sensitivity analysis. *Source*: Adapted from Aspen Plus®

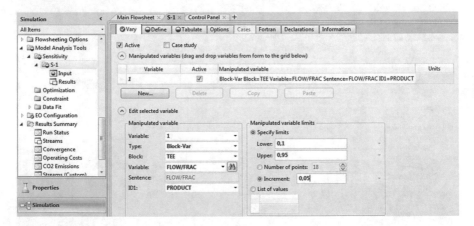

Fig. 1.42 Specification of the manipulated variable for the sensitivity analysis. *Source*: Adapted from Aspen Plus®

will appear. In this case a variable with name T is defined for the temperature in stream 1, while a variable Q is defined for the heat duty of the phase separator V-100. It should be noted that the variables definition can be made using different alternatives according to the type of variable. To do that, Aspen Plus® filters the variables in categories as blocks, streams, properties, etc. In the same way a section to reference the variable to the specifications in the flow sheet is given (Fig. 1.42).

Finally, according to the names assigned to the response variables, the tab *Tabulate* permits the specification of the order in which the results will be presented. By default, the first column is used to present the information of the manipulated variable, and so columns 2 and 3 are used to consign the information of

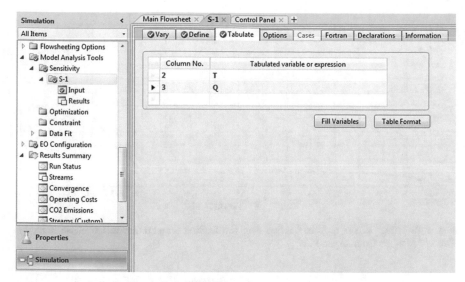

Fig. 1.43 Tabulate tab for the results report. *Source*: Adapted from Aspen Plus®

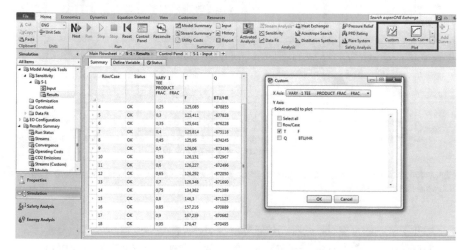

Fig. 1.44 Results for the sensitivity analysis and options of the *Plot* menu. *Source*: Adapted from Aspen Plus®

variables T and Q, as shown in Fig. 1.43. Now the simulation is run and the review of the results can be performed.

In the folder *Model Analysis Tool*, where the sensitivity analysis was defined, the option *Results* containing the resulting data table can be found (Fig. 1.44). These tables can be plotted with the option *Plot* in the tools bar. To do this the following steps must be done:

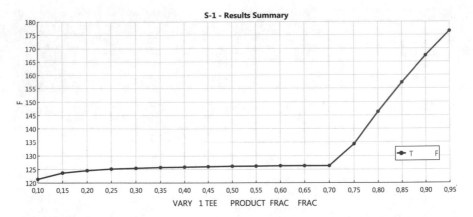

Fig. 1.45 Effect of the recycle fraction over the feeding temperature in the phase separator. *Source*: Adapted from Aspen Plus®

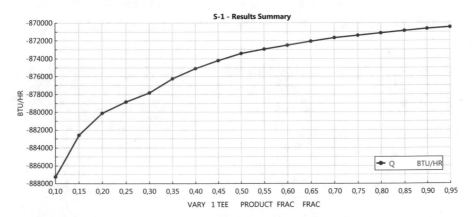

Fig. 1.46 Effect of the recycle fraction over the removed heat from the phase separator. *Source*: Adapted from Aspen Plus®

- Go to Plot menu at the upper right side of the screen and choose *Custom*.
- Mark the VARY 1 as the *X-Axis* variable and select T as the *Y-Axis* variable, and then press *OK*.
- Now a Window like that illustrated in Fig. 1.45 will appear.

Once the plots are generated for both of the variables of interest, the titles and other presentation features of the figure can be changed right-clicking in the plot area and choosing the option *Properties*. In Figs. 1.45 and 1.46 the obtained results are presented.

In Fig. 1.45 can be seen that the most remarkable effects over the feed temperature to the phase separator are obtained when very low or very high recycle fractions of the bottom stream are used. When a high fraction of the bottoms stream is recycled (corresponding to a low fraction extracted as product), the feed temperature decreases as consequence to a higher heat transfer. In the same way, for low recycles the feeding temperature increases importantly. In the case of energy consumption there is no direct effect to be evidenced when the recycle fraction is changed. It is true that, when a higher fluid quantity is recycled, a lower temperature is observed at the entrance of the separator, and therefore a lower cooling duty in required in the separator. However, the flow to be processed in the phase separator increases and, thus, the consumed energy do it as well. In consequence, it could be say in a preliminary way that a product stream corresponding to a fraction of 0.8 (Figs. 1.45 and 1.46) should be used in order to have a relatively low energy consumption with a feeding temperature close to 125 °F. In that way, a minimal flow must be processed in the phase separator, the desired precooling can be achieved and a larger flow of product is obtained. A more rigorous evaluation must include cost functions associated to the equipment sizes, the processed quantities and the total products flow.

1.7.2 Sensitivity Analysis in Aspen HYSYS®

Now the same sensitivity analysis over the heat duty and temperature of the phase separator varying the split fraction in the flow splitter will be performed using Aspen HYSYS®. To do that the simulation developed in Sect. 1.6.2 will be employed. Once the system recycle is closed, all the units of the process are calculated.

Now follow in *Home* menu, under the option *Analysis* press the *Case Studies* button, where the sensitivity analysis is defined by creating the *Case Study 1*. Add the three variables involved:

- Temperature of stream 1
- Heat duty of the separator *V-100* (Stream *QSep*)
- Flow fraction of stream *Product*

For it first select the object; for instance, stream 1, later the variable *Temperature*, and then do the same procedure for each one of the variables. A window like the one shown in Fig. 1.47 will be displayed to select the variables and then a window with the specified variables will appear (Fig. 1.48). The type of variables (independent or dependent) and the variation interval of the firsts can be defined in lower part of the screen.

Regarding the variables, the temperature and heat duty are the dependent variables and the flow ratio is the independent variable. As it was mentioned in the lower part it is necessary to specify the variation interval of the independent variables. For this case the same interval implemented in the sensitivity analysis with Aspen Plus® is taken. Figure 1.48 shows how such information must be specified in the mentioned window. Next, in the *Results* tab, press the *Run* button

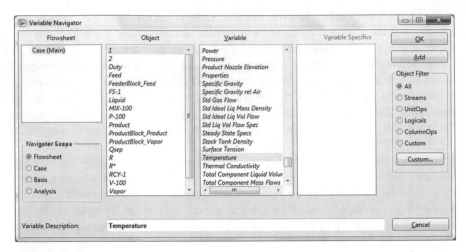

Fig. 1.47 Window for the variable selection from *Case Studies* in Aspen HYSYS®. *Source*: Adapted from Aspen HYSYS®

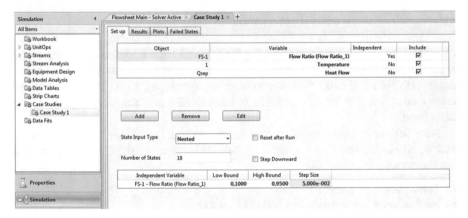

Fig. 1.48 Set up window of the *Case Study* in Aspen HYSYS®. *Source*: Adapted from Aspen HYSYS®

to execute the analysis of the case study (Fig. 1.49) and then select the way in which the results should be reported.

One of the following options can be chosen:

- *Table*: selecting this option, the data will be reported in a table where in the rows are the dependent and independent variables and in the columns the calculated scenario.
- *Transpose Table*: with this option the scenarios will be presented in the rows and the variables in the columns.
- *Results Plot*: selecting this option, the results will be presented graphically with one or two dependent variables in the Y-Axis and the independent variable in the X-Axis.

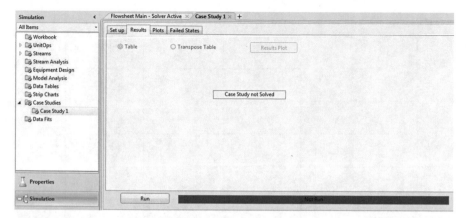

Fig. 1.49 Results tab of the *Case Study* in Aspen HYSYS®. *Source*: Adapted from Aspen HYSYS®

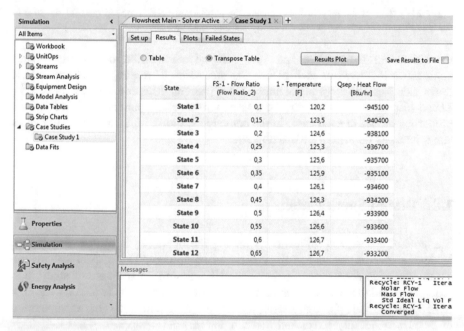

Fig. 1.50 *Results* window in the case of study in Aspen HYSYS®. *Source*: Adapted from Aspen HYSYS®

In Figs. 1.50 and 1.51 the results provided by the simulator are presented. Observe that in the lower part it is possible to change the visualization features of the results. In Fig. 1.51 can be noticed that the results are very similar to the one shown in Figs. 1.45 and 1.46. In both simulators the mass and energy balances are similar since the same thermodynamic property package was selected. In general

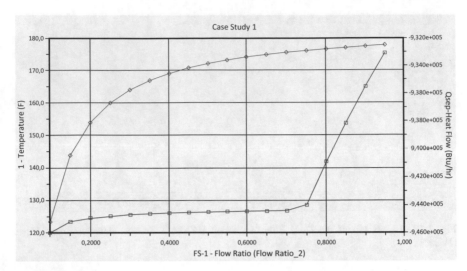

Fig. 1.51 Effect of the recycle fraction over the temperature and removed heat for the case of study in Aspen HYSYS®. *Source*: Adapted from Aspen HYSYS®

the calculated temperatures and the components distribution in the streams approach a lot between each other. The quantities of removed heat and stream flows calculated in both simulators do not present differences larger than 5 %. Those disparities can be attributed to the interaction parameters used by the properties model in each case, to the convergence algorithms and the parameters and tolerances defined in each simulator.

1.8   Design Specifications

The goal of a design specification is finding the operating conditions required to achieve a specific demand of the process or equipment. These specifications are made once it is necessary to refine or detail a design. In the case of study, it is required to find the operating temperature at which it is possible to obtain a top flow of 850 lb/h from the phase separator. It is important to remember that a temperature of 5 °C was initially fixed, and that allowed a flow of approx. 4760 lb/h. This permits the supposition that the temperature required to obtain the new flow is lower.

To do the design specification in Aspen Plus® it is necessary to go to the data browser, in which the navigation tree with all the folders corresponding to the simulation will be displayed. If the calculation was successfully performed, results should be available and marked in blue in the corresponding folders. In the navigation tree the folder *Flowsheeting Options* is chosen. There the information of the manipulated and the response variable can be entered. In the same way that in

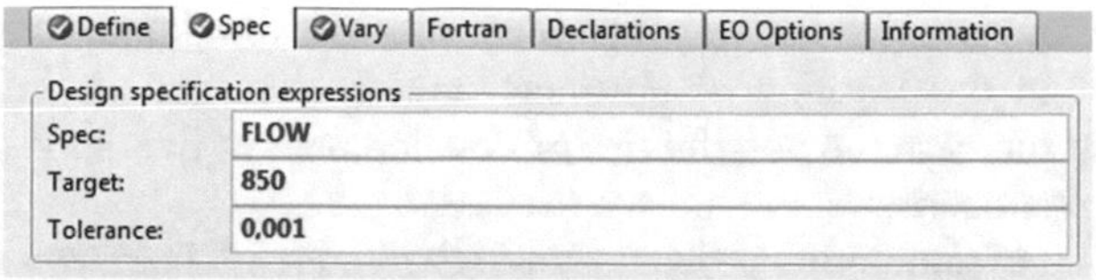

| Define | Spec | Vary | Fortran | Declarations | EO Options | Information |

Design specification expressions

Spec:	FLOW
Target:	850
Tolerance:	0,001

Fig. 1.52 Design Specification for the phase separator. *Source*: Adapted from Aspen Plus®

Main Flowsheet × | S-1 - Results × | Control Panel × | S-1 - Input × | S-1 - Results Summary - Plot × | **Results Summary - Streams** × | +

Material | Heat | Load | Work | Vol.% Curves | Wt. % Curves | Petroleum | Polymers | Solids

Display: All streams ▾   Format: GEN_E ▾   Stream Table   Copy All

	2 ▾	4 ▾	FEED ▾	LIQUID ▾	PRODUCT ▾	R ▾	VAPOR ▾
Temperature F	23,9	122,8	185	23,9	23,9	25,7	23,9
Pressure psia	25	100	100	25	25	100	25
Vapor Frac	0	0,462	1	0	0	0	1
Mole Flow lbmol/hr	133,878	286,962	153,084	267,757	133,878	133,878	19,205
Mass Flow lb/hr	7176,01	15202	8026	14352	7176,01	7176,01	850
Volume Flow cuft/hr	181,338	7485,08	9613,87	362,675	181,338	181,399	3843,23
Enthalpy MMBtu/hr	1,341	4,344	2,995	2,682	1,341	1,349	-0,07
Mole Flow lbmol/hr							
METHA-01	0,233	3,349	3,117	0,465	0,233	0,233	2,884
ETHAN-01	1,327	4,653	3,326	2,655	1,327	1,327	1,998
PROPA-01	11,945	27,819	15,874	23,89	11,945	11,945	3,929
N-BUT-01	13,84	28,808	14,968	27,68	13,84	13,84	1,128
1-BUT-01	19,173	40,133	20,96	38,346	19,173	19,173	1,787
1:3-B-01	87,36	182,199	94,839	174,72	87,36	87,36	7,479

Fig. 1.53 Results summary after performing the design spec calculation. *Source*: Adapted from Aspen Plus®

the sensitivity analysis, there are two tabs named *Define* and *Vary*. Additionally the tab *Spec* can be found, in which the exact value of the response variable and the calculation tolerance can be defined. For this exercise, the mass flow of the stream *Vapor*, leaving the phase separator *V-100*, is named *Flow*. The manipulated variable is the temperature of the phase separator that will be varied between 10 and 200 °F. The flow specification is made as shown in Fig. 1.52.

After entering the required information the simulation is run. In the results folder can be seen that the adjusted value for the flow at the top of the equipment was adjusted to 850 lb/h. In the same way, if the folder *Results>Summary>Streams* (Fig. 1.53) can be found that the temperature of the stream Vapor was established in −4.5 °C (23.9 °F), the same value of the operating temperature in the phase separator

1.9 Summary

Chemical process simulation is a fundamental tool in different tasks regarding design, control, and optimization. The chemical and process engineers can develop complex calculations and evaluate different alternatives and operation scenarios in short periods of time. In that way, chemical process simulators have become an essential tool in the learning of chemical engineering, and its knowledge and understanding increase the possibilities of development in the fields of conceptual and detailed engineering.

The solution of simulation problems through a sequential strategy implies using tear streams over which the iterative process is made. The number of tear streams that permits the solution of a system is not necessarily the number of recycles in it. For that reason, the selection of tear streams is a careful process in which multiple variables must be taken into account, as is the number of recycles, its organization, the units involved, the sensitivity of the streams, etc.

1.10   Problems

P1.1 What is the importance of defining adequately the tear streams in a process simulation?

P1.2 What are the main differences between Design and Rating modes? What information is required for each of these modes?

P1.3 For the block diagram shown in the Fig. P1.1:

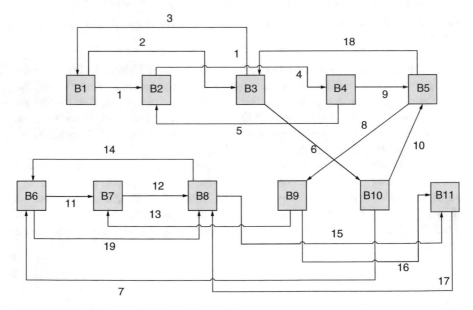

Fig. P1.1 Block diagram for the problem P1.3

a. Do the partitioning process of the flow sheet to find the groups or partitions irreducible.

b. For partitions of more than one unit in **a.**, find the minimum number of tear streams and its position.

P1.4 For the introductory example, do a sensitivity analysis of the work used by the pump P-100 changing the flow fraction that is extracted as product.

P1.5 Determine the flow of benzene, as saturated steam, 1 atm of pressure, which is necessary to mix with a stream of 200 lb mol/h of liquid benzene to increase the temperature from 30 to 60 °F. For an initial estimate, note that the vaporization heat of benzene is 13,200 Btu/lbmol and the specific heat is 0.42 Btu/lb °F.

P1.6 A gas stream with a flow rate of 100 lb mol/h, composed by methane, ethane and propane at 1 atm and 50 °F is fed to a flash separator. The molar composition of the feed mixture is 10 % methane, 20 % ethane and 70 % propane. We wish to determine the possibility to operate the phase separator at 200 psi and 50 °F, assuring the presence of two phases (liquid and vapor). This requires establish the bubble point pressure and the dew point of the mixture.

If possible to operate at 200 psi, determine the compositions of product streams of liquid and vapor, the vaporized fraction and the energy requirements of the operation. The mixture is nonpolar hydrocarbons, it is recommended to use the SRK (Soave Redlich-Kwong) equation of state.

P1.7 For this exercise, the application of a multiple recycle gas compression multistage process is studied. The configuration process together with the main operating conditions of each one of the equipment is shown in the Fig. P1.2. The input stream corresponds to a mole flow rate of 2500 lb mol/h, at 50 °F and 80 psia. The molar composition of the inlet gas is shown in Table P1.1. The goal is to run the simulation and observe the tear streams that the simulator uses to develop the calculation (in the case of Aspen Plus®). Then, select new tear streams, reset the simulation and verify the efficiency of the selection. Finally, make the simulation in Aspen HYSYS® using just two recycle operations to achieve convergence. Can this be done? Justify briefly.

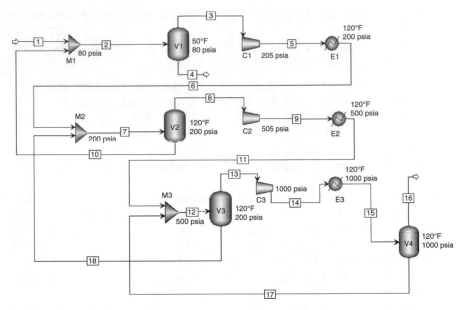

Fig. P1.2 Flow sheet for a process gas compression in multiple stages

Table P1.1 Molar composition of inlet gas. Exercise P1.7

Component	Mole fraction
Nitrogen	0.0069
CO_2	0.0138
Methane	0.4827
Ethane	0.1379
Propane	0.0690
i-butane	0.0621
n-butane	0.0552
i-pentane	0.0483
n-pentane	0.0414
n-hexane	0.0345
n-heptane	0.0276
n-octane	0.0207

References

Babu BV (2004) Process plant simulation. Oxford University Press, New Delhi, Chap. 13
Biegler LT, Grossmann IE, Westerberg AW (1999) Systematic methods of chemical process design. Prentice Hall, San Francisco, Chap. 8

Dimian AC (2003) Integrated design and simulation of chemical processes, 1st edn. Elsevier, Amsterdam, Chap. 3

Schad R (1994) Don't let recycle streams stymie your simulations. Chem Eng Prog 90(12):68–76

Seider WD, Seader JD, Lewin DR (2004) Product and process design principles, 2nd edn. Wiley, Hoboken, Chap. 4

Towler G, Sinnott R (2008) Chemical engineering design, 1st edn. Butterworth-Heinemann, Waltham, Chap. 4

Yee Foo D, Abdul Z, Selvan M, Lynn M (2005) Integrate process simulation and process synthesis. Chem Eng Prog 25–29

Chapter 2
Thermodynamic and Property Models

2.1 Introduction

A thermodynamic model is a set of equations that permit the estimation of pure component and mixture properties. In order to represent chemical processes, their modifications, equipment or new designs, the selection of a thermodynamic model that represents accurately the physical properties of the substances interacting in such process is mandatory (Satyro 2008). Nonetheless, the importance of some properties depends on the goal of the simulation itself. For instance, if the objective is the sizing of heat exchange equipment, transport properties are vital, since these affect the equipment dynamics. Therefore, if there is a substantial error in the modeling of those properties, problems in the performance of the equipment can be evidenced after its sizing, because the real behavior of the apparatus differs from the simulated one (Agarwal et al. 2001a, b; Finlayson 2006).

There are four main groups of thermodynamic models available: the ideal model, the activity coefficients models, the equations of state, and the special methods. The activity coefficients models are especially useful to describe the nonideality in the liquid phase, while the equations of state are used to calculate the nonidcality in the vapor phase. However, under some conditions, the equations of state can be extrapolated to the liquid phase, and the activity coefficients models to the solid phase. Usually both methods are employed to determine the fugacity in the liquid phase.

Particularly, the simulation tools Aspen Plus® and Aspen Hysys® count with an assistant to help the user in the proper selection of the thermodynamic model taking into account some parameters proposed by Eric Carlson (1996), and that are commented in a following section.

The different strategies to calculate physical properties are briefly described below.

© Springer International Publishing Switzerland 2016
I.D.G. Chaves et al., *Process Analysis and Simulation in Chemical Engineering*,
DOI 10.1007/978-3-319-14812-0_2

2.2   Ideal Model

The ideal model is the most elemental of all. Since it does not take the possible molecular interactions into account, it only requires pure component information and the composition of the mixture.

When there is phase equilibrium, the fugacity equality criteria between the phases must be satisfied, in such a way that:

$$f_L^1 = f_G^1 \tag{2.1}$$

Breaking down each term, the following equation is obtained:

$$P^{sat}x_1 = Py_1 \tag{2.2}$$

where the saturation pressure (vapor pressure to achieve vapor–liquid equilibrium) can be calculated using correlations or through experimental data.

As it is well known, most of the substances at normal conditions of pressure and temperature do not have an ideal behavior in any of the both phases, reason why it is common to find important deviations that invalidate the usage of this model.

2.3   Equations of State

The equations of state are models that correspond to PVT (Pressure–Volume–Temperature) correlations that allow to estimate the properties of a pure substance or of a mixture using the Maxwell Relations; in the case of mixtures, mixing rules design for that purpose are employed.

The most important equations of state are the cubic equations, since they combine mathematical simplicity with a good approximation of the theoretical values of the estimated properties.

Some of the most widely used equations of state are: (Dimian and Bildea 2014; Orbey and Stanley 1998).

2.3.1   Redlich–Kwong

Introduced in 1949, the Redlich–Kwong (RK) equation was a substantial improvement with respect to other equation of its time. It is still focus of attention due to its relative simple expression.

Although it is better than the Van der Waals equation, it does not provide good results for the liquid phase, and therefore it cannot be used for vapor–liquid equilibria calculations. Nonetheless, in such cases it can be employed jointly with concrete expressions for the liquid phase.

The Redlich–Kwong equation is adequate to determine vapor phase properties when the ratio of the pressure and the critical pressure is lower than half the ratio of the temperature and the critical temperature.

The corresponding model is presented as follows:

$$P = \frac{RT}{V_m - b} - \frac{a}{\sqrt{T}V_m(V_m + b)} \tag{2.3}$$

$$a = \frac{0.42748R^2T_c^{2.5}}{P_c} \tag{2.4}$$

$$b = \frac{0.08664RT_c}{P_c} \tag{2.5}$$

where:

$$R = 8.314 \text{ J/mol K}$$

$$V_m = \text{molar volume}$$

2.3.2   Soave–Redlich–Kwong

In 1972, Soave replaced the term $a/\sqrt{T}$ of the Redlich–Kwong equation with the expression $\alpha(T, \omega)$, function of the temperature, and the acentric factor. The α function was conceived to fit with the vapor pressure data of hydrocarbons; this equation describes accurately the behavior of those substances. It can be used to correctly represent both liquid and vapor phases.

The model with the corresponding correction is shown below:

$$P = \frac{RT}{V_m - b} - \frac{a\alpha}{V_m(V_m + b)} \tag{2.6}$$

$$a = \frac{0.42748R^2T_c^{2.5}}{P_c} \tag{2.7}$$

$$b = \frac{0.08664RT_c}{P_c} \tag{2.8}$$

$$\alpha = \left\{1 + \left(0.48508 + 1.55171\omega - 0.15613\omega^2\right)\left(1 - T_r^{0.5}\right)\right\}^2 \tag{2.9}$$

$$T_r = \frac{T}{T_c} \tag{2.10}$$

The advantages of using this model in a simulation package (Aspen Technology Inc 2006) are:

- It is a modification of the Redlich–Kwong model
- Results similar to the Peng–Robinson (PR) model but with a lower development than in Aspen Hysys® can be obtained
- Special treatment for key components
- Wide Data Bank of binary parameters

2.3.3 Peng–Robinson

The Peng–Robinson equation was developed in 1976 to accomplish the following objectives:

- The parameters had to be extrapolated with respect to the acentric factor and the critical properties
- The model should be reasonably accurate close to the critical point, particularly for calculations of compressibility factor and the liquid density
- The mixing rules should not employ more than a parameter regarding the binary interactions that should be independent of pressure, temperature, and composition
- The equation should be applicable for all the calculations of all the properties of the fluids in processes involving the use of natural gas

Generally the Peng–Robinson equation provides results similar to those of the Soave equation, although it is better to predict the densities of many components in the liquid phase, especially those that are nonpolar.

The previously mentioned model is presented as follows:

$$P = \frac{RT}{V_m - b} - \frac{a\alpha}{V_m^2 + 2abV_m - b^2} \tag{2.11}$$

$$a = \frac{0.45724R^2T_c^2}{P_c} \tag{2.12}$$

$$b = \frac{0.07780RT_c}{P_c} \tag{2.13}$$

$$\alpha = \left\{1 + \left(0.37464 + 1.5422\omega - 0.26992\omega^2\right)\left(1 - T_r^{0.5}\right)\right\}^2 \tag{2.14}$$

The advantages of using this model in a simulation package (Aspen Technology Inc 2006) are:

It is the most developed model in Aspen Hysys®

- High precision in a wide range of temperature and pressure
- Special treatment for key components
- Wide Data Bank of binary parameters

Note that the factors a, b and α are similar to those proposed by Soave–Redlich–Kwong (SRK), but correcting the constants to improve its precision.

2.4 Activity Coefficient Models

The activity coefficient models surged in face of the inability of the ideal model to represent adequately the behavior of mixtures. These models are based in the evaluation of the excess free Gibbs energy.

Between the activity coefficient models that can be found in the packages of process simulation are the following (Agarwal et al. 2001a, b; Dimian and Bildea 2014; Aspen Technology Inc 2006):

2.4.1 Van Laar Model

The Van Laar Model was developed based on the Van der Waals equation for regular solutions. It has the particularity of being a simple model, with few parameters, allowing an acceptable fitting for engineering applications. Additionally, it can be considered as a pioneer model in this topic and the base for later research in the field.

The model for a binary system is presented below:

$$\frac{G^{\text{ex}}}{x_1 x_2 RT} = \frac{A_{12} A_{21}}{A_{12} x_1 + A_{21} x_2} \tag{2.15}$$

Based on the above, the expressions for the activity coefficients can be deduced as follows:

$$\ln \gamma_1 = A_{12} \left(1 + \frac{A_{12} x_1}{A_{21} x_1} \right) \tag{2.16}$$

$$\ln \gamma_1 = A_{21} \left(1 + \frac{A_{21} x_2}{A_{12} x_1} \right)^{-2} \tag{2.17}$$

However, the parameters of this model are not temperature dependent, something corrected in the following models. In the formulation the presence of two-liquid phases can be included.

2.4.2 Wilson Model

In 1964, Wilson was one of the first contributors in the modern development of the liquid phase coefficients including the concept of *Local composition*, working with binary systems.

This concept is due to the interaction between pair of substances, reason why it is necessary to evaluate the energy involved in the interaction of the molecules through the following expression:

$$\frac{x_{12}}{x_{11}} = \frac{x_2 \exp(-\gamma_{12}RT)}{x_1 \exp(-\gamma_{11}/RT)} \tag{2.18}$$

Such energies are determined when there is an excess of substance 1 in the presence of substance 2 and vice versa. The possible combinations of those energies are designated as $\lambda_{11}, \lambda_{22}, \lambda_{12}$ and λ_{21}. Later, Wilson defined that the constants of its model should be function of the mentioned energies, and proposed the following relations:

$$A_{12} = \frac{v_{2L}}{v_{1L}} \exp\left(-\frac{\lambda_{12} - \lambda_{11}}{RT}\right) \tag{2.19}$$

$$A_{21} = \frac{v_{1L}}{v_{2L}} \exp\left(-\frac{\lambda_{21} - \lambda_{22}}{RT}\right) \tag{2.20}$$

$$\lambda_{11} \neq \lambda_{22} \text{ and } \lambda_{12} = \lambda_{21}. \tag{With}$$

In the same way, the equations for the estimation of the activity coefficients were obtained:

$$\ln \gamma_1 = -\ln(x_1 + x_2 A_{12}) + x_2 \left(\frac{A_{12}}{x_1 + x_2 A_{12}} - \frac{A_{21}}{x_2 + x_1 A_{21}}\right) \tag{2.21}$$

$$\ln \gamma_2 = -\ln(x_2 + x_1 A_{21}) - x_1 \left(\frac{A_{12}}{x_1 + x_2 A_{12}} - \frac{A_{21}}{x_2 + x_1 A_{21}}\right) \tag{2.22}$$

This model, that involves the energy implicated in the interactions, indirectly includes the temperature effect in the molecular phenomena responsible for the phase equilibrium. Nevertheless, the model of Wilson does not take into account the presence of two-liquid phases, reason why it is recommended to use other model for systems reporting immiscibility.

2.4.3   NRTL (Nonrandom Two Liquids)

The NRTL model was developed by Renon and Prausnitz in 1968. It is an extension of the local composition concept that takes into account the no randomness of the interactions. The corresponding expression for the excess free Gibbs energy is:

$$\frac{G^{ex}}{RT} = x_1 x_2 \left(\frac{\tau_{21} G_{21}}{x_1 + x_2 G_{21}} - \frac{\tau_{12} G_{12}}{x_2 + x_1 G_{12}} \right) \tag{2.23}$$

The terms τ_{ij} represent the differences between the interaction energies, $\tau_{ij} = g_{ij} - g_{ii}/RT$. The G_{ij} parameters allow for the no randomness, with help of the parameter α in the following form:

$$G_{ji} = \exp \left(-\alpha \tau_{ji} \right) \tag{2.24}$$

The α parameter can be considered as adjustable; however, it is better to adjust it according to the following suggested values (Table 2.4):

The equations that allow the calculations of the activity coefficients are the following:

$$\ln \gamma_1 = x_2^2 \left[\frac{\tau_{21} G_{21}^2}{(x_1 + x_2 G_{21})^2} + \frac{\tau_{12} G_{12}^2}{(x_2 + x_1 G_{12})^2} \right] \tag{2.25}$$

$$\ln \gamma_2 = x_1^2 \left[\frac{\tau_{12} G_{12}^2}{(x_2 + x_1 G_{12})^2} + \frac{\tau_{21} G_{21}^2}{(x_1 + x_2 G_{21})^2} \right] \tag{2.26}$$

In this way, the NRTL model requires three parameters for a binary system.

2.4.4   UNIQUAC

UNIQUAC comes from UNIversal QUAsi-Chemical, and was developed by Abrams and Prausnitz in 1975. While for the NRTL and Wilson model information about the local volume fraction is required, the UNIQUAC method uses the fraction of local surface area as parameter θ_{ij}.

Each molecule is characterized by two experimental parameters: r, the relative number of segments in the molecule (volume parameter), and q, the relative surface area (surface parameter). The values of these parameters were mainly obtained using mechanical statistics. Other special feature is that for mixtures with alcohols it is possible to include a parameter q' in order to improve the accuracy of the calculations in a third phase.

Table 2.1 Suggested values for the α parameter in the NRTL model

Value	System
0.2	Saturated hydrocarbons with nonassociated polar substances
0.3	Nonpolar components, also for water and nonassociated polar components
0.4	Saturated hydrocarbons and CFS
0.47	Alcohols and other nonpolar self-associated compounds

Source: Adapted from Aspen Technology Inc. (2006)

Table 2.2 Property methods available in Aspen Polymer Plus®

Model	Application
Van Krevelen	Thermodynamic properties using group contribution methods
Tait	Molar volume calculation
Mark–Houwink	Viscosity calculation

Source: Adapted from Aspen Technology Inc. (2001)

Table 2.3 Activity coefficient models available in Aspen Polymers Plus®

Model	Application
Polymer NRTL	It is an extension of the NRTL model that includes the interaction parameters between the segments
	It adjusts well in copolymer systems
Flory–Huggins	This model is well recognized because of its representation of the nonideality of polymeric systems
Polymer UNIFAC-UNIFAC-FV	This predictive model extends the UNIFAC model to polymeric systems considering the involved segments

Source: Adapted from Aspen Technology Inc. (2001)

Table 2.4 Activity coefficient models available in Aspen Polymers Plus®

Model	Application
Polymer ideal gas	This method is employed with other equations of state to calculate thermodynamic properties
Sanchez–Lacombe	This method is widely known, and is based on the Lattice theory adapted for polymers
Polymer PSRK	An extension of the traditional PSRK model
SAFT	It is the traditional rigorous method based on the perturbed-chain theory
PC-SAFT	A modification of the SAFT model

Source: Adapted from Aspen Technology Inc. (2001)

The calculation of the excess free Gibbs energy is executed through two contributions: a *combinatorial* term that represents the influence of structural parameters as size (parameter r) and shape (parameter q), and a *residual* term, that considers the interaction energy among the segments. In the case of a binary system, the expression for the excess free Gibbs energy is the following:

- Combinatorial term:

$$\frac{G^{ex}}{RT} = x_1\ln\left(\frac{\varnothing_1}{x_1}\right) + x_2\ln\left(\frac{\varnothing_2}{x_2}\right) + \frac{z}{2}\left(q_1x_1\ln\left(\frac{\theta_1}{\varnothing_1}\right) + q_2x_2\ln\left(\frac{\theta_2}{\varnothing_2}\right)\right) \quad (2.27)$$

- Residual Term:

$$\frac{G^{ex}}{RT} = -q_1x_1\ln(\theta_1 + \theta_2\tau_{21}) - q_2x_2\ln(\theta_2 + \theta_1\tau_{12}) \quad (2.28)$$

The parameters involved in the equation are explained as follows:

- Average segment fraction (ϕ_i):

$$\phi_i = \frac{x_1r_1}{x_1r_1 + x_2r_2} \quad (2.29)$$

- Average surface area fraction (θ_i):

$$\theta_i = \frac{x_1q_1}{x_1q_1 + x_2q_2} \quad (2.30)$$

- Binary interaction energy (τ_{ij}):

$$\tau_{ij} = \exp\left(-\frac{u_{ji} - u_{ii}}{RT}\right) \quad (2.31)$$

- Coordination number (z) that is equal to 10.

The UNIQUAC model is equally effective than the model of Wilson. Additionally, as added value, it permits, during the calculations, the determination of the situations in which liquid–liquid–vapor equilibrium is reached, it means, the partial miscibility in the liquid phase. In this case, it is possible that additional information of the system can be required. The quality of the data of binary interaction is crucial in order to obtain sufficiently accurate results; in that sense a lot of attention must be paid during parameter regression.

2.4.5   UNIFAC

UNIFAC comes from UNIversal Functional Activity Coefficient, and it is one of the predictive models existing. It is an extension of the UNIQUAC model in which the parameters are estimated using the group contribution method. This means, that the functional groups and the bounds composing the molecule are considered to estimate how a substance behaves in the presence of another one.

In consequence, the equations of this model have the same form of the UNIQUAC model, and the parameters are calculated using the following formulas:

- Molecular volume and area parameters of the combinatorial term:

$$r_i = \sum_k v_k^i R_K \tag{2.32}$$

and

$$q_i = \sum_k v_k^i Q_K \tag{2.33}$$

where v_k^i is the number of functional groups k in the molecule i and R_k and Q_k are the parameters of the functional group k.

- The residual term is replaced by:

$$\ln\gamma_i^R = \sum_k v_k^i \left(\ln\Gamma_k - \ln\Gamma_k^i\right) \tag{2.34}$$

where Γ_k is the activity coefficient of the functional group k for the current mixture, and Γ_k^i is the residual activity coefficient of the functional group k in a reference mixture.

2.5 Special Models

Regularly, special models are not of free access. Petrochemical industry generates very accurate correlations for the calculation of vapor–liquid, liquid–liquid and vapor–liquid–liquid equilibrium for hydrocarbon mixtures. However, they protect such information to avoid that their competitors could use it.

Other industries, such as the polymeric industry, also develop specific model for certain applications with a very good precision (Sada et al. 1975).

In this section, a summary of the special models available in Aspen Plus® as in Aspen Hysys® is presented.

2.5.1 Polymeric Systems

The polymeric industry is one of the fastest growing due to the innovation possibilities and its large spectrum of products with varying properties, designed to meet several requirements.

In order to perform simulations regarding polymerization, Aspen Tech® counts with and add-in for the Aspen Plus® and Aspen HYSYS® environments, namely,

Aspen Polymer Plus® (Aspen Technology Inc 2001), which has a very complete data bank with numerous polymers of industrial application, as well as the related reactants and the intermediates involved in the corresponding reaction mechanism. Additionally, it includes a property methods package to calculate the required physical properties.

Such methods are explained briefly below.

Aspen Tech® manages three types of property methods for the calculation of physical properties; the first one determines properties such as volume, viscosity, etc., while the remaining two are activity coefficient models and equations of state.

2.5.1.1 Property Methods

A table showing the most important methods and their application is presented next (Table 2.5):

2.5.1.2 Activity Coefficient Models

These methods are basically the same explained previously. Nonetheless, for polymeric systems, parameters for the polymer–polymer, segment–polymer, and segment–segment interactions must be taken into account. In Table 2.3 the available methods and its application are enlisted.

2.5.1.3 Equations of State

For the case of equations of state something similar to the previously discussed happens: they are the same models described before, but now considering the interactions with intermediates in the polymeric reactions, if present. In Table 2.4 the main available models and their application are summarized.

Table 2.5 Property methods available in Aspen Plus® for electrolytic systems

Model	Application
Clarke aqueous electrolyte volumen	Molar volume
Jones–Dole	Viscosity
Riedel	Thermal conductivity
Nernst–Hartley	Diffusivity
Onsager–Samaras	Surface tension

Source: Adapted from Aspen Technology Inc. (2006)

2.5.2 Electrolytic System

There is a wide range of electrolytic systems in the chemical industry, since many of the substances employed have the tendency to dissociate due to the presence of water, and many times such property is fundamental for its function in the process (for instance: acids and bases).

In consequence, it is of big importance to know the available models to represent these systems, because it is meaningful to analyze and optimize processes including ionic substances, complex formation, salt precipitation, weak and strong acids, and weak and strong bases. Such systems present important features that must be taken into account for their proper description and simulation. Between those characteristics it is worth to highlight the following:

- The chemistry involved in the dilution of the components must be considered
- There are multiple species in liquid phase
- The real and the apparent composition differ largely from each other
- There is a high nonideality in the liquid phase
- The properties are the function of the ions presence and its concentration

To determine the properties of these systems, it is precise to select between the real and the apparent approaches, since a correction can be made if the apparent composition is known to calculate the properties of the real solution. For example, a way to calculate the fugacity coefficients is presented as follows:

$$\phi_i^{a,l} = \phi_i^{r,l} \frac{x_i^r}{x_i^a} \tag{2.35}$$

where

$\phi_i^{a,l}$ = Apparent fugacity coefficient for component i

$\phi_i^{r,l}$ = Real fugacity coefficient for component i

x_i^a = Apparent composition of component i

x_i^r = Real composition of component i

It is worth to highlight that similar relations for the other thermodynamic properties exist.

This method has some advantages, because the calculations are only made over the apparent components, simplifying the calculation; however, there is a loss of precision that can generate important errors in some systems. For example, to determine the vapor–liquid equilibrium of an electrolytic system, the deviation is not important because, by definition, the real components are dissociations of the apparent components of the system, and considering that these are low volatile, they do not take part in such equilibrium and this approach leads to a very adequate approximation.

2.5.2.1 Property Methods

Other relevant aspect, for the polymers case, is the determination of some global properties that are very useful in the calculation of unit operations. In Table 2.5 a summary of the available methods for the calculation of electrolytic systems is shown.

2.5.2.2 Activity Coefficient Models

For the simulation of electrolytic systems, there are three thermodynamic properties that are fundamental for the calculation of phase equilibrium and chemical equilibrium: activity coefficients, enthalpy and free Gibbs energy. The enthalpy is required for the solution of the energy balances, and the other two properties are related with the estimation of flows, compositions, and phase stability for those systems.

The study of the electrolytic thermodynamics has generated a big quantity of semi-empiric models for excess free Gibbs energy calculations that permit the correlation and prediction of activity coefficient by ions, average activity coefficients, and activity coefficients for molecules and solvents. The mainly used models are electrolytic NRTL, Pitzer equation, and the Bromley–Pitzer model.

2.5.2.3 Electrolytic NRTL Model

The NRTL model is a versatile model for the calculation of the activity coefficients. This modification permits the study of the contributions of each one of the effects in a separate way and the calculation of the activity coefficient of each component in the mixture. When the concentration of electrolytes is zero, this system becomes the basic NRTL model, previously explained.

The electrolytic NRTL model was originally proposed by Chen, and it was addressed to aqueous solutions; later it was extended to mixtures involving other solvents. This model takes as reference state an infinite aqueous solution, and employs the developments of Pitzer–Debye–Hückel and Born to calculate the interactions between the species.

$$\frac{G_m^{*E}}{RT} = \frac{G_m^{*E,\text{PDH}}}{RT} + \frac{G_m^{*E,\text{Born}}}{RT} + \frac{G_m^{*E,\text{lc}}}{RT} \qquad (2.36)$$

This leads to

$$\ln\gamma^* = \ln\gamma^{*,\text{PDH}} + \ln\gamma^{*,\text{Born}} + \ln\gamma^{*,\text{lc}} \qquad (2.37)$$

where

$\quad G_m^{*E} =$ Free Gibbs energy of component m

$G_m^{*E,\text{PDH}}$ = Free Gibbs energy calculated using the Pitzer–Debye–Hückel formula

$G_m^{*E,\text{Born}}$ = Free Gibbs energy calculated with the Born model

$G_m^{*E,\text{lc}}$ = Free Gibbs energy calculated by local contribution

The term calculated with the Pitzer–Debye–Hückel formula considers the long distance interactions, and can be determined as follows:

$$\frac{G_m^{*E,\text{PDH}}}{RT} = -\left(\sum_k x_k\right)\left(\frac{1000}{M_B}\right)^{1/2}\left(\frac{4A_\varphi I_x}{\rho}\right)\ln\left(1 - \rho I_x^{1/2}\right) \tag{2.38}$$

with

$$A_\varphi = \frac{1}{3}\left(\frac{2\pi N_A d}{1000}\right)^{1/2}\left(\frac{Q_e^2}{\epsilon_w kT}\right)^{3/2} \tag{2.39}$$

$$I_x = \frac{1}{2}\left(\sum_i x_i z_i^2\right) \tag{2.40}$$

where

M_B = Molecular weight of solvent B

A_φ = Debye–Hückel Parameter

N_A = Avogadro's number

d = Density of the solvent

Q_E = Electron charge

ϵ_w = Water's dielectric constant

k = Boltzmann's constant

l_x = Ionic strength (mole fraction units)

z_i = Charge number of ion i

ρ = Parameter of the "closest approximation"

Or directly from the expression:

$$\ln\gamma^{*,\text{PDH}} = -\left(\frac{1000}{M_B}\right)^{1/2}A_\varphi\left[\left(\frac{2z_i^2}{\rho}\right)\ln\left(1 + \rho I_x^{1/2}\right) + \frac{z_i^2 I_x^{1/2} - 2I_x^{3/2}}{1 + \rho I_x^{1/2}}\right] \tag{2.41}$$

The term calculated with the Born model has into account how the activity coefficient is influenced from the reference state (infinite aqueous dilution) until the condition of the system. It is calculated in the following form:

$$\frac{G_m^{*E,\text{Born}}}{RT} = \frac{Q_e^2}{2kT}\left(\frac{1}{\varepsilon} - \frac{1}{\varepsilon_W}\right)\left(\frac{\sum_i x_i z_i^2}{r_i}\right)10^{-2} \tag{2.42}$$

where

r_i = Born radius

Or directly with the expression:

$$\ln\gamma^{*,\text{Born}} = \frac{Q_e^2}{2kT}\left(\frac{1}{\varepsilon} - \frac{1}{\varepsilon_W}\right)\left(\frac{z_i^2}{r_i}\right)10^{-2} \tag{2.43}$$

Finally, the term calculated by local contribution is estimated as explained in the regular NRTL model, and considers the binary interactions among the species present in the system.

2.5.2.4 Pitzer Model

The Pitzer model was developed based on the Guggenheim model, which has good accuracy at low concentration of electrolytes, but differs strongly at high concentrations (>0.1 M). The Pitzer model corrected that fact without using higher order equations.

The model can be used for concentrations up to 6 molal, but cannot be used for mixtures of solvents and electrolytes. This model is based on an approximation of the Debye–Hückel theory.

The general equation of the Pitzer Model is the following:

$$\frac{G^E}{n_\text{w}RT} = f(1) + \sum_i\sum_j\lambda_{ij}(I)m_im_j + \sum_i\sum_j\mu_{ij}(I)m_im_j \tag{2.44}$$

with

$$m_i = \frac{x_i}{x_\text{w}}\left(\frac{M_\text{w}}{1000}\right) = \frac{n_i}{n_\text{w}} \tag{2.45}$$

where

G^E = Excess free Gibbs energy

n_w = Mass of water in kilograms

m_i = Molality of component i

x_i = Molar composition of component i

x_w = Molar composition of water

M_w = Water molecular weight

n_i = Moles of component i

The function (I) is the electrostatic term and represents the electrostatic strengths of long distance. The equation has two parameters: λ_{ij} and μ_{ij}. The first one corresponds to the second virial coefficient that takes into account the short distance interactions of the components, while the second one accounts for the interactions between the present solutes. For ion–ion interactions, the second virial coefficient depends on the ionic strength. The dependence of μ_{ij} with the ionic strength can be neglected. These parameters are symmetrical, it means, $\lambda_{ij} = \lambda_{ji}$.

2.5.2.5 Bromley–Pitzer Model

The Bromley–Pitzer model is a simplified case of the Pitzer model, which is much more inaccurate. It uses the interaction parameters of the Bromley model. It cannot be used for systems with a different solvent than water.

$$\frac{G^E}{n_w RT} = f(1) + \sum_i \sum_j B_{ij}(I)m_i m_j + \sum_i \sum_j \theta_{ij}(I)m_i m_j \qquad (2.46)$$

Where

$$B_{ij} = f\left(\beta_{ij}^{(0)}, \beta_{ij}^{(1)}, \beta_{ij}^{(2)}, \beta_{ij}^{(3)}\right) \qquad (2.47)$$

2.6 Integration of the Activity Models with Equations of the State

Since the approach of the equations of the state is the proper phase and the nonideality of this one, and that the activity coefficients model we represent in no ideality of the liquid phase, it is possible to perform on a coupled calculation that allows to take advantage of the benefits that each one of these methods offer.

In Aspen Plus® some integrated methods are presented and are shown in Table 2.6.

To select one of these methods it is important to take into account all the restrictions that the previously described methods present and to evaluate if the mixture meets them, to guarantee that the approximation is adequate. For instance, the Hayden-O'Connell model represents the dimerization in the vapor phase, as is the case of the acetic acid and the propionic acid. The same procure as for model selection can be carried out for this case (Agarwal et al. 2001a, b, Part I–II).

2.7 Selection of Thermodynamic Model

Most of the user manuals of the simulation packages have one chapter that makes reference to the importance of thermodynamics in the process simulation and how it rules the obtained results (Agarwal et al. 2001, Part I). However, these chapters generally are not taken into account, fact that has to be changed. Here at the station tree based on one found on the literature reference is presented (Carlson 1996).

Table 2.6 Some integrated models available in Aspen Plus®

Thermodynamic model	Description
NRTL-RK	NRTL model for the liquid phase and Redlich–Kwong for the vapor phase
NRTL-HOC	NRTL model for the liquid phase and Hayden–O'Conell for the vapor phase
UNIQ-RK	UNIQUAC model for the liquid phase and Redlich–Kwong for the vapor phase
UNIQ-HOC	UNIQUAC model for the liquid phase and Hayden–O'Conell for the vapor phase

Table 2.7 Experimental data of vapor–liquid equilibrium for the system methanol (1) Water (2) (Concentration basis: mole fraction; $P = 95.3$ kPa) (Soujanya et al. 2010)

T (K)	x_1	y_1
371.45	0	0
361.2	0.1	0.381845
354.85	0.2	0.566476
350.4	0.3	0.6755
347.1	0.4	0.748836
344.55	0.5	0.803301
342.45	0.6	0.847367
340.65	0.7	0.885926
339.05	0.8	0.922327
337.55	0.9	0.959379
336.1	1	1

Source: Adapted from (Soujanya and Satyvathi, 2010)

Table 2.8 Regulated volume composition for natural gas in Colombia

Compound	Concentrations (% vol)	
	Minimal	Maximal
Methane	74.0	98.0
Ethane	0.25	12.5
Propane	0.02	5.4
i-Butane	0.0	1.5
n-Butane	0.0	1.5
i-Pentane	0.0	0.6
n-Pentane	0.0	0.4
Hexane	0.0	0.4
CO_2	0.0	5.5
O_2	0.0	0.5

Source: Ministry of Mines and Energy of Colombia (2010)

In consequence, the proper thermodynamic model is vital since it defines is the real situation accurately described or not. As the engineer can use the simulators to describe how certain substances interact with each other under changes of

Table 2.9 Molar
composition of the well La
Creciente (Guajira,
Colombia)

Compound	Composition (%mol)
Methane	97.9779
N_2	1.5337
CO_2	0.1261
Ethane	0.262
Propane	0.051
i-Butane	0.0199
n-Butane	0.0079
i-Pentane	0.0067
n-Pentane	0.0002
n-Hexane	0.0146
Total	100.0

temperature and pressure, heat and other thermodynamic properties, it is precise that the given simulation represents the most loyally possible real operating conditions.

2.7.1 Selection of the Property Model

The thermodynamic model is also named property package, because it is aspects of equations used for the determination of physical properties of the components to simulate reference (Carlson 1996).

To perform an adequate selection, their various parameters to take into account are proposed, and each engineer has its own criteria. With the objective of performing an adequate and responsible selection, many articles and books about this issue have been published. The more recognized medical is the one developed by the Aspen technology symbol engineer, Eric Carlson, who summarizes some main parameters for the selection in several decision trees that are shown as the following reference (Agarwal et al. 2001, Part I).

In Fig. 2.1 the first criteria to take into account for the selection of the thermodynamic models are indicated. The polarity is an important aspect, since it determines the type of molecular interaction that can take place between the substances. It can be deduced that if the polarity is high, the interaction is strong. The polar path must be selected, even if just one of the components is polar. The second parameter taken into account depends on the first one; it means, that the mixture is polar (or at least one of the components) it has to be considered as the substance is not an electrolyte. This parameter is relevant because the electrolytic mixtures are composed of ions obtained from salts.

His behavior generates the necessity of incorporating calculation routines in ionic equilibrium, when these are available. The applications of these type of

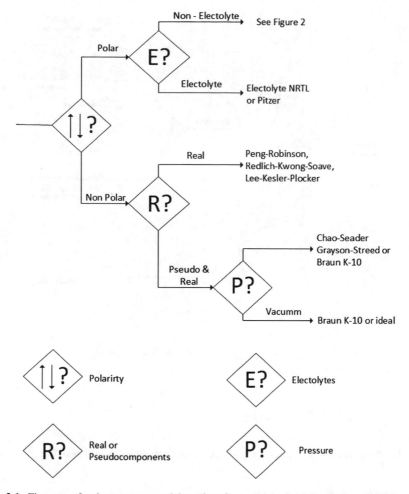

Fig. 2.1 First steps for the property model section. Source: Adapted from Carlson (1996)

mixtures are several: Ash wash, neutralizations, acid production, and salt precipitation.

For no-polar substances, the existence of pseudocomponents has to be considered. These are employed in very complex nonpolar mixtures, as is the case of petroleum, in which, since some components cannot to be identified, and representation of a set of components is generated, generally classified by similar properties, like boiling point, to reduce the number of them.

The properties of these pseudocomponents are obtained like the average of the component's properties.

In Fig. 2.2, subsequent decision tree for a polar mixture of nonelectrolytes is present.

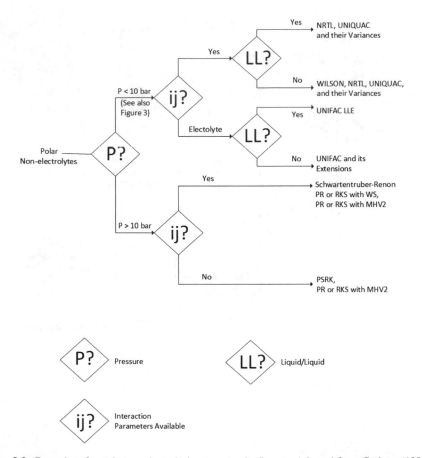

Fig. 2.2 Procedure for polar nonelectrolytic compounds. Source: Adapted from Carlson (1996)

In Fig. 2.3, a decision tree adequate for parameters related with the presence of polymers in the simulation is present.

Five main tasks

In general, five main tasks are mentioned for the proper representation of the physical properties:

1. Select an adequate physical property method
2. Validate the physical properties
3. Properly describe the components that are not present in the database and the missing parameters
4. Obtain and use experimental data
5. Estimate any missing parameter

This process may not be sequential, and to some degree can be concurrent. Additionally to the correct selection of the model, it has to be correctly implemented, because there are situations, like in the case of the partial molar

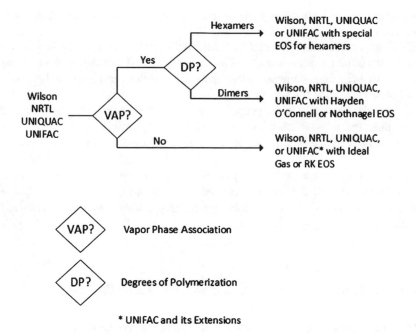

Fig. 2.3 Options for the calculations of the vapor phase with activity coefficient model [Carlson 1996]. *Source*: Adapted from Carlson (1996)

properties that the model does not contemplate and that may be of the importance for the modeling of an operation or process.

2.7.2 Selection of the Properties Model

Previously the selection of the model was discussed, and some remaining aspects are analyzed as follows.

2.7.3 Validate the Physical Properties

The first step in any simulation project is to verify that the properties are representative. For that matter, tables and graphics with the calculated properties must be generated by the simulator for each one of the components (independently of whether they are or not in the database) and are compared with experimental information and reliable bibliographic sources.

This testing is also important to observe their behavior of physical properties such as density, heat capacity, enthalpy, etc., with changes in pressure, temperature, and composition. Its tendency when the values are extrapolated can also be observed.

Similarly, such validation can be made with generated data for liquid–liquid equilibrium (LLE), liquid–vapor equilibrium (VLE), and vapor–liquid–liquid equilibrium (VLLE). Some simulators can generate residue curves for analyses of the distillation operations for ternary mixtures.

If some discrepancies in the properties are found, the pure substance properties can be observed to establish which component is responsible for the error.

The number of phases must be also properly selected to estimate the adequate equilibrium for each situation. For instance, if a distillation column used for the separation of a mixture presenting partial miscibility wants to be modeled, it is precise to specify whether or not two-liquid phase and one vapor phase can take place, so the simulator incorporates the ternary equilibrium in its calculation routines.

2.7.4 Describe Additional Components to the Database

All the simulators provide the possibility of including components that are not in database or at parameters that are missing. For these situations the following aspects must be taken into account:

- Is the component in the quantities? If not, can it be neglected from the simulation?
- Does the component take part of the liquid–vapor equilibrium (VLE)?
- Is the component volatile?
- Is the component polar or nonpolar?
- In case of chemical reaction, that this component compromise the reaction?
- Which properties have to be accurately described for the simulation project?

These questions may help to decide which parameters are important or not for the simulation goal. If these parameters are not available and cannot be found in any bibliographic source, a regression of experimental data can be performed to estimate them. However, it always has to be checked if the parameters are available, since it is dangerous to assume that they are, because the simulator does not provide any warning message.

Some properties are essential for the simulation, as they are the molecular weight, the vapor pressure, the ideal gas heat capacity, among others. Some additional parameters may be required depending on the selected model. Such information appears in the user manuals of the simulators.

Some parameters can be supposed taking into account, criteria corresponding to the component nature. For instance, if a component is not volatile and its vapor pressure is not specified, because the constants of the Antoine equation are not available:

$$\ln P = A + \frac{B}{T + C} \tag{2.48}$$

In that case the following values can be assumed: values, since in that way, being an exponential equation, its vapor pressure is practically equal to zero, and this would adequately represent the behavior of the component.

If for any reason a component is not found in the database, make sure you look for synonyms. In order to be sure, if the browsing is complex, look for the chemical formula.

2.7.5 Obtain and Use Experimental Data

When some important properties are not available, cannot be supposed or a binary parameter regression is required, a search of information must be made. Such information can be found in different sources, as collections of parameter data, specialized journals, manuals or own experimental information.

Since the majority of the streams in a simulation are mixtures, the importance of the pure substance properties should not be underestimated, since they are the basis for the calculation of the mixture properties.

The recommended order for the search of information is the following:

- Critically evaluated sources
- Not evaluated sources
- Experimental data
- Estimation techniques

It should also be taken into account that in most of the cases the parameters taken from different sources do not match, and the best option is to consider the performed experimentation and its accuracy degree as a parameter for the source selection.

Estimate any missing parameter

The regression of data is the most powerful tool of engineering, although it is not the best option. Many simulators include modules for the data regression. The most common examples are the regression of experimental data of liquid–liquid equilibrium (LLE) and vapor–liquid equilibrium (VLE).

As they are many existing components, it is almost impossible to have the binary parameters of all the couples. In many cases it is necessary to obtain them from the literature or employ reliable experimental data to regress them with a good estimation.

To perform a good regression, a proper method must be chosen, as well as a correct objective function, and admissible standard deviation for the data and some adequate initial values for the parameters.

The incomplete information should be handled carefully. Frequently the simulator reports the parameters, but if anyone is missing and the user does not notice that, the simulator selects are routine and estimates them, being possible that the obtained values are incorrect.

2.8   Example of Property Model Selection

When thermodynamic Model is selected it is key, when possible, to compare the generated data by the model with reliable experimental information. In that way it can be observed if the data represents properly or not the model. In the case that it does not fit properly the model, another one can be selected or, if the deviation is too small, the parameters can be fitted using the experimental data.

For this exercise the experimental data is reported by Soujanya et al. (2010) for the system methanol (1) water (2) at the pressure of 95.3 kPa. The compositions of methanol are presented as follows (Table 2.10):

Now, the selection of the thermodynamic model to represent the system can be made according to the decision tree proposed by E. Carlson 1996. Again, the first criteria has to do with the polarity of the compounds; therefore the path *Polar* is chosen. Following this path, the selection between electrolytes and *nonelectrolytes* must be made; for the case of the study nonelectrolytes would be chosen. Later, in Fig. 2.4, it can be observed that a selection parameter is the pressure, for which $P < 10$ bar would be chosen when this process is performed. This allows for the conclusion that the system can be properly represented by the following two models: NRTL or UNIQUAC. Now Aspen properties$^®$ is used to generate the data with the model NRTL, to illustrate the use of these tools.

First the program is open. It can be found in the following path Start>All programs>Aspen Tech>Process Modeling V8.6>Aspen Properties>Aspen Properties Desktop V8.6. A window similar to the one shown by Aspen Plus$^®$ appears. Select *File>New>Blank and recent>Blank Case*.

Then Click on User with *metric units* and *Create*.

In the following screen the components of the system are selected: methanol and water. Go to the button in the bottom part with the name *Find*; later type the component name or its formula. To select the substance, it must be clicked on it and later in the button *Add selected compounds*. Once all the substances are selected, close the window using the button *Close*. Later, go to the *Setup* menu (Fig. 2.5).

Now a window where the title of the project can be included appears. Below, where says *Valid phases*, *Vapor–Liquid* must be selected since the data for the liquid–vapor equilibrium (VLE) are required to compare them with experimental data. Press the button *Next*.

The following step is the specification of the selected thermodynamic model: NRTL. To sign it in go to the option *Method Filter* and select *ALL*. In *Base Method* select *NRTL*. Press the button *Next*. Then verify that the parameters of the model are complete (that there are no empty cells).

Table 2.10 Molar flows of the natural gas from the well La Creciente to be introduced in Aspen Plus®

Component	Molar flow (lbmol/h)
Methane	9797.79
N_2	153.37
CO_2	12.61
Ethane	26.2
Propane	5.1
i-Butane	1.99
n-Butane	0.79
i-Pentane	0.67
n-Pentane	0.02
n-Hexane	1.46
Total	10,000

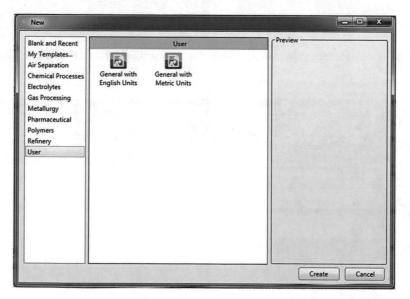

Fig. 2.4 Screen for the template selection in Aspen Properties®. Source: Adapted from Aspen Properties®

Finally, a message box asking if you want the simulator to perform the corresponding calculation routine *Run Property Analysis* appears. Click on *OK*. A window showing the following message appears: *Aspen Properties Setup completed* (Fig. 2.6).

Once the calculation is completed, select the tab *Home*, then the menu *Analysis* and *Binary*. A window useful for the construction of a phase diagram appears. Information for the later comparison with experimental data can be extracted from it (Fig. 2.7).

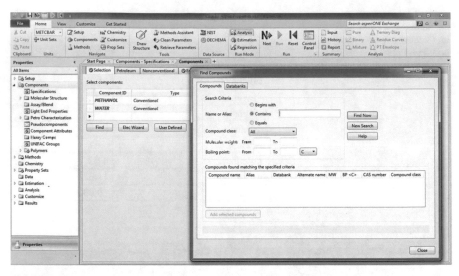

Fig. 2.5 Screen for the components selection in Aspen Properties®. *Source*: Adapted from Aspen Properties®

Fig. 2.6 Screen of the calculation engine in Aspen Properties®. *Source*: Adapted from Aspen Properties®

Where *Analysis type* appears, select T_{xy}. Verify that under *Composition* the molar fraction of *Methanol* is selected in order to compare the values with the available experimental data. In the same way, in the section *Valid phases* the option *Vapor–Liquid* must be selected. Introduce the pressure of the system

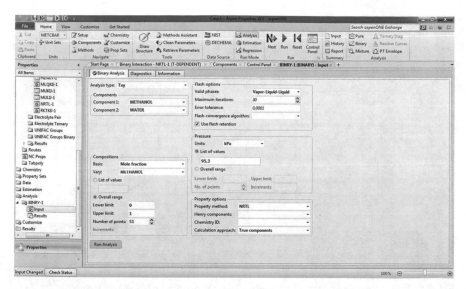

Fig. 2.7 Screen of binary analysis in Aspen Properties®. *Source*: Adapted from Aspen Properties®

(95.3 kPa). Click on *Run Analysis*. After the data is generated, then appears the Graph and the calculation must be re-executed to obtain the corresponding reports (Fig. 2.8).

The following generated diagram appears using the thermodynamic model NRTL. In the left Menu it can be selected *Analysis* and the option *BINRY-1 in Results* the same equilibrium information appears but in a tabular form, which can be exported to a spreadsheet (Fig. 2.9).

Now the experimental data are included in the same diagram to observe if the model accurately represents the experimental behavior of the mixture. For that matter, export the table and compare them with experimental data using a spreadsheet type Microsoft Excel® (Fig. 2.10).

With the observed tendency and its closeness to the experimental data, it can be concluded that the model represents the system properly. Nonetheless, at low methanol concentrations, the liquid phase does not behave as the model predicts, reason why, to obtain a better approximation, a parameter adjustment is advised.

It is recommended to the reader to carry on this exercise using the UNIQUAC model, to compare the results with the experimental data and to analyze which of the models offers a better representation of the system.

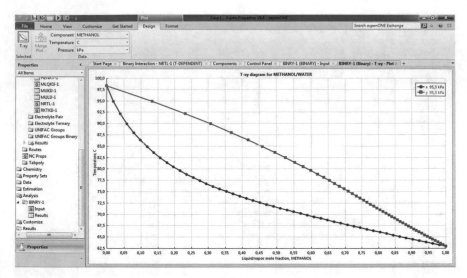

Fig. 2.8 Phase diagram in Aspen Properties®. Source: Adapted from Aspen Properties®

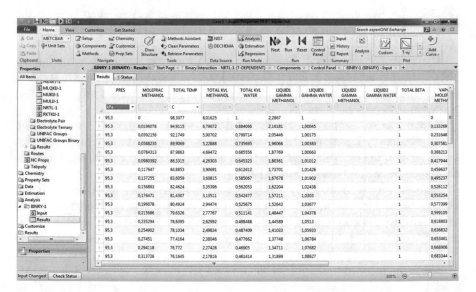

Fig. 2.9 Tabulated phase diagram generated in Aspen properties®. *Source*: Adapted from Aspen Properties®

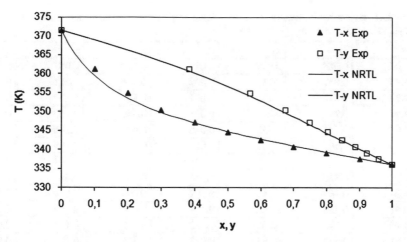

Fig. 2.10 Comparison between the data generated with the NRTL model and the experimental data ($P = 95.3$ kPa). *Source*: Authors

2.9 Example of Phase Diagram

For the Oil and Gas industry there are tools that allow an adequate thermodynamic description of the state of both liquid mixtures (crude oil) and gaseous mixtures (natural gas), as well of both phases coexisting. This analysis is of vital importance for the refining of oil and gas, and for the characterization of the source well.

The phase diagrams are essential to observe the behavior of the natural gas and oil. For purposes of this example the natural gas specifications reported by Ecopetrol S.A.[1] were taken (Table 2.11).

In these conditions the parameters reported by Promigas S.A.[2] of the natural gas extracted of the well Ballenas are reported in Table 2.9.

This natural gas was treated considering the absence of water and hydrogen sulfide These compounds must be removed from the gas before its commercialization due to their corrosivity, their acidity and because these substances diminish the calorific power of the gas, reason why this must be treated.

For the corresponding treatment, the phase diagram is very useful, since the dew and bubble lines determine the changes liquid–vapor required for the separation of the undesired compounds.

---

[1] http://www.ecopetrol.com.co

[2] http ://www.promigas.com

Table 2.11 Experimental data of the vapor–liquid equilibrium for the system ethanol (1) water (2) cyclohexane (3) at 1 atm (Gomis et al. 2005)

T (K)	x_1	x_2	x_3	y_1	y_2	y_3
346.06	0.915	0.053	0.032	0.752	0.05	0.198
345.95	0.851	0.111	0.038	0.776	0.111	0.113
344.77	0.813	0.156	0.031	0.65	0.127	0.223
344.67	0.77	0.204	0.026	0.649	0.163	0.188
344.93	0.724	0.258	0.018	0.631	0.185	0.184
344.83	0.683	0.301	0.016	0.609	0.209	0.182
344	0.623	0.357	0.02	0.525	0.194	0.281
340.98	0.819	0.075	0.106	0.556	0.035	0.409
340.24	0.785	0.113	0.102	0.498	0.089	0.413
339.47	0.738	0.162	0.1	0.446	0.106	0.448
338.76	0.672	0.205	0.123	0.432	0.119	0.449
337.81	0.662	0.263	0.075	0.355	0.127	0.518
338.25	0.741	0.066	0.193	0.456	0.061	0.483
337.37	0.7	0.116	0.184	0.412	0.091	0.497
336.63	0.655	0.155	0.19	0.48	0.132	0.388
337.8	0.715	0.063	0.222	0.448	0.061	0.491
337	0.647	0.111	0.242	0.383	0.09	0.527
337.43	0.673	0.062	0.265	0.429	0.064	0.507
335.78	0.59	0.227	0.183	0.32	0.166	0.514
335.6	0.489	0.164	0.347	0.317	0.168	0.515
335.56	0.47	0.156	0.374	0.318	0.168	0.514
335.93	0.567	0.134	0.299	0.348	0.144	0.508
336.28	0.593	0.12	0.287	0.362	0.133	0.505
336.04	0.589	0.161	0.25	0.347	0.149	0.504
339.37	0.01	0.004	0.986	0.129	0.258	0.613
336.65	0.031	0.005	0.964	0.235	0.22	0.545
339.64	0.004	0.002	0.994	0.13	0.319	0.551

2.9.1 Simulation in Aspen HYSYS®

For the construction of such flow diagram, Aspen HYSYS® counts with a tool called *Envelope Utility*, that can be found under the route in menu *Stream Analysis>Envelope* in the home bar of the simulator. Start a simulation in Aspen HYSYSR entering the components reported in Table 2.9 and using the property model Peng–Robinson represents adequately the natural gas mixtures. Enter in the simulation environment.

Create a material stream with the name *Gas La Creciente* and enter the information reported in Table 2.9. Regarding the operating conditions, any value can be entered since it is not relevant for the analysis.

Now go to *Stream Analysis>Envelope* and select it, as shown in Fig. 2.11.

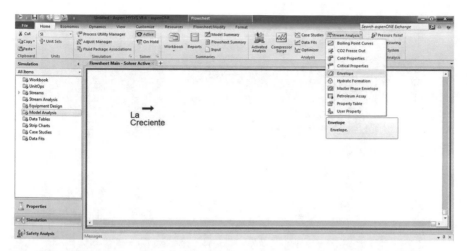

Fig. 2.11 *Utilities* Window in Aspen HYSYS®. *Source*: Adapted from Aspen HYSYS®

Now click on the *Envelope* button to display the main window of the tool and select the stream to analyze and press *Add* (Fig. 2.12).

In the *Name* box a name can be entered when the tool is employed several times for different streams. For this case enter *La Creciente*.

To enter the stream over which the analysis will be made, click on the *Select stream* button and select the stream *Gas La Creciente*. When the stream in entered and the adequate property package is selected, all the required information is completely defined, reason why the tool performs the calculation automatically and reports the status of the calculation in green at the lower part of the window. The information reported in Fig. 2.13 should appear.

With the calculation performed by the simulator, the critical temperature and pressure of the two-phase region can be estimated. They correspond to the *Two-Phase Critical* temperature and *Two-Phase critical* pressure boxes. As the gas does not present liquid–liquid immiscibility under any operating condition, no values are reported in the other boxes.

In the *Maxima* section, two values of high importance for the phase diagram analysis appear: *Crincondentherm* and *Criconderbar*, which represent two points in the phase diagram. *Cricondentherm* represents the maximal temperature in the region when two phases exist, while *Cricondenbar* represent the maximal pressure of such region.

In the tab *Performance* the phase diagram for the selected stream can be observed. In the lower right part the type of graphic to visualize can be selected. For the natural gas application, the P–T diagram can be more widely employed. In Fig. 2.14 the phase diagram visualized in Aspen HYSYS® is presented.

Additionally, the data can be visualized in a tabular form to import them lately into a spreadsheet. For that matter select the option Table in the upper left part of the *Performance Tab*. In this window the data corresponding to the selection made in the previous window in the *Envelope Type* section. In Fig. 2.15 the displayed

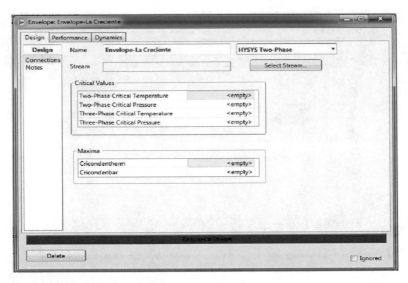

Fig. 2.12 Main Window of the *Envelope Utility* tool. *Source*: Adapted from Aspen HYSYS®

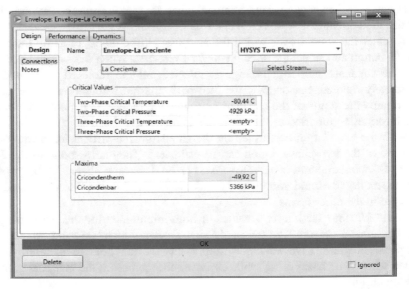

Fig. 2.13 Specification Window of the stream La Creciente. *Source*: adapted from Aspen HYSYS®

window is shown. In this window the data corresponding to each one of the lines conforming the diagram can be selected: Dew point lines (Dew Pt), bubble point lines (Bubble Pt), isobaric lines (Isobar 1), Isotherm lines (Isotherm 1), constant quality lines (Quality 1), and hydrate formation lines (Hydrate).

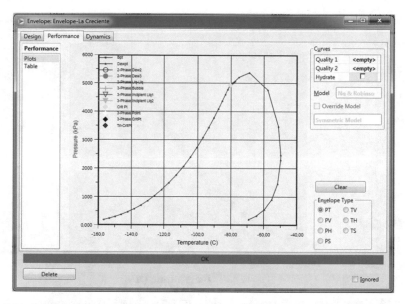

Fig. 2.14 *Performance* tab in the *Envelope Utility* tool. *Source*: Adapted from Aspen HYSYS®

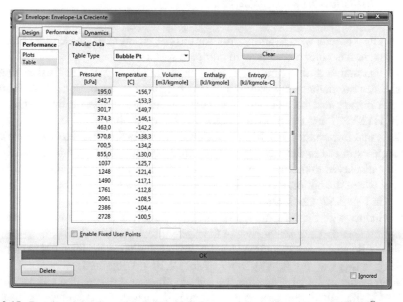

Fig. 2.15 Results of the *Envelope Utility* tool. *Source*: Adapted from Aspen HYSYS®

With the data available in this window it is possible to build a phase diagram using a spreadsheet. Such construction is presented in Fig. 2.16.

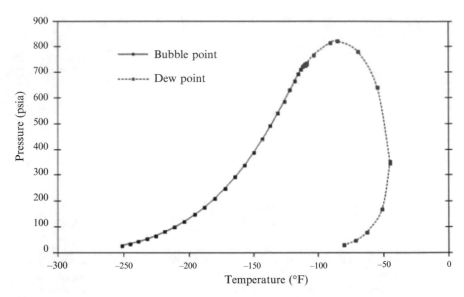

Fig. 2.16 Phase diagram for the gas La Creciente (Guajira, Colombia). *Source*: Authors

2.9.2 Simulation in Aspen Plus®

Aspen Plus® counts with a similar tool for the construction of phase diagrams. The same diagram is constructed now to compare results.

For that matter a new simulation must be started in Aspen Plus® using English units. Since no modules are required for the calculation, the option *Analysis* in the cell left menu must be selected. The components are the same ones entered in Aspen HYSYS®, and the property method is Peng–Robinson.

Once that information has been introduced, an analysis under the route *Analysis* that appears in red in the navigation tree, must be entered. To achieve this, in the window displayed under the mentioned route, Click on the button *New*. A window appears where the option PTENVELOPE must be selected in the *Select type* cell, as shown in Fig. 2.17. Click *OK*.

A window with the information of the gas to analyze is displayed. For it select the option *Specify component flow* in the *System* section (Fig. 2.18).

Since a flow of component must be entered, a calculation basis of 10,000 lbmol/h is taken; in this way the flows given in Table 2.10 are obtained.

Now, below in the tab *Envelope* the initial values of pressure and temperature to be used in the calculation can be defined and, if wished, the calculations inside of the envelope. For that matter, in the *Additional vapor fractions* table, the values to be included in the calculation can be specified.

Once the above is made, the *Analysis* tool is completely specified and the calculation can be performed using the button *Next*. The results can be found under the route *Analysis>PT-1>Results*, and are shown in Fig. 2.19.

Fig. 2.17   Main window of the *Analysis* tool in Aspen Plus®. *Source*: Adapted from Aspen Plus®

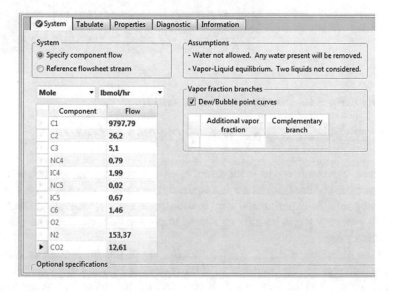

Fig. 2.18   Window for the data input in the PTENVELOPE tool in Aspen Plus®. *Source*: Adapted from Aspen Plus®

2.9.3   Results Comparison

As can be seen in Figs. 2.15 and 2.19, the results provided by both simulators are very similar, since the same property model was used. This tool constitutes one of the many that are available for the design of mixture separation systems or to analyze the behavior of some streams under specific operating conditions.

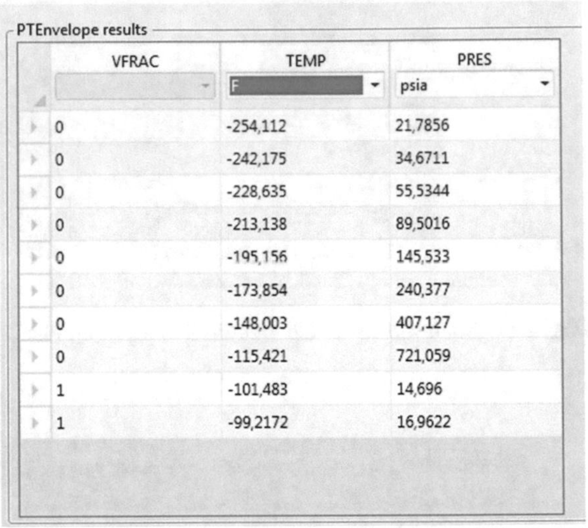

Fig. 2.19 Results of the
PTENVELOPE tool in
Aspen Plus®. *Source*:
Adapted from Aspen Plus®

VFRAC	TEMP	PRES
	F	psia
0	-254,112	21,7856
0	-242,175	34,6711
0	-228,635	55,5344
0	-213,138	89,5016
0	-195,156	145,533
0	-173,854	240,377
0	-148,003	407,127
0	-115,421	721,059
1	-101,483	14,696
1	-99,2172	16,9622

PTEnvelope results

2.10 Example of Parameter Adjustment

2.10.1 Example Using an Activity Coefficient Model

When a thermodynamic model to represent in an accurate way the real behavior of a mixture with a minimal deviation is required, the parameter regression for that model using reliable experimental data is usually necessary. Generally, the model represents precisely a system, although some important discrepancies can exist in a composition range.

With the objective of explaining the functioning of this procedure, an exercise using the data reported by Gomis et al. 2005 is developed using the regression tool of Aspen Plus®. The equilibrium data reported in the example, for the system Ethanol (1) Water(2) Cyclohexane(3), are presented in Table 2.11.

First the program is open. This can be found under the route: *Start>All programs>AspenTech>Process Modeling V8.6>Aspen Plus>Aspen Plus V8.6*. Next a window similar to the one shown by Aspen Plus® at the start appears. Select *Installed Template* and then select. Then *General with Metric Units*, in this case is going to obtain a Data regression. In that way, Aspen Plus® can be used in a similar way than Aspen Properties® (Fig. 2.20).

Next in the left menu in the Setup a window will appear where a title for the project can be assigned. Lower, in *Valid Phase* choose *Vapor–Liquid*, since the goal is to obtain data of vapor–liquid equilibrium (VLE) to compare them with experimental data. Press *Next* (Fig. 2.21).

Later, the components of the system are selected: ethanol, water, and cyclohexane. For this there is a button in the lower part that says *Find*; introduce the component name or its formula. To choose the substance, select it and click on

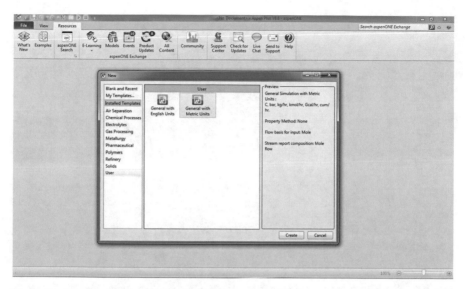

Fig. 2.20 Selection Screen of the template in Aspen Plus®. *Source*: Adapted from Aspen Plus®

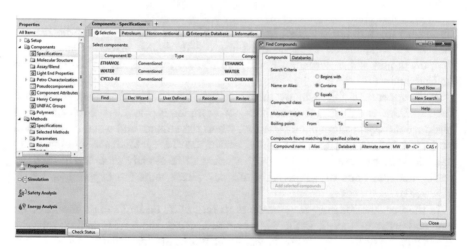

Fig. 2.21 Components input screen in Aspen Plus®. *Source*: Adapted from Aspen Plus®

the button *Add*. Once all the substances were selected, close the window using the button *Close*. Then click on *Next*.

Now the selected thermodynamic model, NRTL, is specified. To enter it go to *Method Filter* and select *All*. In Base *method* select NRTL. Press *Next*. Then verify that the parameters of the model are complete (that no boxes are empty).

In the main menu on *Run Type* section, select *Regression*, now in the left menu will appear *Data*, create a new case, and select type *Mixture*. After that select category *Phase equilibrium* and in *Data type* select *TXY*. Then the components of

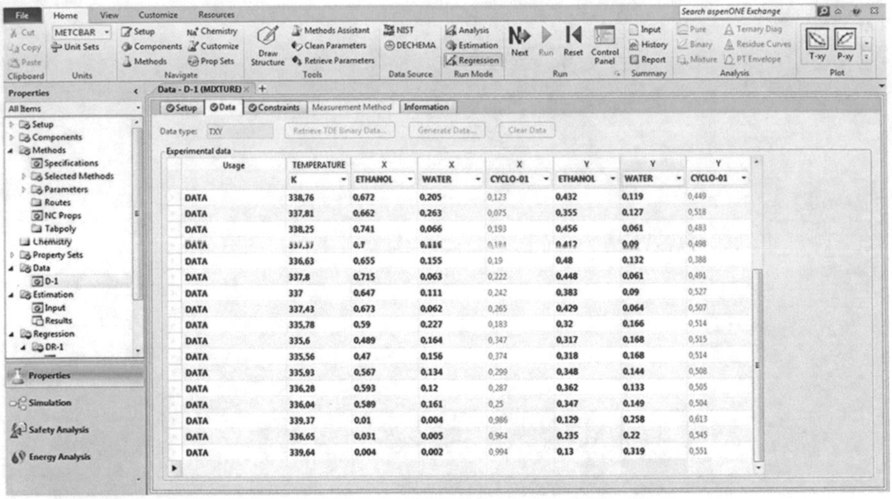

Fig. 2.22 Window for the input of experimental data of the equilibrium in Aspen Plus®. *Source*: Adapted from Aspen Plus®

the known mixture are selected; in this case all three components of the list are chosen. For that matter click on the button with double arrow or select each component and add it normally. Finally, introduce the pressure of the system in the lower part (101.3 kPa).

In the upper part of the window there are three tabs: *Setup*, *data* and *Constraints*. The tab *Data* is still in red since it is not completely specified; click on it and introduce the equilibrium data. Note that when the ethanol and water composition are introduced, the composition corresponding to the cyclohexane automatically gets blocked, because being a ternary mixture it is already completely defined. Verify that the units of temperature are consistent with the experimental data (Fig. 2.22).

In the tree, the category *Regression* is in red; click on it. Then do the same on *Next* and, later, in *Accept*. In the window that appears, the NRTL method must be selected in the section *Property options*. After it the introduced data must be added; for it, in the *Data Set* list, click and add the data package D-1. In the upper part, click on *Parameters*.

As the objective is the regression of the parameters of the NRTL model, it must be taken into account how many parameters have the mentioned model. In the regression screen 12 parameters must be registered, since the model has three types of parameters (A_{ij}, B_{ij}, α_{ij}) and there are three components. In Aspen Plus® the A_{ij} parameters are denoted with the number 1 and the b_{ij} parameters with the number 2.

To register the parameters the following must be specified: in *Type* select *Binary parameter*, in *Name/Element* NRTL must be selected and next to it enter the number 1. Then the three compounds in all possible combinations must be considered, creating one parameter for each combination with number 1 (Ethanol–Water,

Ethanol–Cyclohexane, Water–Ethanol, Cyclohexane–Ethanol, etc.). After it repeat the procedure with the number 2, to represent the B_{ij} parameters of the model. Then click on *Next*.

At this moment two notifications appear, one indicating that everything is properly defined and other asking if the user wants to proceed with the calculations. Then other notifications ask which regressions must be executed and which not. It must be ensured that the regression R-1 is executed.

Once the corresponding calculations are carried on, it is asked to the user if the parameters in the database should be replaced with those obtained from the experimental data. Click on the button *No*, and then verify the reliability of the data used in the regression in order to guarantee a good parameter selection.

In the *Regression* Tab is the option *Results*, where the results of the data regression can be observed, as well as some statistic variables useful to determine the accuracy or the calculation of the new parameters.

In Table 2.12 the values obtained from the regression are reported.

It is recommended to the reader to perform this exercise to regress parameters with the UNIQUAC model and compare with the obtained values. It must be taken into account the number of parameters that such model requires. For more information about the model, Sect. 2.4 can be revised.

2.10.2   Example Using an Equation of State

As the procedure for parameters regression using activity coefficient models was previously explained, now the same procedure is made for an equation of state, which will allow comparing the models and establishing differences and similitude of significance. With the objective of performing this analysis, the data reported for the system Carbon Dioxide (1) Pentane (2) at a temperature of 220 °F is used, employing the Peng–Robinson equation. The data is reported in Table 2.13.

As follows, the differences with respect to the procedure previously described are presented. Do not forget to specify Data Regression as the simulator mode in order to be able to enter the experimental data.

The selected thermodynamic model is chosen (Peng–Robinson). For that matter, go to *Method Filter* and select *All* in *Base Method* select *PENG-ROB*. Click on *Next* Then verify that the parameters of the model are complete (all boxes must contain information).

In the main menu on *Run Type* section, select *Regression*, now in the left menu will appear *Data*, create a new case, and select type *Mixture*, choose the option *Phase equilibrium*, and in the section *Data Type* select PXY. Then the components that are part of the mixture are selected; in this case, the two components of the list. For it, click on the button with double arrow or select each component and add it normally. Finally introduce the temperature of the system in the lower part of the window (220 °F).

Table 2.12 Values obtained from the interaction parameters regression for the NRTL model

Parameter	Component i	Component j	Value (SI)	Standard deviation
NRTL/1	Ethanol	Water	6.35808813	0.67695745
NRTL/1	Ethanol	Cyclo-01	1.404	0
NRTL/1	Water	Cyclo-01	41.236362	171.099372
NRTL/1	Water	Ethanol	−2.2764816	33.6439038
NRTL/1	Cyclo-01	Ethanol	8.56948383	0.29148555
NRTL/1	Cyclo-01	Water	42.9329209	57.6523955
NRTL/2	Ethanol	Water	−2215.62	0
NRTL/2	Water	Cyclo-01	−9628.6239	58090.1967
NRTL/2	Ethanol	Cyclo-01	−84.006605	72.0847761
NRTL/2	Water	Ethanol	1385.7611	11597.7001
NRTL/2	Cyclo-01	Water	−12818.847	19524.2423
NRTL/2	Cyclo-01	Ethanol	−1926.6822	0

Source: Authors

Table 2.13 Data reported for the system carbon dioxide (1) pentane (2) at a temperature of 220 °F

P (psia)	x_1	y_1
94	0	0
132	0.0119	0.2568
214	0.0482	0.5092
341	0.1115	0.6668
486	0.1797	0.741
653	0.2548	0.7854
850	0.3452	0.8094
1055	0.4367	0.8162
1286	0.5601	0.8
1397	0.6447	0.7674

Source: Authors

Click on the *Data tab* and introduce the equilibrium data. Note that when the composition of the carbon dioxide is introduced, the one of the pentane gets blocked, because being it a binary mixture, it is already completely defined. Verify that the units of pressure are consistent with the experimental data.

As the goal is the regression of the parameters of the Peng–Robinson equation, it must be taken into account how many parameters such model has. Nonetheless, for this example, only the binary interaction parameter K_{ij} is regressed.

To register the parameters, the following must be specified: in *Type* select *Binary Parameter*; in *Name/Element*, *PRKBV* must be selected and next to it the number 2 must be introduced. Then consider as component i the carbon dioxide and the Pentane as component j. Click on *Next*.

2.10.3   Comparison and Results Analysis

It can be observed that the process for the two property models is basically the same. However, it is important to know the model to establish which parameters does it have, what does each of the mean, and which would be required to regress to adjust the system of interest.

It can be observed that the UNIQUAC model better represents the equilibrium Ethanol–Water–Cyclohexane after the parameter regression is made, overcoming the prediction and the adjustment of those achieved with the NRTL model.

2.11   Hypothetical Components

Although, there is a huge quantity of compounds which properties are registered in the database of simulators, there are also cases where some component properties are not reported and, additionally, there is a lack of experimental information to perform a regression, being necessary the usage of hypothetical components. These constitute a powerful tool that, using different methods, allows the prediction of compound properties using reduced information such as chemical structure, molecular weight, critical temperature and pressure, among others. In that way, the hypothetical compound can be used in simulation and a first approach to the behavior of the system can be obtained.

This tool is also very useful in the analysis of multicomponent mixtures. In this case, the hypothetical components can be used to predict multicomponent mixtures with relatively similar properties. An example of this application is the modeling of oil fractions, case in which a huge quantity of heavy components exists in the mixture in very low proportions. Such components, for simulation purposes, can be modeled as a hypothetical component that clusters all of them.

2.11.1   Usage in Aspen HYSYS®

With the objective to illustrate the usage of this tool, the modeling of the properties of the following component is proposed (Fig. 2.23).

Additionally in order to compare the accuracy of the data predicted by the simulator, it is necessary to count with some experimental data for this substance (Table 2.14).

In order to start the simulation in Aspen HYSYS® a new file must be opened, and then a new list of components is added. In the left menu the *Hypothetical* option must be selected; in this way a window as the following appears (Fig. 2.24):

In the main menu options appear: *Hypo Manager*, which allows the administration of components or group of hypothetical components created, in *Component*

Fig. 2.23 Chemical
structure of 1,3-Thiazole

Table 2.14 Some important reported properties of 1,3-Thiazole (Linstrom and Mallard 2010)

Variable	Value	Source
Normal boiling point	390.7 K	Weast and Grasselli 1989
Melting point	239.58 K	Meyer and Metzger 1966
Triple point	239.53 K	Soulie, Goursot et ál. 1969
C_p Liquid at 298.15 K	121.00 J/mol K	Soulie, Goursot et ál. 1969

Source: Adapted from Linstrom and Mallard (2010)

Fig. 2.24 Hypothetical components window in Aspen HYSYS®. *Source*: Adapted from Aspen
HYSYS®

List a *Hypo Component*, which allows the fast creation of a hypothetical component
(this option comes with the default configuration for the creation of a hydrocarbon);
and the *Quick Create* a *Solid Component*, which allows the fast creation of a solid
component. In this case, the *Hypo Manager* option is selected, and the following
menu is displayed (Fig. 2.25):

This menu is very similar to the components list (in fact, it can be accessed
without passing through the components list); there hypothetical components can
be added and edited from the hypothetical components lists. Adding a list of
hypothetical components, the following window displays (Fig. 2.26):

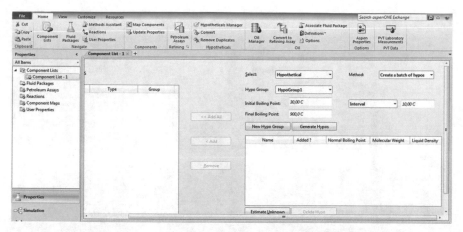

Fig. 2.25 *Hypo Manager* in Aspen HYSYS®. *Source*: Adapted from Aspen HYSYS®

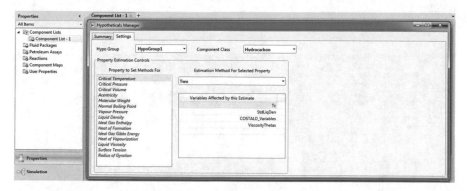

Fig. 2.26 Group of hypothetical components. *Source*: Adapted from Aspen HYSYS®

In this window it is possible to select and modify options like method of properties estimation of the hypothetical components and the class of component to be calculated. In this window the option *Add Hypo* is selected; a new component is generated, and its data will be modified later going to Component added and double click.

Then a window (Fig. 2.27) is displayed with parameters such as family, chemical formula, ID number, among others. Additionally, in this menu, it is possible to use the UNIFAC method to predict values based on the chemical structure of the compound. On the other hand, on the lower tabs, the user can access different options that allow the modification of the values of boiling temperature, critical variables, and specific heat capacity dependent on temperature, among others. Now the option *Structure Builder* is accessed (Fig. 2.28).

In this menu, the different groups of the UNIFAC method can be added to perform the predictive calculation. After this window is closed, the option Estimate

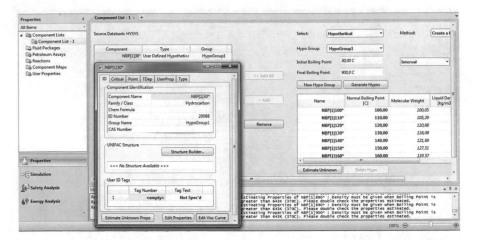

Fig. 2.27 Menu of the hypothetical component in Aspen HYSYS®. *Source*: Adapted from Aspen HYSYS®

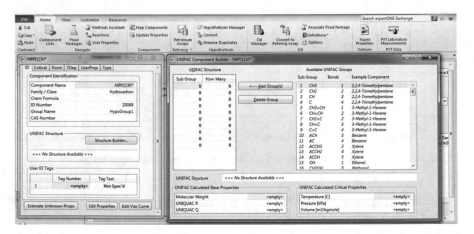

Fig. 2.28 *Structure Builder* menu based on the UNIFAC model. *Source*: Adapted from Aspen HYSYS®

Unknown Props is selected, option with which the simulator calculates the not specified properties for the hypothetical component.

Later the values generated by the simulator must be compared with reliable experimental data to validate the usage of the hypothetical components and its usage in a simulation project.

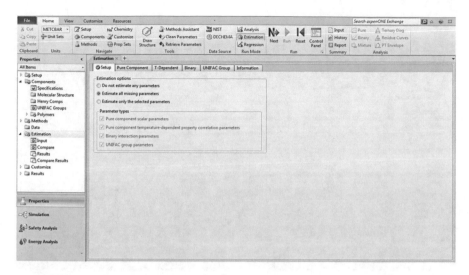

Fig. 2.29 Estimation window in Aspen Plus®. *Source*: Adapted from Aspen Plus®

2.11.2 Usage in Aspen Plus®

Aspen Plus® counts with a powerful tool for the properties estimation, tool based on the group estimation method of Joback. An illustration of how this tool is used is given as follows.

As first step it is necessary to create a new file in Aspen Plus®, and the *Estimation* calculation type is selected in the main menu; then it is necessary to specify the name of the component which properties are going to be estimated, in this case the 1,3-Thiazole. Later, in the estimation window, the simulation is set to estimate all the nonexistent parameters of the component (Fig. 2.29).

Then it is necessary to go to the Molecular Structure tab, where the molecular structure of the component which properties will be estimated is configured; the molecular structure of the component must be set in such a way that it matches with the structure of 1,3-Thiazole (Figs. 2.23 and 2.30).

There are other forms that introduce a structure in order to estimate its properties; among these it is possible to access to a file with the extension *.mol, which contains the molecule information. These types of files can be created using different specialized programs or they can be also downloaded from different data banks, such as the NIST *Chemistry Webbook*.

To import the *.mol file it is necessary to go to the Structure tab and select the *Import*.

Structure option; in this way a browser window will appear to find the *.mol file and import it. After the _le is imported, it is required to calculate the molecule bounds; in that way the simulator adapts the *.mol structure in such a way that the properties can be evaluated and estimated.

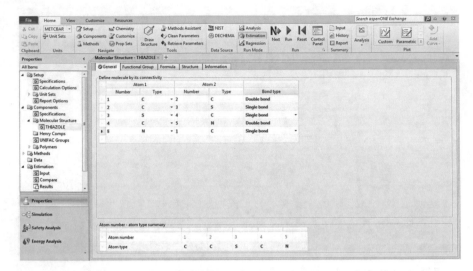

Fig. 2.30 Molecular structure of 1,3-Thiazole entered in Aspen Plus®. *Source*: Adapted from Aspen Plus®

After the acceptance, the execution of the estimation is carried on. In the control panel it is possible that a serial of warnings appears, this due to the different methods used for the estimation doesn't count with the parameters for some bounds. After it, the results can be consulted under *Estimation>Results*. There a set of calculated properties for the component will appear, as well as the method used for the estimation (Figs. 2.31 and 2.32).

2.12   Summary

Several aspects must be taken into account to guarantee the proper selection of the properties model. For instance, the selection of the right number of phases allows the obtainment of coherent results when interpreting the results of a model. In consequence, the selection of the adequate model and the use of the known information allow an approach to reality, necessary for the development of a good simulation project.

The nonideal model does not work in most of the applications due to the fact that the molecular interactions between the components responsible of the nonidealities of the thermodynamic systems are not taken into account.

On the other side, the state equations usually used are cubic equations, which combine the simplicity of the calculation with a proper approximation of the properties of most of the substances.

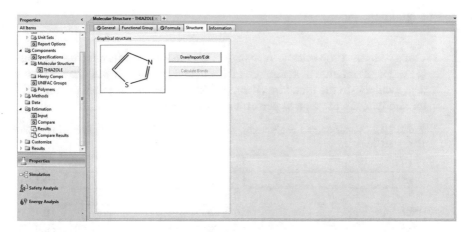

Fig. 2.31 Window for the *.mol file import in Aspen Plus®. *Source*: Adapted from Aspen Plus®

PropertyName	Parameter	Estimated value	Units	Method
MOLECULAR WEIGHT	MW	85,1296		FORMULA
NORMAL BOILING POINT	TB	387,84	K	JOBACK
CRITICAL TEMPERATURE	TC	629,049	K	JOBACK
CRITICAL PRESSURE	PC	6,65302e+06	N/SQM	JOBACK
CRITICAL VOLUME	VC	0,2125	CUM/KMOL	JOBACK
CRITICAL COMPRES.FAC	ZC	0,270313		DEFINITI
IDEAL GAS CP AT 300 K		69908,6	J/KMOL-K	JOBACK
AT 500 K		105485	J/KMOL-K	JOBACK
AT 1000 K		150830	J/KMOL-K	JOBACK
STD. HT.OF FORMATION	DHFORM	1,6918e+08	J/KMOL	JOBACK
STD.FREE ENERGY FORM	DGFORM	1,9547e+08	J/KMOL	JOBACK
VAPOR PRESSURE AT TB		101301	N/SQM	RIEDEL
AT 0.9*TC		3,15495e+06	N/SQM	RIEDEL
AT TC		6,65302e+06	N/SQM	RIEDEL
ACENTRIC FACTOR	OMEGA	0,235302		DEFINITI
HEAT OF VAP AT TB	DHVLB	3,56543e+07	J/KMOL	DEFINITI
LIQUID MOL VOL AT TB	VB	0,0781433	CUM/KMOL	GUNN-YAM
SOLUBILITY PARAMETER	DELTA	22930	(J/CUM)**.5	DEFINITI
UNIQUAC R PARAMETER	GMUQR	2,72973		BONDI
UNIQUAC Q PARAMETER	GMUQQ	1,816		BONDI

Component: THIAZOLE   Formula: C3H3NS

Fig. 2.32 Results of the estimation in Aspen Plus®. *Source*: Adapted from Aspen Plus®

An integrated method takes advantage from an activity coefficient method and the state equations. Nonetheless, it must be verified that the method represents the mixture properly.

2.13   Problems

P2.1 Using Aspen Properties®, calculate the binary equilibrium for ethanol (1) water (2) mixture at a pressure of 1 atm and compare it with experimental data available in the literature (Perry 1992). Remember to select properly the thermodynamic package to perform a reliable analysis.

Table P2.1 Vapor–Liquid equilibrium data for ethanol (1) water (2) system at 1 atm (Perry 1992)

T (K)	x_1	y_1
95.5	0.0190	0.1700
89	0.0721	0.3891
86.7	0.0966	0.4375
85.3	0.1238	0.4704
84.1	0.1661	0.5089
82.7	0.2337	0.5445
82.3	0.2608	0.5580
81.5	0.3273	0.5826
80.7	0.3965	0.6122
79.8	0.5079	0.6564
79.7	0.5198	0.6599
79.3	0.5732	0.6841
78.74	0.6763	0.7385
78.41	0.7472	0.7315
78.15	0.8943	0.8943

P2.2 Using Aspen Properties® generate the phase equilibrium for the system 2-propanol (1) water (2) using the Wilson model and compare it with the following experimental information (Perry 1992).

Table P2.2 Vapor–Liquid equilibrium data for the 2-propanol (1) water (2) system at 45 °C

P (kPa)	x_1	y_1
15.252	0.0462	0.3936
17.412	0.0957	0.4818
18.505	0.01751	0.5211
19.132	0.2815	0.5455
19.838	0.4778	0.5981
20.078	0.6046	0.6411
19.985	0.7694	0.7242
19.585	0.8589	0.8026

P2.3 Based on the decision tree (Carlson 1996) shown in the chapter, select the models recommended for the following systems:

(a) Ethane–Pentane–Water

(b) Ethyl acetate–Ethanol–Water–Acetic acid
(c) Methyl chloride–Ethylene glycol
(d) Hydrogen–Oxygen–Water
(e) Polyethylene–Ethylene

P2.4 What disadvantages can be presented by Wilson model for a mixture of polar substances?

P2.5 What other choice you have to do calculations of pure substances as water, ammonia, and some refrigerants?

P2.6 What characteristics should have a group of components to be entered as a pseudocomponent?

P2.7 On Promigas SA web site it is possible to find information about the composition of some natural gases of Colombian wells. With the information reported in Table P2.3, construct the P–T phase diagram using Peng–Robinson equation as property package.

Table P2.3 Molar composition of natural gas from the well of Guepaje (Córdoba, Colombia)

Component	Composition (mol %)
Methane	96.4387
N_2	2.6901
CO_2	0.1293
Ethane	0.637
Propane	0.0007
i-Butane	0.0507
n-Butane	0.0005
i-Pentane	0.0157
n-Pentane	0.0156
n-Hexane	0.0217
Total	100

P2.8 Enter the component "6-aminohexanoic acid," also known as "6-aminocaproic acid" which is not found in the database using experimental data. Compare the boiling point calculated with the reported in the literature.

P2.9 Input the component Sucrose found in the database using experimental data, and compare the results with the data available in the simulator.

References

Abrams DS and Prausnitz JM (1975) Statistical Thermodynamics of Liquid Mixtures: A New Expression for the Excess Gibbs Energy of Partly or Completely Miscible Systems. AIChE Journal 21(1):116–128

Agarwal R, Li Y-K, Santollani O, Satyro MA (2001a) Uncovering the realities of simulation part I. Chem Eng Prog 97(5):42–52

Agarwal R, Li Y-K, Santollani O, Satyro MA (2001b) Uncovering the realities of simulation part II. Chem Eng Prog 97(6):64–72

Aspen Technology Inc (2006) Aspen process engineering webinar. Aspen HYSYS Property Packages

Aspen Technology Inc (2001) Aspen polymers plus 11.1. user's guide. Aspen Technology Inc, Cambridge

Carlson EC (1996) Don't Gamble with Physical properties for simulations. Chem Eng Prog 92:35–46

Dimian AC, Bildea C (2014) Integrated design and simulation of chemical processes. Comput Aided Chem Eng 35:201–251

Finlayson B (2006) Equations of state. In: Finlayson B (ed) Introduction to chemical engineering computing. Wiley-Interscience, New York, pp 7–29

Gomis V, Font A, Pedraza R, Saquete MD (2005) Isobaric vapor–liquid and vapor–liquid–liquid equilibrium data for the system water + ethanol + cyclohexane. Fluid Phase Equilib 235(1):7–10

Linstrom P and Mallard W (2010) NIST Chemistry WebBook, NIST Standard Reference Database Number 69

Meyer R. Metzger J (1966) C. R. Acad. Sci. Paris, Ser. C, 263:1333

Orbey H, Stanley S (1998) Modeling vapor-liquid equilibria: cubic equations of state and their mixing rules, 1st edn. Cambridge University Press, Cambridge

Perry RH (1992) Perry's chemical engineers' handbook. McGraw-Hill, New York

Renon H and Prausnitz JM (1968) Local composition in thermodynamic excessfunctions for liquid mixtures. AIChE Journal 14:135–44

Sada E, Morisue T, Miyahara K (1975) Salt effects on vapor-liquid equilibrium of tetrahydrofuran-water system. J Chem Eng Data 20(3):283–287

Satyro MA (2008) Thermodynamics and the simulation engineer. Chem Prod Process Model 3(1):1–41

Soulie MA, Goursot P, Peneloux A, Metzger J (1969) Proprietes Thermochimiques du Thiazole. J. Chim. Phys. Phys. Chim. Biol. 66:603–610

Soujanya J, Satyavathi B, Vittal Prasad TE (2010) Experimental (vapour + liquid) equilibrium data of (methanol + water), (water + glycerol) and (methanol + glycerol) systems at atmospheric and sub-atmospheric pressures. J Chem Thermodyn 42(5):621–624

Weast RC and Grasselli JG, ed(s). (1989) CRC Handbook of Data on Organic Compounds, 2nd Editon, CRC Press, Inc., Boca Raton, FL, 1

Chapter 3
Fluid Handling Equipment

3.1 Introduction

Many operations in chemical engineering involve fluids, either liquids or gases. It is therefore important to know the different options that allow transport equipment and conditioning fluid within a process. In any process plant there are pumps and piping networks gas compression systems. Fluid handling has a variety of applications, from water injection into an oil well to the steam distribution and other services in a chemical plant. Fundamental knowledge of fluid mechanics is important to understand the design of systems and equipment mentioned above, since they depend on the operation of the other processes, and similarly ensures that the flow and pressure are appropriate. It is also important to know the required power to carry out transport operations, since this information not only affects the operation from the technical point of view; it translates directly into associated costs within the operation. Process simulation is relevant for these systems. For this reason, this chapter will illustrate the various options that have simulators to calculate fluid handling equipment.

3.2 General Aspects

Before introducing the calculation tools with which simulators feature, you need to remember the principles that are involved behind the calculation itself. This ensures a proper understanding of the type of calculation performed by the simulator, and likewise introduces some criteria to evaluate the results and propose design alternatives.

© Springer International Publishing Switzerland 2016
I.D.G. Chaves et al., *Process Analysis and Simulation in Chemical Engineering*,
DOI 10.1007/978-3-319-14812-0_3

3.2.1   Background

The regime by which the fluids moves generally classified as follows: laminar flow, referring to a flow pattern without significant mixing of the particles, and turbulent flow, which is a random flow pattern and particle randomly mixed. To differentiate it is required to introduce the definition of Reynolds number:

$$Re = \frac{VL}{\vartheta} \tag{3.1}$$

where V and L correspond to the speed and length characteristics of the fluid, and θ to the kinematic viscosity of the fluid. For Reynolds numbers less than 2000 laminar flow regime is present; for higher Reynolds 10,000 turbulent flow, and between these two values is known as transition regime. Applying a fluid energy balance moving along a flow line leading to obtain an expression known as Bernoulli's equation:

$$\frac{V_1^2}{2} + \frac{p_1}{\rho} + gh_1 = \frac{V_2^2}{2} + \frac{p_2}{\rho} + gh_2 \tag{3.2}$$

where V_1 and V_2 are the velocities of the fluid at two specified points, p_i fluid pressure, fluid density ρ_i and fluid height h_i. This equation of Newton's second law should not be confused with the energy equation. Several simplifications are also included:

- Inviscid flow, i.e., without shear
- Continuous flow
- Constant density
- Balance along a streamline

Bernoulli's equation is widely used in studies of fluids. However, care must be taken because this equation only complies with the restrictions mentioned above. Also not to be confused with the energy equation, since the Bernoulli results comes from applying Newton's second law on a differential of fluid.

Aside from making a force balance on a particle of fluid, it is possible to perform an energy balance on the same particle. Omitting the mathematical development of this, the energy equation is:

$$H_P + \frac{V_1^2}{2g} + \frac{p_1}{\rho g} + z_1 = H_T + \frac{V_2^2}{2g} + \frac{p_2}{\rho g} + z_2 + h_L \tag{3.3}$$

where H_P and H_T correspond to the added energy and removal system, respectively, and h_L represents the energy losses in the system. The losses are given by accessories and by friction with the tube; thus simply brings flowing pressure losses, and if some accessories are added, these losses increase. Loss calculation due to flow is made based on the following equation:

$$h_L = f\frac{L}{D}\frac{V^2}{2g} \tag{3.4}$$

where L is the path length through the fluid, D is the pipe diameter, and f is the friction factor, which is obtained by Moody diagram or other experimental correlations. Moreover, in the case of pressure losses due to accessories the equation has the following form:

$$h_L = K\frac{V^2}{2g} \tag{3.5}$$

where K corresponds to an empirical factor depending on the type of accessory used. All these correlations can be easily found in any text on fluid mechanics. Meanwhile, the simulators are based on the differential expression of the energy balance to calculate losses through a pipe.

$$\frac{dp}{dL} = \rho_m g \sin\theta + \left(\frac{dp}{dL}\right)_f + \rho_m v\frac{dv}{dL} \tag{3.6}$$

The first term on the right side of Eqn. (3.6) corresponds to the pressure gradient which is as a result of gravity, where ρ_m is the fluid density and the inclination angle θ of the pipe. The second term corresponds to the losses caused by fluid friction. The last term represents the angular acceleration component on the pressure drop, and is proportional to the change in fluid velocity v. Depending on the situation, either of these terms can be more representative when calculating the pressure drops through piping (Potter and Wiggert 2006). To perform this calculation, a very accurate model should be used for predicting the fluid density, as this is involved in the calculation of all terms of the equation. If there is more than one phase, the simulator calculates the density from the fluid holdup in the pipe, using the same correlation which depends on the operating conditions. Similarly to the holdup calculation, the simulator uses different correlations to determine the quantity of the friction term. In this case there is no correlation that successfully meets all conditions, since many of these have been determined for very specific situations. The following section illustrates some of the most important correlations available in process simulators.

3.2.2   Piping

Different correlations and calculation models, available in process simulators, are used to simulate piping systems. However, not all of these correlations allow the user to obtain the same results, and will not work for all systems. Although principles governing the mechanics and fluid dynamics are the same, different correlations have been developed in order to obtain more accurate results when modeling specific systems. The methods reported in Table 3.1 have been developed to calculate the pressure drop for two-phase flow. Among the specific conditions under which it is recommended to use them, some of these models that have been developed exclusively for horizontal flow or vertical flow or both. Also some of these methods are calculated based on the flow rate, so that the stream flow regime is determined, and according to this result, using a flow map, the appropiate model for calculation is selected. Similarly, some models allow calculation of fluid holdup for two phases and others do it to make a homogeneous mixture.

These correlations have been developed for two-phase flow. When only one-phase flow is present, the Darcy equation is used to predict the line-pressure drop. It is also possible to know the heat transfer by estimations that use information about the diameter and material of the pipe along with internal and external conditions. Some of the models (Aspen Technology Inc. 2005b Operations Guide) in Table 3.1 are discussed in more depth in the following sections. How they were developed and the most important considerations to keep in mind when using them are also shown.

Table 3.1 Models used for the calculation of pipes (Aspen Technology Inc. 2005b, Operations Guide)

Model	Horizontal flow	Vertical flow	Liquid holdup	Flow map
Aziz, Govier, and Fogarasi	No	Yes	Yes	Yes
Baxendell and Thomas	Carefully	Yes	No	No
Beggs and Brill	Yes	Yes	Yes	Yes
Duns and Ros	No	Yes	Yes	Yes
Gregory, Aziz, Mandhane	Yes	No	Yes	Yes
Hagedorn and Brown	No	Yes	Yes	No
Htfs homogeneous	Yes	Yes	No	No
Htfs liquid slip	Yes	Yes	Yes	No
Olgas 2000	Yes	Yes	Yes	Yes
Orkisewski	No	Yes	Yes	Yes
Poettman and Carpenter	No	Yes	No	No
Tacite hydrodynamic module	Yes	Yes	Yes	Yes
Tulsa	No	Yes	Yes	Yes

3.2.2.1 Aziz, Govier and Fogarasi (Aziz and Govier 1972)

This model considers that the flow rate is independent of the phase viscosity and pipe diameter. However, it is proportional to the cube root of the gas density. From this statement, the superficial gas and liquid velocities are modified according to a map of flow regimes. Once the system is determined, the proper correlation is selected to calculate the pressure gradient and other parameters as the flow rate.

3.2.2.2 Baxendell and Thomas (Baxendell 1961)

The Baxendell and Thomas model is the result of an extension of the Poettman and Carpenter model, which is intended to include much larger flows. It is based on a homogeneous pattern in which the friction factor for the two phases is calculated using a pilot correlation. It is appropriate to use this model in situations of horizontal flow; this is in addition to the vertical flows considered by Poettman and Carpenter model. Likewise, this correlation does not take into account the different flow regimes which can occur in a given time, and it is assumed that the pressure gradient is independent of viscosity.

3.2.2.3 Beggs and Brill (1973)

It is the method used by default in modules in Aspen HYSYS® pipes, due to the versatility and the calculation restrictions, unlike several of the other models which do not correspond to systems with very specific conditions.

The method is based on the work developed for air–water mixture under different conditions; this module can be applied to inclined flow conditions.

When using Beggs and Brill correlation, the flow rate is determined using the Froude number and content of incoming liquid. The flow map (Fig. 3.1) is based on horizontal flow systems; regimens depend on inlet liquid fraction, i.e., the stream quality, and accordingly define four regimens, namely segregated, intermittently distributed and transition. Each of these systems has its own characteristics, as shown below:

- Segregated flow: corresponds to a stratified and annular flow.
- Intermittent flow: rich in viscous liquid.
- Flow transition presents an intermediate behavior between the segregated flow and intermittent flow.
- Distributed flow: bubbles and mist.

Once the flow rate is determined, it proceeds to calculate the fluid holdup to a horizontal pipe through appropriate correlation. A correction factor is applied if there is any inclination. Then, with the predicted holdup friction factor for the two phases and finally, the pressure gradient is calculated.

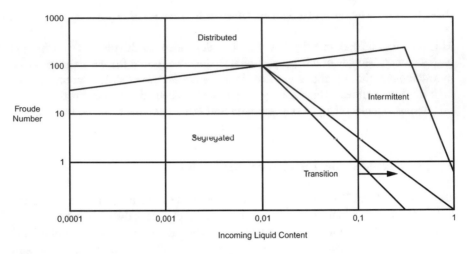

Fig. 3.1 Flow map for Beggs and Brill method

3.2.2.4 Duns and Ros (1963)

Duns and Ros model is based on a laboratory research about the vertical flow of air, oil, and water systems. The model uses three flow sections, namely:

- Region I: liquid phase is continuous.
- Region II: liquid and gas phases alternate.
- Region III: gas phase is continuous.

According to the region in which the flow is found, the proper correlation is used. The regions are distinguished according to two dimensionless groups, one corresponding to the gas velocity and other to the liquid content. All correlations for this model are defined in terms of these dimensionless groups.

3.2.2.5 Gregory Aziz Mandhane Pressure Gradient (Gregory and Mandhane 1975)

An elaborated model to predict the pressure drop for two-phase flow is used. It makes use of a flow map to set the flow rate of a stream. After knowing the flow rate, this method uses the adequate correlations to estimate parameters of the pipe.

3.2.2.6 Hagedorn and Brown (Hagedorn 1965)

The systems data on which the construction of this model was made for upward flow of air or air–water mixtures or crude. Pressure drop is calculated using a derivative factor Moody chart, using a Reynolds number for two phases, which in case of single phase flow should be reduced to the Reynolds number for this phase. This model developed a curve in which the void fraction is related to the same dimensional parameters for Duns and Ros model.

3.2.2.7 HTFS Models (Aspen Technology Inc. 2009)

The two HTFS models share calculation method for calculating the pressure gradient. However they differ in the method for calculating the gradient of static pressure. For the homogeneous model, the void fraction is assumed to be a homogeneous fraction, while for the other model for the fraction is calculated. These modules have been validated for horizontal and vertical flow in both directions.

3.2.2.8 OLGAS 2000 (Aspen Technology Inc. 2005b, Operation Guide)

This model allows flow calculation in two or three phases. It is one of the most powerful models at the time of simulating transport operations in the oil and gas industry models. For calculations, OLGAS 2000 has models for every four schemes, namely stratified annular and dispersed flow like mud. The multiphase flow is a complex process between the different phases involved considering fluid properties, geometry, and reservoir conditions, the well, pipeline and processing plant. To model this, OLGAS model has the ability to handle up to three-phase flow, and different substances such as oil, gas, and sand, among others. OLGAS predicts the pressure gradient, fluid holdup, and flow rate. It has been designed from trials involving horizontal flow to vertical and inclined flows.

3.2.2.9 Orkisewski (1967)

Orkisewski correlation is a relationship for vertical upward flow based on various methods. In these four flow regions, different correlations are used according to the rules defined. Among the correlations used by Orkisewski, the Duns and Ros correlation is included.

3.2.2.10 Poettman and Carpenter (Poettmann 1952)

This model assumes that the contribution of the acceleration term in the overall pressure drop is negligible and the pressure drop can be calculated using a homogeneous model. Also assumes that the pressure drop can be estimated using a homogeneous two-phase density. In the model, a correction factor for calculating friction factor based on experimental data for gas ascending wells under varying conditions is proposed. This model assumes that the pressure gradient is independent of viscosity.

3.2.2.11   Tacite Hydrodynamic Mode (Aspen Technology Inc 2005b, Operation Guide)

This module is mainly used to simulate multiphase flow, and for the design and control of oil and gas pipelines. The module provides two options for the calculation: one for gas–liquid and another based on the correlation of Zuber-Findlay. In this module three regions of flow are identified:

- Stratified: the model assumes a time balance between the phases present in the pipe section.
- Flashing this type of flow is solved as a problem of two regions. The gas phase is considered as stratified flow, and the flow of liquid is dispersed.
- Sparse: This is a special case of the previous region.

3.2.2.12   Tulsa (Aspen Technology Inc 2005b, Operation Guide)

This model proposes a mechanical approach to predict the flow, pressure drop, and fluid holdup in an upward vertical flow for two phases. There five regions are included, each with its specific correlations, mostly based on existing models such as Aziz and Hagedorn method. Tulsa model has been tested in more than two thousand wells and, in general, it is recognized to get very accurate results.

3.2.3   Pumps

From the energy balance equation, there are two terms corresponding to the added energy and system withdraws. This energy, expressed in the equation as height, belonging to a given by a pump or a turbine work taken. This makes it possible to know the work used in both cases by the following expressions:

For turbines:

$$\dot{W}_T = \dot{m}gH_T\eta_T \tag{3.7}$$

For pumps:

$$\dot{W}_P = \frac{\dot{m}gH_P}{\eta} \tag{3.8}$$

where W_P corresponds to the work and η to the efficiency. Similarly, there is an additional parameter for pump calculation known as the net positive suction head (NPSH). This value is related to the difference between the pressure at a given point and the vapor pressure of the liquid, and is of vital importance because if the fluid pressure at a point is found to be less than the vapor pressure, a cavitation phenomenon occurs. NPSH is defined in two different ways. NPSH required is related to the minimum value to avoid cavitation given by the manufacturer and the NPSH available, which depends on the liquid and the characteristics of the system in which it will operate (Kenneth 1989). These parameters can be calculated with the following equations:

For NPSH required:

$$NPSH_r = H_z + \frac{V^2}{2g} \tag{3.9}$$

where H_z is the minimum at the rotor inlet pressure and v the velocity.

For NPSH available:

$$NPSH_d = \frac{P_a}{\gamma} - H_a - P_{ds} - \frac{P_v}{\gamma} \tag{3.10}$$

Where P_a is the suction pressure, the suction height H_a, P_{ds} pressure drop in the suction line and P_v the vapor pressure at pumping temperature. To avoid cavitation, the NPSH available must be greater than required. In process design, NPSH available is calculated and with a factor of 2 ft (according to design practice), the NPSH required is specified.

3.2.4 Compressors and Expanders

Compressors are equipment used to increase the pressure of a gas stream, normally a large flow rate at low pressure for centrifugal compressors. According to the information entered in the simulator, compressor module can calculate the properties of any stream or compressor efficiency. These operations are also used to simulate some kind of pumps, for situations in which the conditions are close to the critical point where the incompressible fluid becomes compressible. This is because the algorithm for calculating these modules takes into account the compressibility of the liquid and allows a more rigorous calculation. Moreover, there are models for expanders. These work contrary to the compressors, since they take a stream of gas at high pressure and get a stream of low-pressure gas and high speed as output (Greene 1992). The expansion process involves a change in the internal energy of the gas kinetic energy, energy that can be converted into work. Several methods to calculate these units have been developed, based on information available to calculate, such as current condition and the possibility of a characteristic curve. Usually the solution of these units is given in terms of flow, the pressure change, the used power and efficiency. In Table 3.2 the most common strategies for the calculation of these units are shown.

Both units are governed by the same thermodynamic principles. The only difference is the flow direction of the energy stream: compression requires energy,

Table 3.2 Compressors and expanders calculation strategies (Aspen Technology Inc. 2005a; Aspen Technology Inc. 2005b)

Curves are available	Curves are NOT available
• Flow rate and inlet pressure are known	• Flow rate and inlet pressure are known
• Specify outlet pressure	• Operating speed is specified
• Specify efficiency (adiabatic or polytropic)	• Simulator calculates efficiency and head
• Required energy, outlet temperature, and efficiency are calculated	• Outlet pressure, temperature, and applied duty are calculated
• Flow rate and inlet pressure are known	• Flow rate, inlet pressure, and efficiency are known
• Efficiency and required energy are specified	• Simulator interpolates curves to determine operating speed and head
• Outlet pressure, temperature, and efficiency are calculated	• Simulator calculates outlet pressure, temperature, and applied duty

and expansion releases energy. For a compressor, the isentropic efficiency is defined as the ratio between the energy required to perform the isentropic process (ideal) with respect to the used energy.

$$\text{Efficiency}(\%) = \frac{\text{Required energy}_{\text{Isoentropic}}}{\text{Required energy}_{\text{Real}}} \times 100\% \qquad (3.11)$$

In the case of the expander, efficiency is the ratio between the energy actually produced compared to the energy that would be released if it were an isentropic process.

$$\text{Efficiency}(\%) = \frac{\text{Release energy}_{\text{Real}}}{\text{Used energy}_{\text{Isoentropic}}} \times 100\% \qquad (3.12)$$

When performing adiabatic calculation, simulators used strictly isentropic line from inlet to outlet pressure, wherein an enthalpy value is obtained. With this value and the known efficiency, the simulator calculates the corresponding outlet enthalpy.

3.3 Modules Available in Aspen Plus®

Aspen Plus® has different modules for these fluid handling operations; all of them are in the *Pressure Changers* tab. Different modules are summarized in Table 3.3.

Table 3.3 Fluid handling modules available in Aspen Plus®

icon	Name	Description
Pump	Pump	Modifies the pressure of a liquid stream and provides results on the required power. Usually it is used as pump or hydraulic turbine
Compr	Compr	Make changes in pressure of a gas stream and provides results on the required power. It is generally used as a compressor or turbine
MCompr	MCompr	Allows changes of a gas flow pressure in several steps with intermediate cooling and the possibility of removing the condensed liquids. It is used to simulate multistage compressors
Valve	Valve	Models the pressure drop through a valve. It can be used to rating valves within the vendor internal database
Pipe	Pipe	Modeling the pressure drop through a pipeline. This device corresponds to a single diameter tube with possible accessories
Pipeline	Pipeline	Modeling the pressure drop through a pipe. This module allows you to use pipes of different diameters and elevations among with a range of accessories

3.4 Modules Available in Aspen HYSYS®

Aspen HYSYS® fluid handling modules are just below the modules for heat transfer equipment in the *Object Palette*. These have different options compared to Aspen Plus® modules. It is worth noting that Aspen HYSYS® does not have multistage compressors, but have a specific module for gas pipes (Table 3.4).

Table 3.4   Fluid handling modules available in Aspen HYSYS®

Icon	Name	Description
	Pump	Increases the pressure of a liquid stream. According to the input data, one can calculate output pressure, temperature, or efficiency
	Expander	It is used to reduce the pressure of a gas stream to produce work. It allows obtaining results on the work accomplished by this operation or the pressure drop requires obtaining a certain amount of work
	Compressor	Compressor makes it possible to calculate and/or provides results on the output pressure, efficiency, and work. You can calculate two types of compressors by modifying the method of solution and the respective input parameters
	Gas pipe	Calculates pipes for compressible flows. It is primarily used for transient vapor calculations
	Pipe segment	It is used to calculate pipe with one or more stages. Allows input different pipe segments and, as input data, automatically selects the conditions in which it must operate
	Valve	Develop a material and energy balance on the valve. Assume that the valve has an isenthalpic operation. Allows valve sizing
	Relief valve	Makes it possible to model relief valves; is primarily used in the dynamic environment

3.5   Gas Handling Introductory Example

3.5.1   Problem Description

Currently, natural gas is one of the most used energy sources worldwide, due to its advantages over other energy sources. Natural gas transport to the processing plant is vital because without this the plant could not process the obtained gas to produce end products. Therefore, it should ensure sufficient supplies for processing and analyzing, by simulation, the performance of these systems. This simulation is intended to illustrate a transmission natural gas besides piping modules and some of the different options that exist in them. The simulation system comprises a network of pipes distributed from wellheads to the processing plant including scrubbers to remove condensates that will be formed considering pressure and temperature changes in transport process.

In Table 3.5 gas composition data to be used is reported. In Fig. 3.2 the piping network is shown with elevations reported in meters.

The distances and configurations are reported in Table 3.6.

The operating conditions for every well are reported in Table 3.7. All lines have no insulation. In the junction points, scrubbers are installed to remove liquids produced in piping network.

In the outlet well lines, a separator is installed to remove liquids and free water.

Table 3.5 Natural gas composition analysis

Component	Weight %
Carbon dioxide	11.31
Nitrogen	0.59
Methane	60.83
Ethane	6.59
Propane	5.23
i-Butane	2.46
n-Butane	3.53
i-Pentane	1.36
n-Pentane	2.06
n-Hexane	1.25
n-Heptane	2.76
n-Octane	0.56
n-Nonane	0.65
Decane	0.25
Undecanes+	0.57
Total	100.00

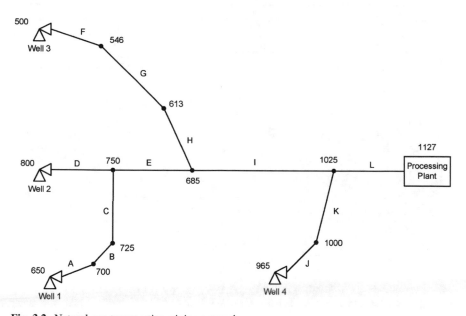

Fig. 3.2 Natural gas transporting piping network

3.5.2   *Simulation in Aspen HYSYS*®

The simulation objective is to illustrate the piping transporting network mentioned above. The next figure shows the simulation flow diagram to be carried out (Fig. 3.3).

Table 3.6 Piping network configuration

Segment	Diameter (in.)	Schedule	Length (m)
A	6	80	2100
B	6	80	1015
C	6	80	550
D	6	80	800
E	8	80	1536
F	6	80	412
G	6	80	698
H	6	80	1526
I	10	80	1752
J	8	80	125
K	8	80	456
L	10	80	2582

Table 3.7 Well operating conditions

Well	Pressure (psig)	Temperature (°F)	Std gas flow (MMSCFD)
1.	900	90	25
2.	850	90	35
3.	700	90	20
4.	600	90	30
Total			120

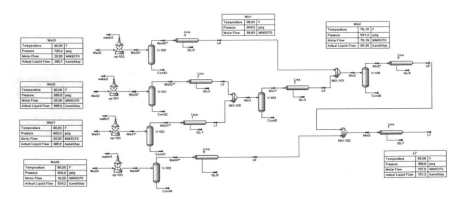

Fig. 3.3 Simulation flow diagram to enter Aspen HYSYS®

For this simulation, the thermodynamic package to be used is *Peng Robinson* and the gas composition is shown in Table 3.5. After defining this data, the simulation environment can be accessed. There, can be defined the process streams corresponding to the four wells shown in Fig. 3.2.

For Wells 3 and 4, the pressure is not known. However, these pressures will be calculated further taking into account that the compositions are equal for all production wells. To saturate the gas with water, it is recommended to use the

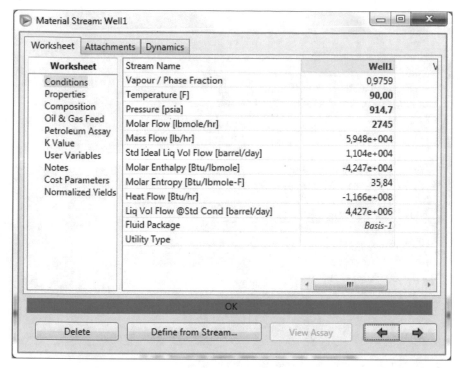

Fig. 3.4 Stream *Well1* specification window in Aspen HYSYS®

Saturator module available in Aspen HYSYS®. This module is not shown in the *Object Palette*, this is why it is necessary to press *F12* button to access the *UnitOps* dialog box which display all modules available in Aspen HYSYS®. In this window on the left side are displayed the different filters in order to search more easily modules. Select in the right side the *Saturate with water* module (Fig. 3.4).

In the module screen, the input stream will correspond to *Well1*, its output stream is *Well1** and water flow, *water1*. Use this scheme for remaining well streams. The module shows a warning message because there wet bulb temperature for a two-phase mixture cannot be calculated, but it is not necessary this calculation for current simulation. Then, it is necessary to saturate the wells streams, *Well2*, *Well3*, and *Well4* with water using the *Saturate with water* module (Fig. 3.5).

In Fig. 3.6 it is shown the specification tab, select as feed stream the *Well1* stream. Create as water stream water1, and as product stream *Well1**. This module calculates the water requirement to saturate the inlet gas at the given temperature and pressure; for this case, it represents the actual outlet condition for every well. The module shows a warning because two phases were found and this module cannot calculate a wet bulb temperature for this system. However, this calculation does not affect the purpose of this simulation.

After installed the *Saturate with water* module for *Well1* stream, please install a *Saturate with water* module for remaining well streams as shown in Fig. 3.7.

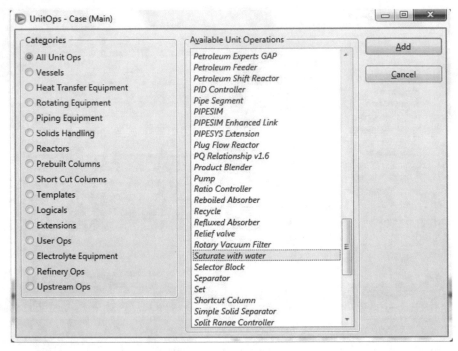

Fig. 3.5 UnitOps selection window in Aspen HYSYS®

Fig. 3.6 Saturate with
water specification window
in Aspen HYSYS®

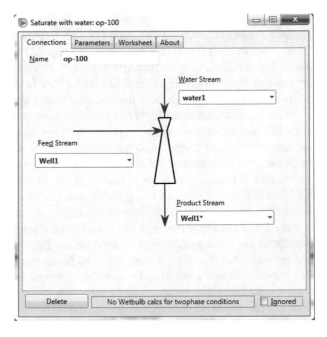

Fig. 3.7 Partial flow
diagram for piping network
in Aspen HYSYS®

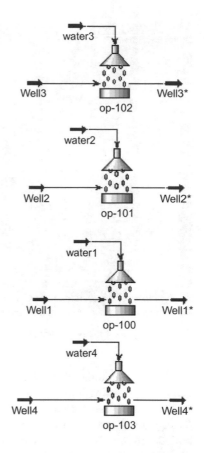

These streams need a separator to remove liquids and free water. In the *Object Palette*, select the *Separator* module. Install it in the flow sheet and select as inlet the *Well1** stream. The vapor outlet is *Well1*** and the liquid outlet *Cond1*. No extra specifications are required because this separator operates adiabatically (Fig. 3.8). Please install a separator for remaining well streams as shown in Fig. 3.9.

The next step is specifying the first piping segments. From the *Object Palette*, select the *Pipe Segment* module and drag it to the flow sheet window. Once the module is selected, name it as *Line 1*. The inlet stream will be *Well1***, the outlet stream *L1* and the energy stream, which represents the energy loss due to pressure drop and heat transfer, *QL-1* (Fig. 3.10).

In the *Rating* tab, the geometry from Table 3.7 and Fig. 3.2 will be entered. Press the button *Append Segment* to add a new segment in this module. In this window, outer and inner diameter can be selected, for this case Schedule 80 pipe is selected with an outer diameter of 6.625 in and an inner diameter of 5.761 in (Fig. 3.11).

The next step is entering the length and the elevation change of each segment. The lengths are reported in Table 3.7. For elevation changes, see Fig. 3.2. In Line 1 module enter the segments A, B, and C from Fig. 3.2 as it is shown in Fig. 3.12. In

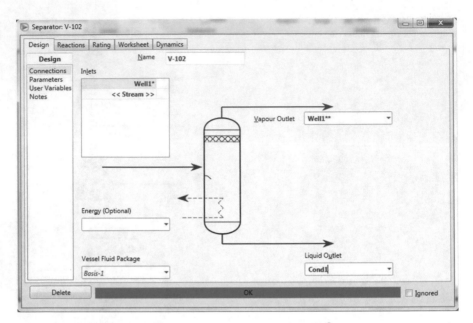

Fig. 3.8 Separator module specification window in Aspen HYSYS®

the Rating tab, in the option Heat Transfer, found in the right section, the parameters for evaluating heat exchange with the ambient must be entered (Fig. 3.13).

Then, heat transfer parameters must be specified in Heat Transfer option in *Rating* tab. According to problem specification, piping has no insulation and is buried 1 m into the ground. Also ambient temperature must be specified. To insert each condition the next options are available:

- *Heat Loss*: Permits to enter a heat flow representing the heat loss across the piping configuration.
- *Overall HTC*: Permits to enter overall heat transfer coefficients to calculate the heat loss considering the inlet and ambient temperatures.
- *Segment HTC*: Permits to define different heat transfer parameters for each segment added.
- *Estimate HTC*: Permits to enter information to calculate heat transfer coefficients and heat loss flow.

Select Estimate HTC option and activate the options: *Include Pipe Wall*, *Include Inner HTC*, and *Include Outer HTC*. For this purpose, an ambient temperature of 85 °F is selected and entered in the corresponding box. The remaining option, *Include Insulation*, is not required considering pipe network does not have insulation. With this information, the module can calculate the pressure drop and the gas outlet temperature. Please insert a *Pipe segment* module for segment D. Viewing the liquid flow present in gas stream is important to evaluate the presence of condensates. It is necessary to visualize some important process variables to take

Fig. 3.9 Partial flow
diagram for piping network
in Aspen HYSYS®

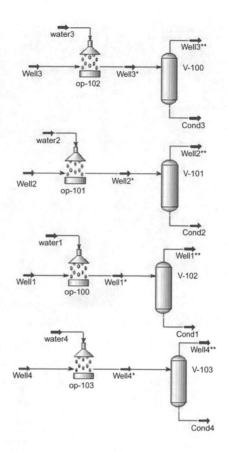

decisions about the design. Aspen HYSYS® has tables to display these variables for
a particular stream or the entire process. In order to view variables for a stream it is
necessary to right-click over the stream *Well1* and select the option *Show Table*. By
default, it is shown Pressure, Temperature, and Flow but if it is necessary, other
variables can be included as follows. Click in the *Add Variable* button and select the
variable *Actual Liquid Flow*. Thus, it shows the liquid flow from Well1 stream
(Fig. 3.14).

Repeat this procedure with *Well2*, *Well3*, and *Well4* streams. The resulting flow
sheet is shown in Fig. 3.15.

The other possible way to see important variables as flow, temperature, and
pressure for the entire flow sheet is using the following commands:

- *Shift + F*: Displays molar flow where it is reported the name of each stream.
- *Shift + T*: Displays temperature where it is reported the name of each stream.
- *Shift + P*: Displays pressure where it is reported the name of each stream.
- *Shift + N*: Displays the name of each stream again.

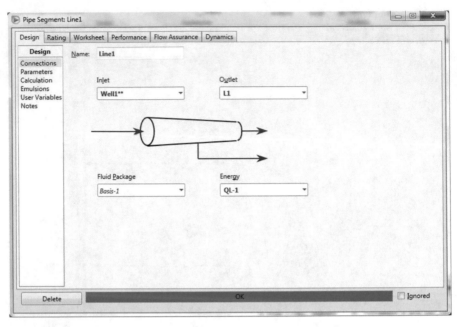

Fig. 3.10 Pipe segment module from Object Palette in Aspen HYSYS®

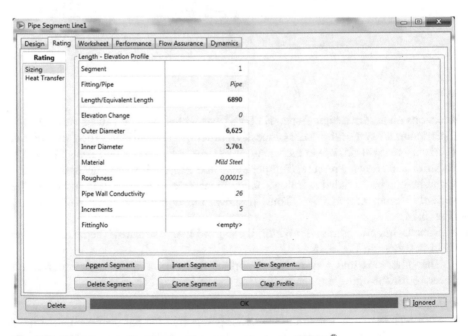

Fig. 3.11 Diameter specification for pipe segment in Aspen HYSYS®

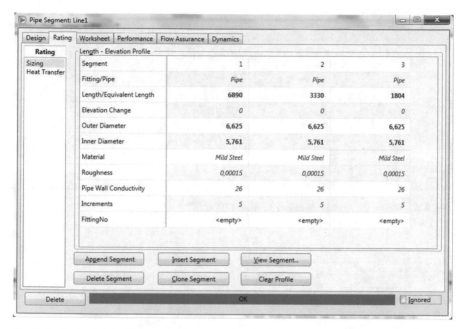

Fig. 3.12 Rating tab in the pipe segment module from Object Palette in Aspen HYSYS®

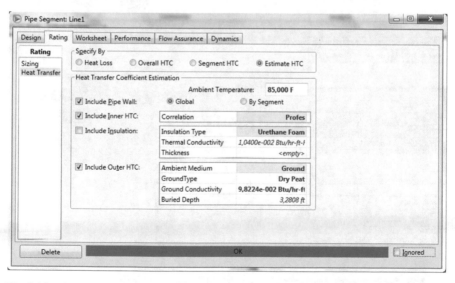

Fig. 3.13 *Heat transfer* information in the *Pipe segment* module from *Object Palette* in Aspen HYSYS®

In Fig. 3.16 is shown an example of these commands showing pressure of each stream.

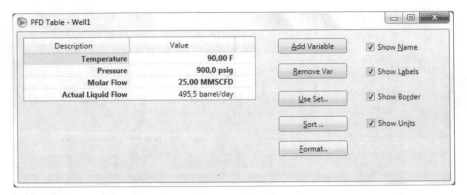

Fig. 3.14 Well 1 stream table window in Aspen HYSYS®

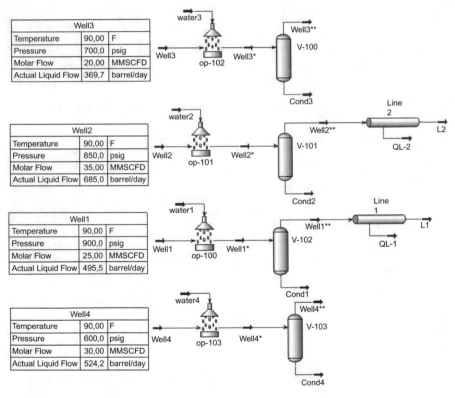

Fig. 3.15 Partial flow diagram for piping network in Aspen HYSYS®

Also there is a visual way to see variables in the simulation flow sheet. In the top right of the Flowsheet/Modify tab there is an icon and a selecting list which correspond to the color scheme box. Click the button, and a window will display like Fig. 3.17.

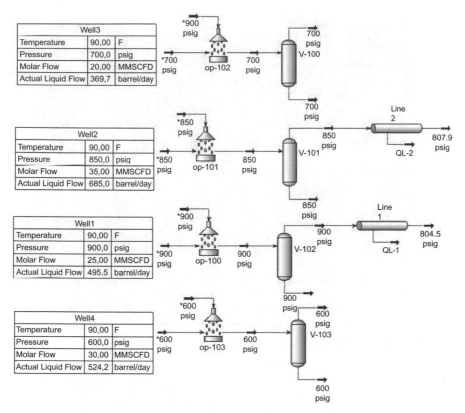

Fig. 3.16 Partial flow diagram for piping network in Aspen HYSYS® showing pressures

In this window, A process variable can be selected and a color scale to use in the simulation flow sheet. For this purpose, in the *Scheme Name* field select *Add New*. In this new window select the variable *Actual Liquid Flow* and click in the *OK* button (Fig. 3.18).

In the next window, it can be defined the color scale that will be used. Considering the liquid contents shown in Fig. 3.16, enter the information reported in Fig. 3.19 and close the window. The flow sheet will look like the Fig. 3.20.

To restore the color scheme to default, select *Default color scheme* option in the selecting list.

The next step in the example is to mix the *L1* and *L2* streams and adding an additional separator. Name the mixture stream as *Mix1*. Create a table for *Mix1** stream including *Actual Liquid Flow*. The resulting flow sheet is shown in Figs. 3.21 and 3.22.

Then, insert *Line 3* and *Line 4* modules with segments E, F, G, and H from Fig. 3.2.

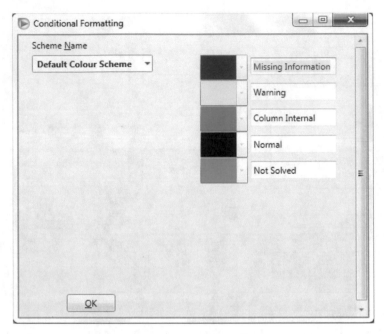

Fig. 3.17 Color scheme dialog window in Aspen HYSYS®

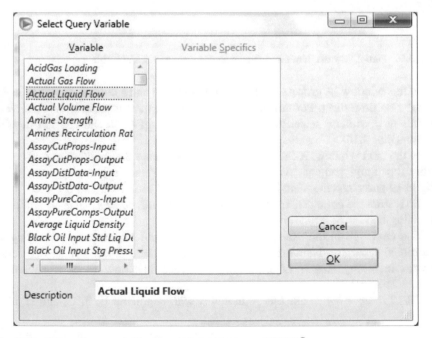

Fig. 3.18 Color scheme variable select window in Aspen HYSYS®

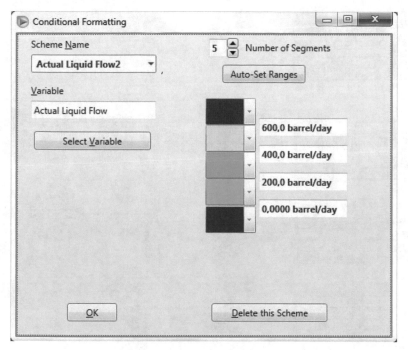

Fig. 3.19   Color scheme scale window in Aspen HYSYS®

The next step is to mix the *L3* and *L4* streams and adding an additional separator. Name the mixture stream as *Mix2*. Create a table for *Mix2** stream including *Actual Liquid Flow*. The resulting flow sheet is shown in Fig. 3.23.

Then, insert *Line 5* and *Line 6* modules with segments I, J, and K from Fig. 3.2.

The next step in the example is to mix the *L5* and *L6* streams and adding an additional separator. Name the mixture stream as *Mix3*. Create a table for *Mix3** stream including *Actual Liquid Flow*. Finally, install the *Line 7* with segment L from Fig. 3.2.

3.5.3   Results Analysis

In this example, six separators are installed to remove liquids form due to pressure drop across the lines. In Table 3.8 these flows are reported.

The saturate with water module is a powerful tool to saturate a gas and perform calculations when liquid condensates are probably formed.

According to pressure drops and flows reported in Table 3.9, it is shown that increasing flow leads to an increase in pressure drop. However, Line6 has a lower pressure drop compared with Line4. This can be explained with the information reported in Table 3.6 where segments J and K sums 581 m (corresponding to the

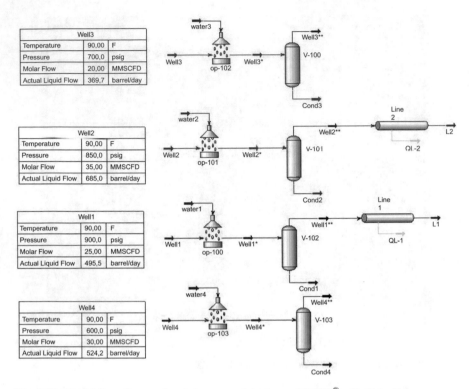

Fig. 3.20 Partial flow diagram for piping network in Aspen HYSYS® with Color Scheme

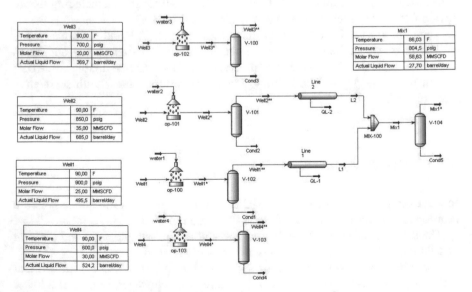

Fig. 3.21 Partial flow diagram for piping network in Aspen HYSYS®

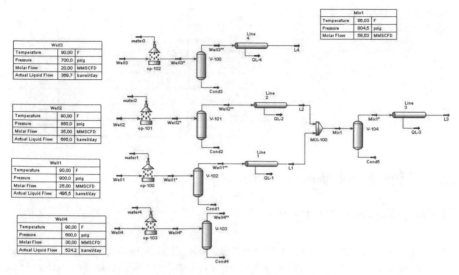

Fig. 3.22 Partial flow diagram for piping network in Aspen HYSYS®

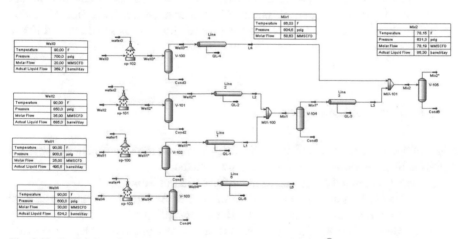

Fig. 3.23 Partial flow diagram for piping network in Aspen HYSYS®

Table 3.8 Results for piping network

Stream	STD flow (bpd)
Cond1	486.7
Cond2	674.1
Cond3	365.5
Cond4	519.3

length for Line6 module) meanwhile segments F, G, and H sums 2636 m. This proves that length has an increasing effect in pressure drop, so both criteria (flow and length) must be taken into account when hydraulic calculations are made.

Table 3.9 Pressure drops for pipe modules

Pipe	STD flow (mmscfd)	Δp (psig)
Line1	24.4	95.5
Line2	34.2	41.7
Line3	58.6	54.6
Line4	19.6	68.7
Line5	78.0	66.4
Line6	29.4	11.0
Line7	107.5	155.1

3.6   Liquid Handling Introductory Example

3.6.1   Problem Description

For any industry is necessary to consider the distribution of services. In the following case an existing piping system to distribute water is evaluated. It is possible to also evaluate the performance of the pumping facilities. Basically, the system is composed by a piping arrangement used to distribute water in different points. All water is pumped by pump *P1* as it can be noted in Fig. 3.24. The technical details and specifications of the process flow sheet are discussed in the next section.

3.6.2   Process Simulation

The only substance involved is water; then, some of the steam tables properties package that comprise Aspen Plus® can be selected.

Stream 1 is defined according to the information in Table 3.10.

The pipe sections are then installed. Some of the sections are installed as a *Pipe* module and another as *Pipeline* module. Before installing the modules is necessary to clarify that when installing each segment, regardless of the module, in the *Flash Options* tab, must be specified that the fluid to pass through the tubes has only one phase, which in this case corresponds to *Liquid Only*. This avoids warnings when running the simulation. The first of the sections belonging to section 1 is installed as a *Pipeline* module. In this module it is possible to enter the pipe geometry as a system of coordinates or nodes and segments with known length and angle. For this case select the *Enter node coordinates* in the *Segment geometry* option (Fig. 3.25).

In *Settings* tab you can also specify alternative calculation. For example, define the calculation sequence (forth or backwards), i.e., if output conditions are calculated with given inlet conditions. Also the model in which the simulator estimates the properties through the pipeline, the heat transfer options, among others. After selecting the option to enter the geometry as a node system, in the *Connectivity* tab

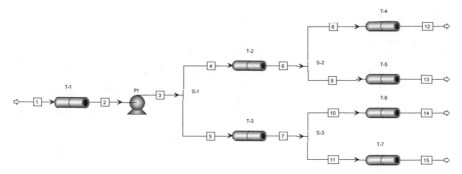

Fig. 3.24 Simulation flow diagram for water distribution system

Table 3.10 Stream
1 specifications for water
distribution system

Cell	Value
Name	1
Flow	0.0072 m^3/s
Temperature	25 °C
Pressure	450 mmH$_2$O
Composition	100 % water

Main Flowsheet ×   T-1 (Pipeline) ×   +

✓ Configuration   ✓ Connectivity   Methods   Property Grid   ✓ Flash Options   Solids Conveying   Information

◉ Fluid flow   ○ Solids conveying

Calculation direction
◉ Calculate outlet pressure
○ Calculate inlet pressure

Segment geometry
◉ Enter node coordinates
○ Enter segment length and angle

Thermal options
○ Do energy balance with
   surroundings
◉ Specify temperature
   profile:

   Constant temperature   ▾

Property calculations
◉ Do flash at each step
○ Interpolate from property grid

Pipeline flow basis
◉ Use inlet stream flow
○ Reference outlet stream flow

Inlet conditions
Pressure: bar   ▾
Temperature: C   ▾

Fig. 3.25 Pipeline model configuration window in Aspen Plus®

can be specified the node coordinates and the line diameters. The specifications are
reported in Table 3.11.

Having specified, section 1 can enter the pump *P1*. This is calculated by a
performance curve. To specify that the calculation method includes a performance
curve should select the appropriate option, as shown in Fig. 3.26. By selecting the

Table 3.11 Segment specifications for first line of piping system

Name	T-1
Inlet stream	1
Outlet stream	2
Node coordinates 1	0;0;2.71
Node coordinates 2	1;0;2.71
Node coordinates 3	1;0.29;2.71
Diameter	2.5 in. (for both lines)

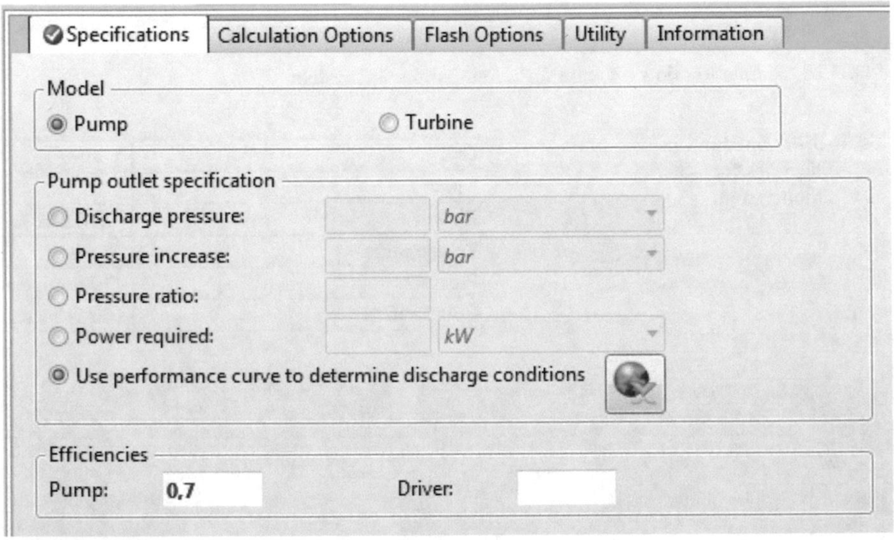

Fig. 3.26 Specifications tab for pump *P1* system

option *Performance Curves*, the possibility of entering the performance curves in the navigation tree is activated. To enter it, select curve format as *Polynomials* considering the data is in a polynomial way, where their performance is variable head (*Head*) and flow variable corresponding to volumetric flow (*Vol Flow*). These specifications are shown in Fig. 3.27.

By having the information of the performance curve, you can enter the same values in the *Curve Data* tab. The curve corresponding to the following polynomial:

$$\text{Head}(m) = 63.22 + 1134 * \text{Flow}\left(\frac{m^3}{s}\right) - 2.142 \times 10^5 \left(\text{Flow}\left(\frac{m^3}{s}\right)\right)^2 \quad (3.13)$$

When values are specified, care should be taken when selecting the units, as this can lead to miscalculations in the simulation.

The outlet stream of the pump passes through a flow splitter which information is reported in Table 3.12.

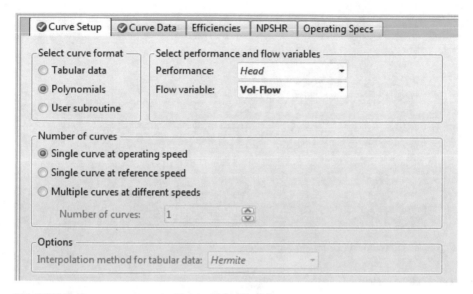

Fig. 3.27 Performance curve specification for pump *P1*

Table 3.12 Splitter S-1 specifications

Splitter	S-1
Inlet stream	3
Outlet stream	4
	5
Flow fraction	0.8050
For stream	4

Table 3.13 Segment specifications for second line of piping system

Name	T-2
Inlet stream	4
Outlet stream	6
Node coordinates 1	0;0;2.71
Node coordinates 2	2;0;2.71
Node coordinates 3	2;0.75;2.71
Diameter	2.5 in. (for both lines)

For section 2, a Pipeline module is also used analogously to section 1 with the parameters presented in Table 3.13.

To specify section 3, a *Pipe* module will be used. This module specifies only one pipe, and it also allows changing calculation alternatives. For this simulation these are not modified. Figure 3.28 shows *Pipe* module interface and the parameters specified are included in Table 3.14.

Now you can specify the two remaining flow splitters. Table 3.15 shows the specifications of the flow splitters.

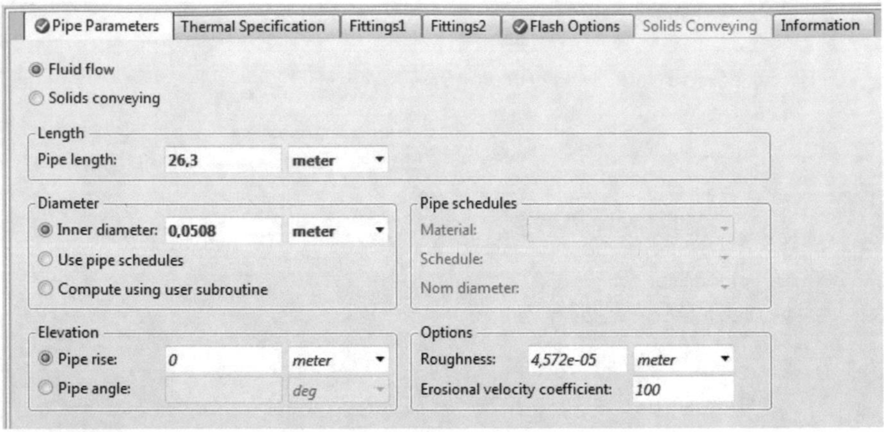

| ⊘ Pipe Parameters | Thermal Specification | Fittings1 | Fittings2 | ⊘ Flash Options | Solids Conveying | Information |

◉ Fluid flow
○ Solids conveying

Length
Pipe length: 26,3   meter   ▾

Diameter	Pipe schedules
◉ Inner diameter: 0,0508 meter ▾	Material:
○ Use pipe schedules	Schedule:
○ Compute using user subroutine	Nom diameter:

Elevation	Options
◉ Pipe rise: 0 meter ▾	Roughness: 4,572e-05 meter ▾
○ Pipe angle: deg	Erosional velocity coefficient: 100

Fig. 3.28 Pipe module interface in Aspen Plus®

Table 3.14 Segment specifications for third line of piping system

Name	T-3
Inlet stream	5
Outlet stream	7
Length (m)	26.30
I.D. (in.)	2
Elevation change (m)	0

Table 3.15 Splitter S-2 and S-3 specifications

Splitter	S-2	S-3
Inlet stream	6	7
Outlet streams	8	10
	9	11
Flow fractions	0.7758	0.6428
For stream	8	11

Table 3.16 Segment specifications for other lines of piping system

Section	T-4	T-5	T-6	T-7
Inlet stream	8	9	10	11
Outlet streams	12	13	14	15
Length (m)	10.73	10.73	14.95	3.9
I.D. (in.)	1.5	0.75	2	1
Elevation change (m)	0	0	0	0

Other pipe sections also correspond to a *Pipe* model, and are specified according to the data reported in Table 3.16.

Thus, the whole simulation is defined and the calculation can be executed. Simulator may show some warnings, the most common related to default phases

Table 3.17 Conditions for streams *12*, *13*, *14* y *15*

Stream	12	13	14	15
Temperature (°C)	25.1	25.1	25.1	25.1
Pressure (bar)	5.43	4384	5893	5832
Flow (m³/h)	16,199	4682	1807	3251

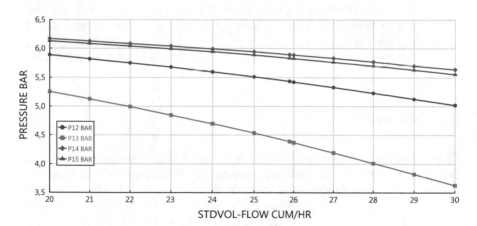

Fig. 3.29 Effect of the flow rate on the outlet stream pressure

considered in pipe calculation corresponding to the vapor–liquid equilibrium. In this case the fluid flowing through the pipe is only liquid. To solve this warning the valid phases should be modified (*Liquid Only*) in the *Flash Options* tab of each module. The conditions of the output streams are reported in Table 3.17.

3.6.3   Results Analysis

Aspen Plus® develops piping calculations similar to those present in Aspen HYSYS® modules. However, Aspen Plus® allows you to configure complex networks in a friendlier manner by the *Pipeline* module, since this can be introduced in pipe networks using node coordinates. To check the effect of the performance curve on the pump and the output currents, a sensitivity study on pressure is carried out by varying the flow rate of the input current to the system. The results are shown in Fig. 3.29.

As shown, the pressure of stream 13 is the most affected. Alternatively, stream 12 is also affected, but not to the extent that it has to stream 13. Streams 14 and 15 have a very similar behavior between them. Reviewing the routes of each of the streams, it may be noticed that stream 14 and 15 pass through pipe sections where the flow: diameter ratio tends to be constant, and checking the Eqn. (3.4), it can be deduced that the pipe diameters is high enough to compensate for the effect of the increased speed and friction factor. Moreover, these two streams handle the lower flows, so the flow of these streams does not increase in the same amount as the others.

For streams 12 and 13 there is a similar trend but it is more evident in the case of stream 13. These two streams have the same relationships that were presented to the stream 14 and 15. However, the trend appears to be more pronounced (especially for the case of stream 13). This is because of the pipe diameters of these streams, which are tight. Being so small as compared with large flows, according to Eqn. (3.4), it is expected that losses will increase more sharply in relation to the flow.

3.7 Summary

Process simulators studied here have very similar modules that allow the calculation of equipment for fluid handling. However, each simulator has different alternatives at the time of introducing data and showing results. On one hand, Aspen Plus® allows a theoretical analysis as well as more specific data entry forms. On the other hand, Aspen HYSYS® is more geared to make a practical calculation without showing intermediate results. In both simulators it is necessary to consider the calculation module and correlations, because, as it was determined, each one counts with advantages and restrictions. In general, these methods provide a very good approximation to reality. In fact, they is now mainly used to calculate oil and gas lines, pipelines, among others. Pump and compressor modules are also used. However, these do not allow a detailed design. These provide rough estimates that in case of the equipment design are a good starting point for detailed calculations. Nonetheless, if very detailed information on the performance of the pump is required, it is possible to make a detailed rating using specialized software.

3.8 Problems

P3.1 Does the pressure drop through a pipe is constant over time? Explain in terms of the Bernoulli equation and energy balances.

P3.2 What is a performance curve of a pump? How is the calculation affected if it is not included?

P3.3 Why are there different correlations for vertical, horizontal, and inclined flow?

P3.4 What advantages does a flow map offer to select adequate correlations?

P3.5 For gas handling system:

- Determine what would happen to double the flow of the well 3 and well 4. Would it help the system as it is implemented? What changes would be necessary? Explain.

- Perform an analysis of temperatures through all modules of tubes. What can be said? If the rate of heat transfer model is changed, it affects the pressure drop?

P3.6 For the water distribution system:

- In case of doubling the feed flow, what pipe should be modified? Explain your answer.
- Similarly or the foregoing analysis, show the effect of the inlet pressure on the outlet streams?

P3.7 Build the simulation of water distribution system in Aspen HYSYS®. Compare with the results presented in the example.

References

Aspen Technology Inc. (2005a) Aspen HYSYS gas gathering. Cambridge

Aspen Technology Inc. (2005b) Aspen HYSYS operations guide. Cambridge

Aspen Technology Inc. (2009) HTFS handbook, vol. 2

Aziz K, Govier GW (1972) Pressure drop in wells producing oil and gas. J Can Pet Technol 11:38–48

Baxendell PB (1961) The calculation of pressure gradients in high-rate flowing wells. J Petrol Tech 13:1023–1028

Beggs HD, Brill JP (1973) A study of two-phase flow in inclined pipes. J Petrol Tech 25:607

Duns H, Ros NC (1963) Vertical flow of gas and liquid mixtures in wells. Proceedings of the 6th World Petroleum Congress, pp. 451–465. Frankfurt, June 1963

Greene RW (1992) Compresores: selección, uso y mantenimiento. McGraw-Hill, Ciudad de México

Gregory GA, Mandhane JM (1975) Some design considerations for two-phase flow in pipes. J Can Pet Technol 14:65–71

Hagedorn AR (1965) Experimental study of pressure gradients occurring during continuous two-phase flow in small-diameter vertical conduits. J Petrol Tech 17:475–484

Kenneth J (1989) Bombas: selección, uso y mantenimiento. McGraw-Hill, Ciudad de México

Orkisewski J (1967) Predicting two-phase pressure drops in vertical pipe. J Petrol Tech 19:829–839

Poettmann FH (1952) The multiphase flow of gas, oil and water through vertical flow strings with application to the design of gas-lift installations. API Drill Prod Pract 257–317

Potter M, Wiggert D (2006) Mecánica de fluidos, 3rd edn. Thomson, Mexico city

Chapter 4
Heat Exchange Equipment and Heat Integration

4.1 Introduction

Heat exchangers play an important role in process engineering. Their main role is adjusting the thermodynamic condition of the input stream to an operation or storage.

In order to perform a thorough equipment design necessary for this operation, one must have sufficient grounds on the heat transfer mechanisms, and also, knowledge about other aspects such as construction standards, materials, and some heuristic rules that allow appropriate selection parameters to get better performance in these operations.

This chapter presents the general aspects required to perform a good heat exchanger design, simulation tools available and an analysis of its implementation. It mainly focuses on shell and tube exchangers due to their wide application, and the various computational tools existing to develop the corresponding mathematical model. However, note that there are several equipment that can be simulated with such tools as it provides modules for coolers, intercoolers, turboexpanders, etc.

To get deeper information on standards, building materials, and the theory of heat transfer could be found in references section suggested at the end of this chapter.

4.2 Types of Programs Available

The simulation of heat transfer equipment has been a widely used tool in recent decades. However, in the beginning there were three independent programs: one company-oriented involved in the equipment manufacturing (design software), other involved in process engineering (evaluation program), and finally another

© Springer International Publishing Switzerland 2016   139
I.D.G. Chaves et al., *Process Analysis and Simulation in Chemical Engineering*,
DOI 10.1007/978-3-319-14812-0_4

who performed calculations in order to predict the heat exchanger performance (simulation program).

These three types of programs were developed independently because the tools were not sufficiently robust, but only for specialized assessment. Subsequently based on calculation routines in design mode, because the computational complexity of such programs is substantially, higher programs were developed. And finally, the three calculation methods were integrated into a single program. A clear example of this strategy is Aspen HTFS + ® (HTFS means Heat Transfer and Fluid Flow Services) that, apart from calculations in the three modes discussed above, uses experimental information to validate results.

The program focused on heat exchangers design may employ different strategies to properly design this equipment. One of them, the factorial method is to sequentially examine all possible geometries; another is to search an evolutionary search quickly routed to the most promising region, i.e., toward the designs that are closer to the optimal, and search directed toward it. These strategies can be used taking into account the above-mentioned construction standards. Over time, the efficiency of the factorial method was improved by making some modifications to identify which calculations are simpler and thus carry out such studies first.

Another very important aspect in heat exchangers design is the possibility of having parameters to optimize the cost. So, the best standard geometry that meets the requirements at the lowest cost is selected. Aspen HTFS + ® performs mechanical and hydraulic calculations that allow a more detailed design considering vibration, noise, and performance problems during operation.

4.3   General Aspects

In process simulators there are a variety of heat exchange modules; however, all use the same basic equations. However, there are variations in equations solving, variables involved and simplifications that can be performed to estimate heat transfer equipment (Butterworth and Cousins, 1976).

Heat exchanger modules are generally classified into process simulators according to the type of substances used in the heat exchanger and according to the phenomena that take place inside (vaporization condensation latent heat exchange only, etc.). This classification can be summarized in the following categories:

- Heat exchanger: both sides in a single phase, and both streams are process streams.
- Heater: just a phase, one process stream, and on the other side a hot utility.
- Cooler: one process stream; cooling is performed with water or air.
- Condenser: a vapor stream is condensed by using air or steam.

- Chiller: similar to the condenser, but in this, condensation takes place at subatmospheric pressure or temperature, so that temperature of the coolant service is other than water or air.
- Reboiler: exchanger is used in distillation column bottoms; you can use a service or other hot process stream.

In general, it is possible to define two calculation routines: *short calculations* and *detailed calculations*. The short calculation is based on estimating the amount of heat added or removed. Moreover, the detailed calculation is based on determining the area, the geometrical parameters, and heat transfer coefficients. Then, equations and principles behind the various calculation routines are illustrated.

4.3.1 Shortcut Calculation (Holman 1999)

Both simulators have this method; is primarily used to calculate thermodynamic condition change only. However, according to the module can be used to access some additional calculations.

In general, this calculation routine does not involve geometric results or film coefficients; however, sometimes used to calculate the correction factor for the MLDT. It can be obtained information about heat load or required exchange area using an energy balance as it is shown below:

$$m_1 C_{p1} \Delta T_1 + m_1 \lambda_1 = m_2 C_{p2} \Delta T_2 + m_2 \lambda_2 \tag{4.1}$$

$$Q = U A F_t \, \text{MLDT} \tag{4.2}$$

$$F_T = \frac{\sqrt{R^2 + 1}}{1 - R} \frac{\ln\left[\frac{1 - P^* R}{1 - P^*}\right]}{\ln\left[\frac{2 - P^*\left(R + 1 - \sqrt{R^2 + 1}\right)}{2 - P^*\left(R + 1 + \sqrt{R^2 + 1}\right)}\right]} \tag{4.3}$$

$$\text{Si } R = 1 \; F_T = \frac{P^* \sqrt{2}}{\left(1 - P^*\right)\ln\left[\frac{2 - P^*\left(2 - \sqrt{2}\right)}{2 - P^*\left(2 + \sqrt{2}\right)}\right]} \tag{4.4}$$

Where F_T is the correction factor for MLDT, R is the ratio of heat capacities and flow and thermal effectiveness P^* per unit area.

$$R = \frac{W_f C_{pf}}{W_c C_{pc}} \tag{4.5}$$

$$P^* = \frac{\left[\frac{1-PR}{1-P}\right]^{\frac{1}{N}} - 1}{\left[\frac{1-PR}{1-P}\right]^{\frac{1}{N}} - R} \qquad (4.6)$$

$$\text{Si } R = 1 \quad P^* = \frac{P}{P - NP + N} \qquad (4.7)$$

where P is the overall heat exchanger effectiveness and N the number of shells in series.

$$P = \frac{\Delta T_{\text{Cold Fluid}}}{T_{\text{Outlet Hot Fluid}} - T_{\text{Inlet Cold Fluid}}} \qquad (4.8)$$

The simulator solves the equations presented only with a minimum of input data. This allows process engineer to use these modules when heat exchange is not important in the simulation analysis, or they will be analyzed in a further stage.

4.3.2   Rigorous Calculation (Holman 1999)

The detailed calculation makes it possible to estimate the film coefficients, pressure drops, F_T correction factor for MLDT, among others, by rigorous methods and detailed maps. According to the calculation mode simulator it can get different results:

- Design Mode: Streams are thermodynamically defined and there are conditions to fulfill. The simulator is responsible for proposing and evaluating alternatives for some geometry to find the optimal equipment that meets the energy requirements.
- Rating mode: Here there is a widely known geometry and flows together with one or more conditions to be met. The simulator is responsible for assessing whether the condition is met and what the minimum parameters to be fulfilled.
- In simulation mode has input streams and clearly specified geometry. The simulator gives results about equipment performance. Also it is allowed to modify certain parameters, as desired.

Before explaining in detail each of these methods is necessary to remember what a heat exchanger shell and tube comprises.

4.3.2.1 Geometry

For effective design, several heuristic rules are available that depend mainly on heat exchangers type and its application. In the case of heat exchangers shell and tube a pretty clear definition of all mechanical components such as further requires: tubes, baffles, shell, shell and cover nozzles, among others. Exchangers of this type are divided into three main sections: front head, back head, and shell.

There are several methods to classify the geometry of the shell and tube exchangers; one of them is by TEMA standards (Mukherjee 1998; TEMA 1997), which are encoded by different letters that specify body and head configurations available. This allows different configurations covering most of the requirements of the operations in the chemical industry, as each configuration allows for a specific behavior of the most representative fluids used industrially. Subsequently the corresponding standards TEMA code is shown.

However, it can carry out a detailed simulation to know the most significant aspects of the geometry of the exchanger; these are grouped into three broad categories: the tubes, shell, and baffles. Below, these categories are illustrated.

Tubes

Before introducing the calculation of the tubes is necessary to illustrate the different configurations that can be found in the different exchangers according to TEMA standard (1997). The tubes can be classified by the mechanical configuration in which they are arranged or the type of service used (Table 4.1).

Mechanically, the classification also includes the head type, because according to the arrangement of the tubes must select the appropriate. The tubes can be found in three forms:

- Fixed tubes: This configuration has fixed the setting, straight pipes on both sides using welded stationary tube sheets. Some TEMA-type exchangers with these tubes are: AEL, BEM, and NEN, among others. The main advantage of this arrangement is its low-cost construction. Also if you have a removable head can be done easy clean only removing the head. Its major disadvantages are maintenance since being fixed to the shell cannot be mechanically clean this part of the tube; so it is very difficult to change a damaged tube (Fig. 4.1).
- U-tubes: as the name suggests, this configuration consists of tubes bent in the form of U. In this configuration requires only a stationary tube sheet; however the cost is equal because it is expensive to produce tubes that form. Thus, the cost of the U-tubes is similar to the cost of the fixed tube. One advantage is that because of its configuration is required to install no expansion joint; similarly it is much easier to select an appropriate head with which to remove the tube bundle thus facilitating cleaning of the shell side. However, cleaning the side of the tubes is very complicated because the U-shape is necessary to use cleaning equipment with moving head. Such tubes should not be used for services with very dirty fluids (Fig. 4.2).

Table 4.1 TEMA classification for heat exchanger equipment (Mukherjee 1998; TEMA, 1997)

TEMA	Header	TEMA	Shell	TEMA	Header
A		E		L	
B		F		M	
C		G		N	
N		H		P	
D		J		S	
		K		T	
		X		U	
				W	

- Floating head: It is the most versatile and also the most expensive configuration. In this design, one of the head is relatively fixed to the shell, while the other end is left free to "float" inside the shell. Thus thermal expansion joints without allowing and likewise mechanically possible to clean both sides of the tubes. This design can be used for applications where both fluids (the tubes and the shell) are dirty. To apply this design TEMA headers S and T are recommended (Table 4.3).

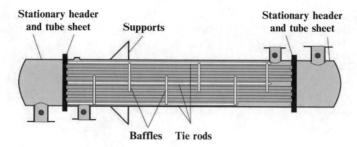

Fig. 4.1 Fixed tube heat exchanger scheme

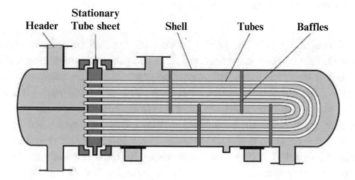

Fig. 4.2 U-tube heat exchanger scheme

Tubes calculation and detailed design is necessary to specify an optimal heat exchanger. In the simulator often not all of them are known, so it is necessary to understand the parameters involved. Hence, if required the designer can make well-founded assumptions and evaluate the results. Required parameters are the following:

- The two streams flow rates, which determine the heat load transfer in accordance with Eqn. (4.1).
- The inlet and outlet temperatures, which likewise affects the calculation of the heat load.
- The operating pressure of both streams. Required especially in the case of gases, since the pressure affects the density and other parameters involved in heat transfer.
- The pressure drop of both streams. It is an extremely important factor. Corresponds generally to liquid, to a value between 0.5 and 0.7 kg/cm^2 (7 and 10 psi) flowing through the shell. In case of viscous liquids this value may be higher. For gases is 0.05–0.2 kg/cm^2 (0.7–2.8 psi), 0.1 kg/cm^2 (1.4 psi) are the most common.
- Fouling factor for both streams. If not known, it can be taken from TEMA standards or from previous experimentation. This factor enables a better modeling of the operation in time.

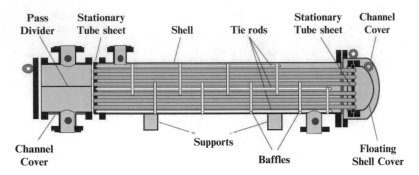

Fig. 4.3 Floating head exchanger scheme

- Physical properties of both fluids: Since independent transfer coefficients are calculated; it is equally important to know the property variations with temperature, and if necessary simplifications or selection of an adequate calculation method is required. The simulators employ large components, so it is important to verify that these properties are as accurate as possible to real.
- Exchanger type: It cannot be defined a priori, but can be defined based on the types discussed above, or the calculation in design mode, if required.
- The pipe diameter: To reduce pressure losses by reducing the use of accessories or increasing pipe diameter. It is used as a criterion for sizing the nozzles.
- The shell maximum size. Determines the maximum tube length, and the possibility of removing the tube bundle. Sometimes it is limited by the available space in the existing plant or planned for such equipment in case of a complete process design space.
- Building materials: These determine both the price and durability of the equipment. Must properly select the material to avoid adverse reactions with fluids running on the equipment.
- Special considerations. Depending on the case, additional considerations are given such as alternative operating scenarios, intermittent operation, etc.

After meeting most of the above specifications, the calculation of heat transfer coefficients can be performed. There are several correlations developed for calculation, and simulators have implemented several of them. Each correlation is subject to certain restrictions and has been developed for optimum performance in specific applications, so it is important to know the correlation used by the simulator to improve or correct the calculations. Similarly, you should consult the restrictions and simplifications of each map to determine the appropriate for each application. If more information is needed, it is highly recommended to use help information provided in the simulators. However, you can apply a basic equation for this calculation, as many of the above correlations have similar forms:

$$Nu = 0.027 \left(Re^{0.8}\right)\left(Pr^{0.33}\right)\left(Re^{0.8}\right)\left(Pr^{0.33}\right) \tag{4.9}$$

And using the definition of each of the parameters is:

$$h = 0.027(DG/\mu)^{0.8}(c\mu/k)^{0.33}(DG/\mu)^{0.8}(c\mu/k)^{0.33}(k/D) \tag{4.10}$$

Knowing this equation it is possible to appreciate the relationship that different properties have with the heat transfer coefficient and to propose alternatives for enhanced or improved equipment performance.

Also, there is calculation of the pressure drop. This is a straightforward calculation, as is done based on the Ergun equation. In this case the pressure loss coefficients are given by configuration of tubes. The simulators include methods for calculating the pressure drop and certain correlations for specific applications.

Shell

The calculation of this part of the exchanger is more complex than tubes, as in the shell that also features cross flow. For shell there are many possible flow arrangements, depending on the shell type, the tube arrangement, and baffle type, among others.

TEMA defines different arrangements based on the flow in the shell: E, F, G, H, J, K, and X. Each of these will be explained in more detail below (See Table 4.1).

- Shell type E: is a single step shell, the fluid enters the shell at one end and exits on the other. This is the most common and simple type of shell.
- Shell type F: This shell consists of a longitudinal baffle, which divides the shell into two steps. The fluid enters one end, reaches the other end of the pass, from there goes to the next and then comes out at the end of the second step. This setting is used in situations where temperature crosses are presented. Here F_T factor usage is implemented to correct the MLD_T calculation.
- Shell type J: corresponds to a shell where there are three nozzles, two on one end and one at the opposite end. In this type of shell you can have two fluid input streams with one output or one input streams with two output.
- Shell type G: is a split flow shell, mainly used in thermosyphon reboilers. There is only one center support and no baffles. This type of shell cannot be used for tube lengths greater than 3 m (10 ft), it exceeds the limit for the supported tube length.
- Shell type H: basically consists of two laterally attached shell type G, so there are two support plates. Also used in thermosyphon reboilers, and one of the main advantages and the shell type G, is the substantial reduction of the pressure drop due to the absence of cross baffles.
- Shell type X: This shell has only crossflow. This fluid enters the top or bottom of the shell, passes through the tubes and out in the opposite side. To ensure the fluid distribution throughout the shell, arrangements are made to the nozzles.

In this type of shell pressure drop is very low, and because of that this type of shell is mainly used for condensing vapor at low pressure vacuum pressures.
- Shell K type: generally used for Kettle type reboilers. It has shell space for steam available.

Another important operation factor of the shell corresponds to the tube arrangement, that is, the way the tube sheet inside the shell is arranged. For this purpose, there are four patterns:

- Triangular (30°): this allows to accommodate more tubes in one place that the square arrangement. Also produces greater turbulence which leads to higher heat transfer coefficients. Usually the step handles is 1.25 times the tube diameter. This does not always allow mechanical cleaning, due to the proximity between tubes, so it is recommended to use the triangular arrangement only for clean fluids.
- Rotated triangular (60°): this does not offer several advantages to the triangular arrangement; therefore it is not as popular.
- Square (90°) is usually used for dirty fluids through shell applications. This allows the mechanical cleaning.
- Rotated square (45°) causes increased turbulence that square arrangement is used in situations where the Reynolds number has a value below 2000.

These arrangements should be selected taking into account the two streams in the heat exchanger. An example of this could be select a fixed head, with very narrow tubes in triangular arrangement, for an operation where the fluid of the shell is quite dirty, so that this does not allow mechanical cleaning of the outside of the tubes, which would prevent heat exchange fouling.

Baffles

With respect to the baffles there are three important variables in their configuration: baffle type, spacing, and cut.

- Baffles type: Baffles have different functions in an exchanger. They are used to support the tubes through the heat exchanger; maintaining the shell with fluid velocity, desired turbulence and prevent malfunction due to the flow-induced vibrations. Two main types of baffles: plate and rod; among the plates have segmented baffles, double-and triple-segmented.
- Spacing: This value corresponds to the center-to-center distance between baffles. It is one of the most important exchanger design parameters. TEMA suggests one fifth of the internal diameter as shell minimum spacing. A very small spacing results in problems of mechanical cleaning and fluid maldistribution. A too large spacing results in a predominantly longitudinal flow, this flow is less efficient than cross flow to heat transfer. It is recommended, from the effects of balancing heat transfer and pressure drop, spacing between baffles types in a range from 0.3 to 0.6 times the internal shell diameter is recommended.

- Baffle Cut: This value corresponds to the height of plate removal, through which the fluid flow of the shell is allowed. This is presented as a percentage of the internal diameter of the shell; this value is normally between 15 and 45 %. For both large and very small cuts, the efficiency of heat transfer is decreased in the shell. If these settings do not allow achieving the desired performance, you can think in different actions, such as increasing vertical segments vertical rather than horizontal baffles type and vice versa.

4.3.3 Calculation Models

There are several ways to develop the detailed calculation of heat exchangers. Either of the simulators used in this text will have several models:

- Simple Weighted
- Simple End Point
- Simple Steady State Rating
- Dynamic Rating
- Rigorous Shell & Tube

Each of these models has different alternatives which makes them ideal for certain types of exchangers; additionally, each take different considerations, although the principles are the same. This is why the method should be carefully selected.

The Weighted and End Point models have an added value, because you can specify whether the heat exchanger losses presented for inclusion in the equipment calculation. It may also specify whether the losses occurring throughout the exchanger or those losses can take place in the ends.

4.3.3.1 Simple End Point Model

This model is based on the basic equation for calculating the heat load in a heat exchanger, where it is part of the overall heat transfer coefficient, the area available for heat exchange and the logarithmic mean temperature.

This model assumes that the overall heat transfer coefficient is always constant, as the specific heat of the both end streams of the equipment. Here it is also assumed that both heat curves are linear. Similarly, assuming that no phase-changes and the specific heat is constant. Thus, for problems where the curves cannot be assumed to be linear, it is not recommended to use this module.

Finally, the parameters needed for calculations using this model are the following: the pressure drops for both sides of the exchanger, the UA product, and exchanger geometry.

4.3.3.2   Simple Weighted Model

The use of this model for systems where heat curves exhibit a nonlinear, such as phase-change systems for pure components on either side of the equipment recommended behavior. With this model, the curves are separated at intervals, and for each of them an energy balance is performed, so that the value LMTD and UA are calculated for each interval and finally summed to obtain the total value of the exchanger.

The Weighted model is available only for counter-current exchangers. It should be noticed also that the calculation of FT correction factor does not take into account the geometry.

The use of the Weighted model requires the following information: the pressure drops on both sides of the exchanger, the UA product, and the specific details for the heating curves. This last parameter includes specifying the number of intervals to be calculated and the assumptions taken into account in the calculation.

4.3.3.3   Simple Steady State Rating Model

This model corresponds to an extension of the model to incorporate an End Point assessment calculation, so that it uses the same assumptions. This model requires very detailed information on the equipment geometry; if you have this information, we recommend running the calculation with this model. It is clear that the model is specifically designed, as its name suggests, assessing steady state exchangers.

4.3.3.4   Dynamic Rating

In this model there are two alternative calculations: basic and detailed. Should have very little information, for example, two temperatures and the UA product or three temperatures, the heat exchanger can be evaluated by the basic model. In case you have much more information available, use detailed model.

This method has the same assumptions in the model for the model End Point; in fact, the basic model is the dynamic adaptation of the latter. However, the basic model can also be used to perform calculations in steady state.

The detailed model is based on the same assumptions of weighted model for steady state, so that this model also divides the heat exchanger into small intervals and performs calculations through them. For using the detailed model information exchanger geometry is required, and as for the case of the basic model, also can be used to calculate steady state.

4.3.3.5   Rigorous Shell & Tube

In this option, the calculation using more detailed methods that are available enable HTFS platform. For this reason, these calculations require a very specific and detailed information and exchanger geometry, but likewise offer the most rigorous results between the methods available for assessing or designing a heat exchanger.

This can be called through certain modules in both simulators. However, HTFS has its own interface to the detailed calculation of equipment, shown in the introductory example in this chapter.

4.4   Modules Available in Aspen Plus®

Aspen Plus® has several modules to calculate heat exchange equipment. These modules allow heat exchange to develop different types of calculation, with various functions, which include settings from the thermodynamic conditions of a stream, detailed calculations of heat exchangers, pressure drops, evaluation and design of heat exchangers, reboilers to calculate Kettle type and condensers, among others.

Aspen Plus® groups these functions in different calculation modules. Brief descriptions of the functions of each of these modules are reported in Table 4.2:

4.5   Modules Available in Aspen HYSYS®

In the upper section of the taskbar, located under the phase separators, are the heat exchanger modules (Fig. 4.4). There are, like in Aspen Plus®, a thermodynamic type exchangers and detailed exchangers; exchangers also include thermodynamic processes for LNG (Liquefied Natural Gas), air coolers and furnaces.

A brief description of the heat exchange modules found in Aspen HYSYS® is presented below (Table 4.3).

4.5.1   Thermodynamic Heat Exchangers

Unlike Aspen Plus®, Aspen HYSYS® divides the thermodynamic heat exchangers in two types: Cooler and Heater, depending on the service provided.

To install this type of exchanger, drag the icon from the object palette to the simulation environment, the material input and output streams, and energy stream that represents the heat given or received to achieve the specified conditions.

Table 4.2 Heat exchanging modules available in Aspen Plus®

Icon	Module name	Description
Heater	Thermodynamic heat exchanger. *Heater*	This heat exchange module is one of the most versatile and so it is best used in Aspen Plus®. This model can be made thermodynamic calculations for a phase, and the calculations include
		• Dew or bubble points
		• Remove or add a specified amount of heat
		• Get subcooling or superheat calculations
		• Determine the amount of heat that is necessary to add or remove to achieve a given vapor fraction
		This module allows specifying the heat load used to calculate from other operations present. Thus, this module can be modeled coolers or heaters (only one side). Wider use is adjusting thermodynamic properties of a given stream
HeatX	LNG thermodynamic heat exchanger. *MHeatX*	This module is a multifluid heat exchanger; one of its applications is for LNG systems configurations. Model allows two streams and exchangers have an outlet for clarified water. This module corresponds to several interconnected Heaters; in this way it is easier to reach the convergence
MHeatX	Shell and tube heat exchanger. *HeatX*	This is one of the most powerful calculation modules for heat exchangers which have Aspen Plus®. Allows calculations of heat exchange between two streams; these calculations can be short or detailed, and additionally allows use of interfaces Hetran, Airborne, and TASC to make calculations and rigorous heat exchanger designs. The main difference between these methods of calculation that provides the module lies in how to calculate the overall heat transfer coefficient
		Methods which do not involve an external interface are:
		• The short method is one in which you must enter the minimum amount of data compared to other methods of calculation. No geometric results or film coefficients are obtained. This calculation method can also specify or calculate the MLDT to the exchanger
		• The method allows detailed assessment calculations and simulation geometry defined exchanger. With this method the pressure drops, the correction factor for the MLDT and film coefficients are calculated
HXFlux	HXFlux module	HXFlux is used for performing calculations between the sink and a heat source, using the convective transfer mechanism. The driving force for convective heat transfer is calculated as a function of the LMDT
		Requires specified variables between: input and output temperatures, heat load, heat transfer coefficient,

(continued)

Table 4.2 (continued)

Icon	Module name	Description
		and heat transfer area. HXFlux calculates the unknown variable and determines the MLDT, using either approximate or rigorous method

Fig. 4.4 *Object Palette* in Aspen HYSYS®

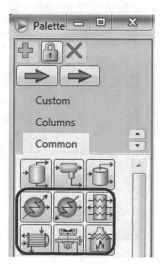

Table 4.3 Heat exchanging modules available in Aspen HYSYS®

Icon	Module name	Description
	Thermodynamic heat exchangers. *Heater, cooler*	Thermodynamic exchangers perform the thermodynamic calculation of the output stream under specification of two of the three thermodynamic specifications of the output stream, i.e., pressure (or pressure drop), temperature or enthalpy (or heat load)
	LNG Thermodynamic heat exchanger. *LNG exchanger*	This module essentially performs the same calculation that heaters–coolers, in terms of which is limited to the thermodynamic calculation of the output streams, but with the peculiarity that one device can handle multiple process streams, both hot and cold. This equipment does not allow the installation of energy streams, which means that the energy transfer occurs only between hot and cold streams process
	Shell & tube heat exchanger. *Heat exchanger*	This module is used in detailed design and detailed assessment of shell and tube exchangers under different calculation methods
	Air-cooled heat exchangers. *Air cooler*	The Air Cooler unit uses an ideal mixture of air as heat transfer medium to heat or cool a process stream

Fig. 4.5 Connectivity for
Cooler heat exchanger
module in Aspen HYSYS®

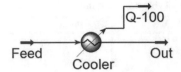

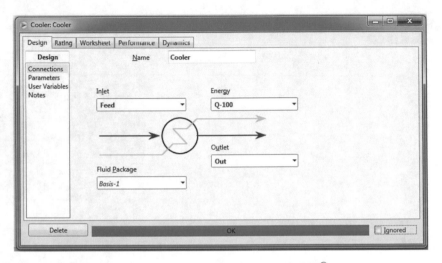

Fig. 4.6 Design tab in the *Cooler* module interface in Aspen HYSYS®

The configuration required for the module is remarkably elementary, requiring only the thermodynamic condition of the process stream at the inlet and the outlet. A possible sequence configuration is presented below.

Once you have connected the streams, as shown in Fig. 4.5, double-click on the exchanger module; in this way the window shown in Fig. 4.6 is opened.

In the upper section of the object palette, located under the phase separators are the heat exchanger modules (Fig. 4.4); there are, like in Aspen Plus®, a thermodynamic type exchangers and detailed exchangers; thermodynamic processes in addition to LNG, Air Cooler exchangers and furnaces are included.

This window displays the input, output, and energy streams, the properties package used in the calculation. Then, the pressure drop is specified in the *Parameters* option (Fig. 4.7).

In *Delta P* box, the pressure drop is specified, or can be specified entering the stream output pressure. Now you need to set the temperature of the output stream; this can be done directly on the *Worksheet* tab in the properties window of the output stream. An alternative is specifying the heat load in the energy stream, as well.

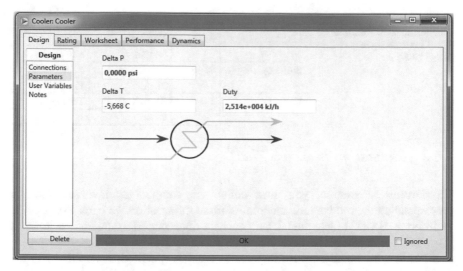

Fig. 4.7 Specifications for the cooler module in Aspen HYSYS®

4.6   Introductory Example

4.6.1   Problem Description

In order to introduce the design and simulation equipment topic for heat exchangers, it is proposed to develop the example 12.1 of Kern's book (1999) to address the simulation packages available. The problem information is presented below:

4.6.1.1   Calculate a *n*-Propanol Horizontal Condenser

It is required a 1–2 horizontal condenser to cool 60,000 lb/h of pure *n*-propanol, coming from a distillation column bottom operating at 15 psi g, with a boiling point of 244 °F. As coolant, water will be used to average 85 °F. Fouling factor of 0.003 is used with an allowable pressure drop of 2 psi for *n*-propanol and 10 psi for water. Due to condenser location, assume that tubes are 8 ft long. Tubes are ¾ in. ED, 16 BWG, arranged in 15/16 in. triangle.

According to Kern 1999, an exchanger with the specifications reported in Table 4.4 is obtained. This water flow 488,000 lb/h is obtained, with a temperature change from 85 to 120 °F. There is also specified that the water flows through the tubes and the *n*-propanol through the shell.

Since Aspen Plus® and Aspen HYSYS® do not have the ability to do detailed designs, the equipment performance is evaluated, and then it is compared with a detailed design from Aspen HTFS +®.

Table 4.4 Geometric
specifications for example
12.1 heat exchanger
(Kern 1999)

Shell		Tubes	
I.D.	31 in.	Number	766
		Length	8 ft
Baffle spacing	31 in.	DE, BWG	¾ in., 16 BWG
		Pitch	15/16 in.
Tubes passes	1	Arrangement	Triangular
		Pass	1

4.6.2   Simulation in Aspen Plus®

To start the simulation, you must define the components involved and the corresponding thermodynamic model. For more information, go back to Chap. 2.

Insert a *HeatX* block model and connect four material streams as follows (Fig. 4.8):

After this, the conditions of feed streams must be specified, as described by the problem. Next step is to define the fluid connections. The calculation mode is set to detail. Hot fluid is connected to the tubes side and a vapor fraction at the outlet of 0 is specified. Additionally the calculation mode is set as rating. In this way the heat exchanger geometry is enabled. Finally, a fouling coefficient of 0.0015 h ft² R/Btu is set for both sides. This is done in the *Film Coefficients* tab, as shown in Fig. 4.9.

The geometry is entered in the *Geometry* tab on the left side of the screen. There you can enter the geometric variables given in the problem; to do this the program has several tabs to specify each of the geometric parameters of the exchanger; *Flanges* refers to one side of the heat exchanger, that is, in the shell can be defined in other pipes, and on the other baffles last nozzles. Thus, you can specify each problem in the simulator, so you should have a configuration as follows (Figs. 4.10, 4.11, 4.12, and 4.13):

With these data already specified, you have a fully defined exchanger to run the simulation. After executed, it can be seen in the *Thermal Results* tab, the results of the exchanger evaluation. In this menu several tabs where you can see both a summary of the exchanger, material balance calculations regarding the pressure drop and the exchanger details (such as the required area, the area today, the LMTD appear, etc.), among others (Fig. 4.14).

4.6.3   Simulation in Aspen HYSYS®

Create an Aspen HYSYS® simulation where the corresponding components and the appropriate thermodynamic model are entered. Later in the simulation environment, install a *Heat Exchanger* module.

The first step in the simulation is to drag the icon for the shell and tube heat exchanger to the simulation area and double-click on it; the window shown in Fig. 4.15 appears.

Fig. 4.8 Connectivity of
HeatX module in Aspen
Plus®

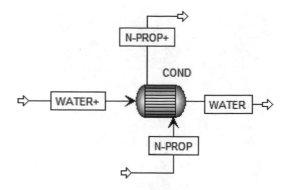

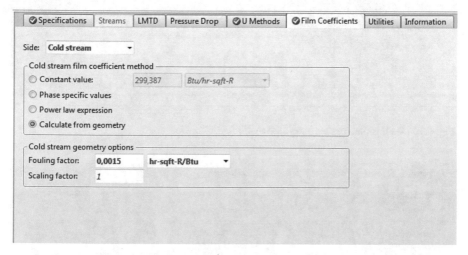

Fig. 4.9 Fouling factor specifications in Aspen Plus®

In this window the process streams are installed, which may come from different flow sheets; the option of selecting different sets of properties for each of the process streams are also provided; obviously this is useful when heat exchange fluids have significantly different characteristics, for example, ethyl acetate cooling using propane.

Once the connection is necessary to fully define the input streams exchanger using data provided by exercise is shown below (Fig. 4.16):

Once defined the feed streams, the selection of model calculation is performed. This is introduced in the *Parameters* section. Then select the desired option from the *Heat Exchanger Model* drop-down list. In the simulator, modes can be selected: *Rating, Design* (*End Point* and *Weighted*), among others; for this case, because the geometry is already indicated by the exercise evaluation mode selected, i.e., *Steady State Rating* (Fig. 4.17).

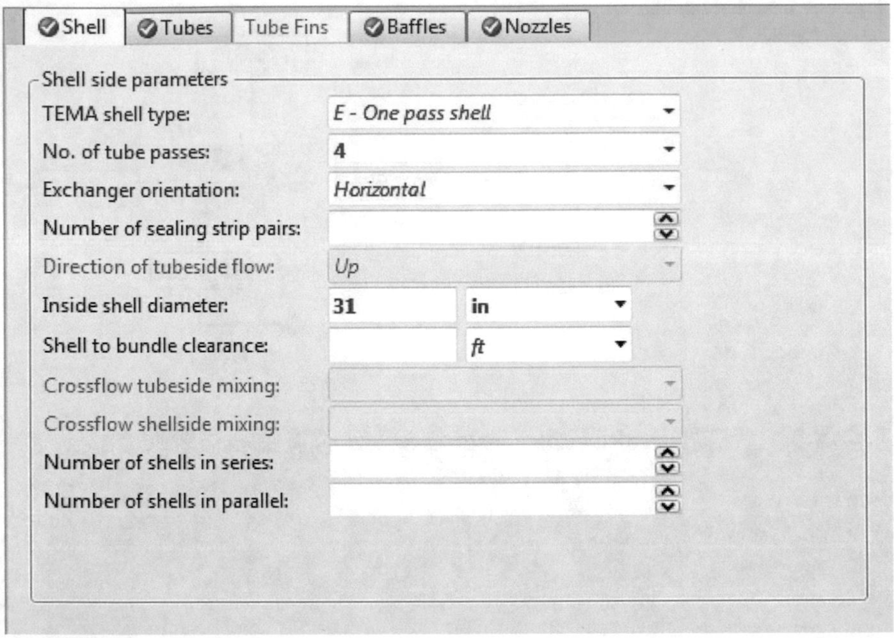

Fig. 4.10   Shell geometry data entry in Aspen Plus®

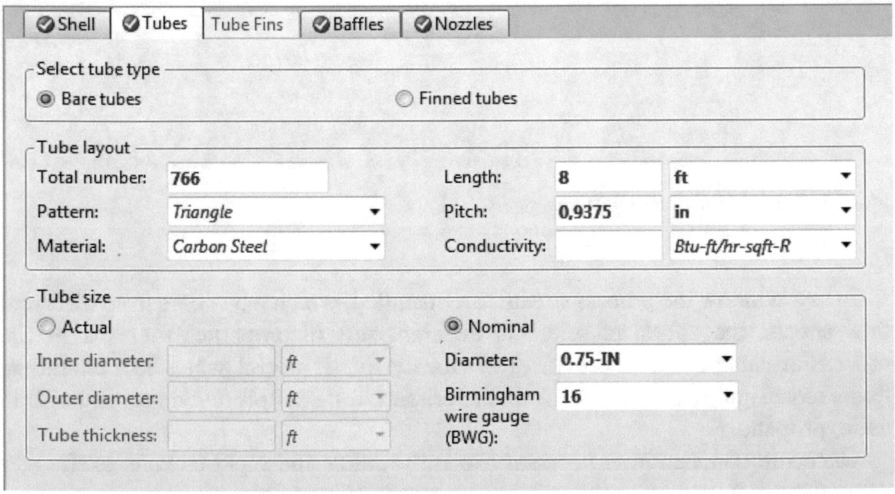

Fig. 4.11   Tubes geometry data entry in Aspen Plus®

Having defined the input streams to the equipment and the method of calculation, you must enter the geometric data of the equipment; this is done in the *Rating* tab (Fig. 4.18).

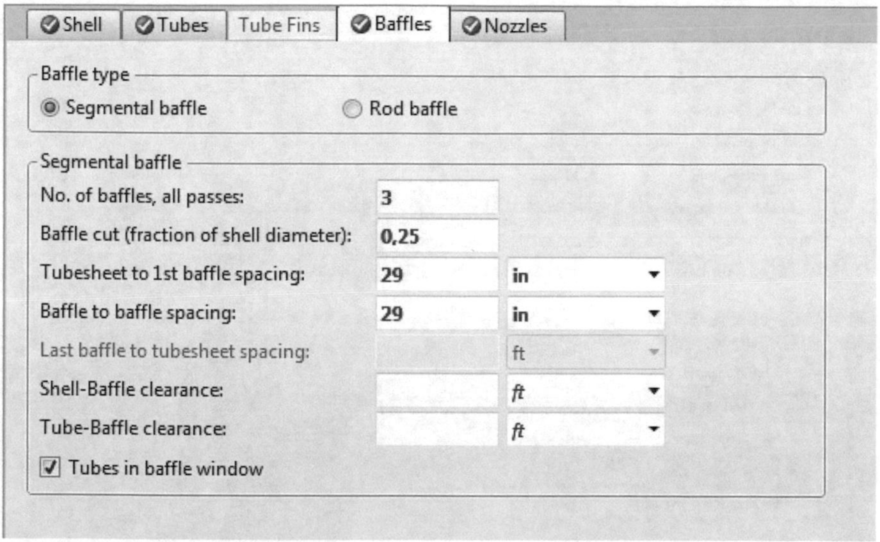

Fig. 4.12 Baffles geometry data entry in Aspen Plus®

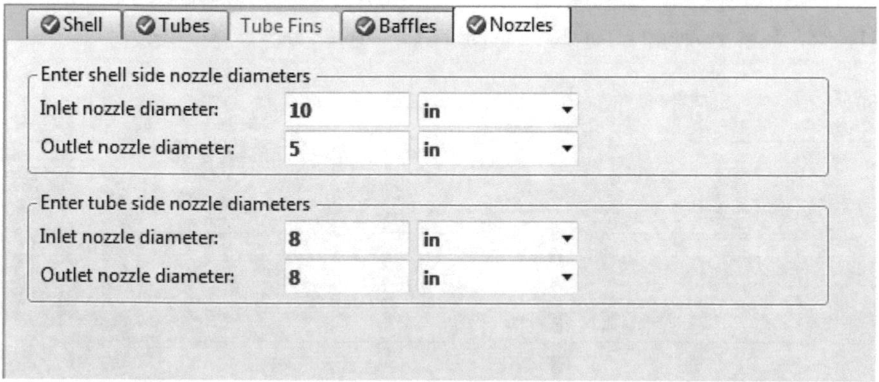

Fig. 4.13 Nozzle geometry data entry in Aspen Plus®

4.6.3.1   Geometric Data Entry

In Fig. 4.18 the geometric data input window in Aspen HYSYS®, which divides the input geometric data, is shown in three types: general, shell, and tube data; selection is changed using the buttons in the top left of the window.

The data entered in each window is:

- *Overall*

   Shell passes
   Shells in series

Summary	Balance	Exchanger Details	Pres Drop/Velocities	Zones	Utility Usage	⊘ Status

Exchanger details

Calculated heat duty:	3,9673e+07	Btu/hr
Required exchanger area:	1203,25	sqft
Actual exchanger area:	1203,23	sqft
Percent over (under) design:	-0,00113051	
Average U (Dirty):	221,993	Btu/hr-sqft-R
Average U (Clean):	371,73	Btu/hr-sqft-R
UA:	267113	Btu/hr-R
LMTD (Corrected):	148,525	F
LMTD correction factor:	0,999845	
Thermal effectiveness:	0,0332793	
Number of transfer units:	0,0343047	
▶ Number of shells in series:		
Number of shells in parallel:	1	

Fig. 4.14 Heat exchanger results for example 12.1 in Aspen Plus®

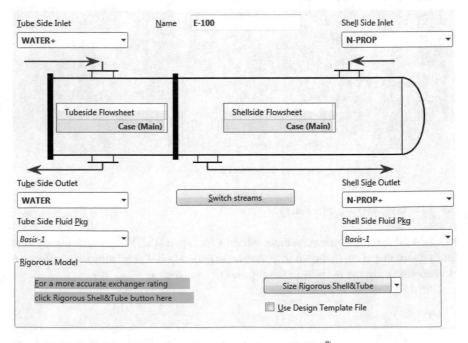

Fig. 4.15 Shell & Tube heat exchanger interface in Aspen HYSYS®

Worksheet	Name	WATER+	WATER	N-PROP	N-PROP+
Conditions	Vapour	0,0000	1,0000	1,0000	0,0000
Properties	Temperature [F]	85,00	193,8	244,0	239,0
Composition	Pressure [psia]	15,00	5,000	29,70	27,70
PF Specs	Molar Flow [lbmole/hr]	879,1	879,1	998,4	998,4
	Mass Flow [lb/hr]	1,584e+004	1,584e+004	6,000e+004	6,000e+004
	Std Ideal Liq Vol Flow [barrel/day]	1087	1087	5110	5110
	Molar Enthalpy [Btu/lbmole]	-1,223e+005	-1,026e+005	-1,065e+005	-1,238e+005
	Molar Entropy [Btu/lbmole-F]	1,835	33,63	44,36	22,18
	Heat Flow [Btu/hr]	-1,076e+008	-9,023e+007	-1,063e+008	-1,236e+008

Fig. 4.16 *Worksheet* window in Shell & tube heat exchanger in Aspen HYSYS®

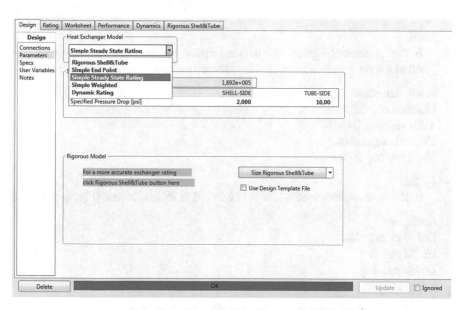

Fig. 4.17 *Parameters* tab in Shell & tube heat exchanger in Aspen HYSYS®

Shells in parallel
Tube passes
Heat exchanger orientation
First tube pass flow direction
Elevation
TEMA type

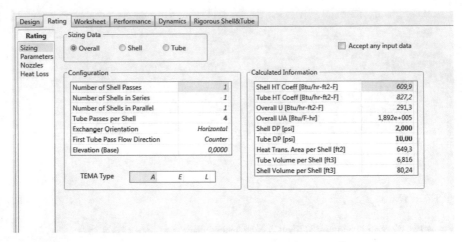

Fig. 4.18 Rating tab in Shell & tube heat exchanger in Aspen HYSYS®

- *Shell*

  In this window the geometric data is entered in two sections; one of them is *Shell and tube bundle data* where is entered the following information.

  Shell diameter
  Number of tubes
  Tube spacing (pitch)
  Tube arrangement
  Shell fouling factor

- *Baffles*

  In the *Shell baffles* section the data entered is shown below (Fig. 4.19):

  Baffle type
  Baffle orientation
  Baffle cut %
  Baffle spacing

- *Tube*

  In this window two sections are found where the data is entered, *Dimensions* and *Tube Properties*, as shown in Fig. 4.20

  In *Dimensions* section
  Diameter (OD)
  Inner diameter (ID)
  Tube thickness
  Tube length

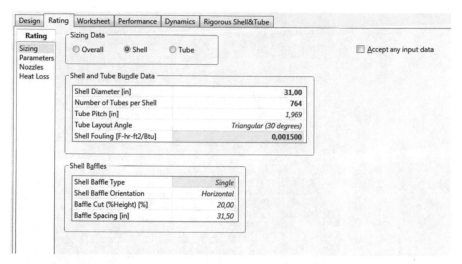

Fig. 4.19 Shell geometric data entry in Shell & tube heat exchanger in Aspen HYSYS®

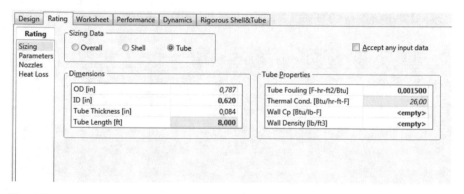

Fig. 4.20 Tube geometric entry in Shell & tube heat exchanger in Aspen HYSYS®

In *Tube Properties* section

Tube fouling factor

Thermal conductivity

Wall Cp

Wall density

Importantly, it is not necessary to know some of the above data, because the simulator uses default values. By entering the values that can be obtained of the problem statement, the state of the simulation passes *OK* indicating that it has completed the calculation of the heat exchanger without warnings.

A summary of the results can be seen in the *Performance* tab (Fig. 4.21), and the results for the temperatures of the streams are displayed on the *Worksheet* tab (Fig. 4.22).

Design	Rating	Worksheet	Performance	Dynamics	Rigorous Shell&Tube

Performance

Performance		
Details		
Plots		
Tables		
Setup		
Error Msg		

Overall Performance

Duty	1,732e+07 Btu/hr
Heat Leak	0,000e-01 Btu/hr
Heat Loss	0,000e-01 Btu/hr
UA	1,70e+05 Btu/F-hr
Min. Approach	64,268 F
LMTD	101,9 F

Detailed Performance

UA Curvature Error	0,0000 Btu/F-hr
Hot Pinch Temp	244,0000 F
Cold Pinch Temp	179,7322 F
Ft Factor	0,992
Uncorrected LMTD	102,677 F

Fig. 4.21 Performance tab in Shell & tube heat exchanger in Aspen HYSYS®

Design	Rating	Worksheet	Performance	Dynamics	Rigorous Shell&Tube

Worksheet	Name	WATER+	WATER	N-PROP	N-PROP+
Conditions	Vapour	0,0000	1,0000	1,0000	0,0000
Properties	Temperature [F]	85,00	179,7	244,0	239,0
Composition	Pressure [psia]	15,00	5,000	29,70	27,70
PF Specs	Molar Flow [lbmole/hr]	884,3	884,3	998,4	998,4
	Mass Flow [lb/hr]	1,593e+004	1,593e+004	6,000e+004	6,000e+004
	Std Ideal Liq Vol Flow [barrel/day]	1093	1093	5110	5110
	Molar Enthalpy [Btu/lbmole]	-1,223e+005	-1,028e+005	-1,065e+005	-1,238e+005
	Molar Entropy [Btu/lbmole-F]	1,835	33,46	44,36	22,18
	Heat Flow [Btu/hr]	-1,082e+008	-9,087e+007	-1,063e+008	-1,236e+008

Fig. 4.22 Worksheet tab in Shell & tube heat exchanger in Aspen HYSYS®

4.6.4 Simulation in Aspen Exchanger Design and Rating®

The proposed approach to this problem involves a preliminary design with the help of Aspen Exchanger Design and Rating® software to become familiar with the use of this tool, and a subsequent comparison with the information given in the same exercise. This comparison can also be valuable if done at hydraulic level.

As a first step, you must open the program. This is located in the following path: *Programs > AspenTech > Exchanger Design and Rating V8.4 > Aspen Exchanger Design and Rating V8.4 > Shell and Tube Exchanger (Shell & Tube)*. A window where the software modules described above are shown, among other shows. Tasc + should be selected because it is desired to design a shell and tube exchanger.

TASC® was originally written for the capacitor design program, but it had no application to phase-change equipment and evaporators. The calculation algorithm

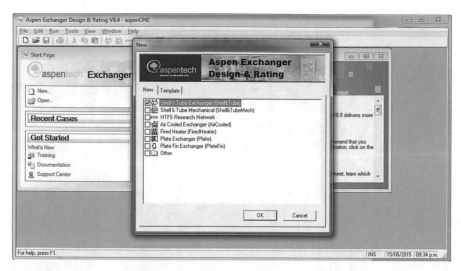

Fig. 4.23 Module selection screen in Aspen Exchanger Design and Rating®

used was relatively simple, with the intention to speed up calculations and reduce the computation time (Fig. 4.23).

Subsequently a tree where two options are deployed: *Input* and *Results*, select the first option. Here are two additional options that are displayed in red. All data reported in that color is missing information required to make the proposed calculation.

Select the *Problem Definition* option and then *Application Options*. In this window you can specify parameters related to model simulation and the fluid on the hot and cold side of the heat exchanger. You can also select the standards used for sizing the heat exchanger. For this example TEMA standards, which are typically applied, are used.

With respect to the simulation mode, there are three options: design, rating, and simulation, which are described below:

Design Mode: is used when it is required to make a design exchanger given operating conditions. The simulator sets the geometrical parameters in order to optimize the costs associated with the installation and operation of equipment. It is the default model in this module Aspen Exchanger Design and Rating®.

Rating mode: is used when a defined geometry, for example a team that seeks to leverage and operating conditions required have. The simulator performs calculations that allow you to calculate what temperature would the process stream, comparing it with the entered conditions and assess whether it meets the requirement (Fig. 4.24).

Simulation Mode: Rating mode is similar to, but the simulator does not perform the comparison between the existing equipment (defined geometry) and operating conditions. It just makes the calculation with the input data and determines the output conditions.

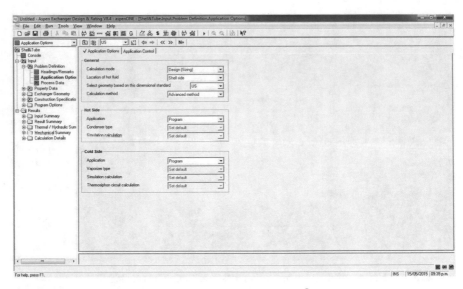

Fig. 4.24 Main screen in Aspen Exchanger Design and Rating®

Table 4.5 Data entry in Aspen Exchanger Design and Rating®

Stream	n-Propanol	Cooling water
Mass flow rate (lb/h)	60,000	–
Inlet temperature (°F)	–	85
Outlet temperature (°F)	–	115
Inlet vapor fraction	1	–
Outlet vapor fraction	0	–
Operating pressure (psia)	29.7	40
Allow. pressure drop (psia)	2	10

In this case, a *Design* mode calculation is performed considering that it is necessary to design the equipment with given operating conditions shown in the example. For each exchanger side (hot and cold) it must be specified the heat transfer conditions. Water as cold fluid must be selected; remember no phase-change in this side. Select *Cold Side* in the *Application* tab and then, *Liquid, no phase-change*.

The next step is to enter the operating information; for this purpose select the *Process Data* tab. In this window it can be specified the stream names, inlet and outlet conditions (as temperature, vapor fraction, pressure, pressure drop, etc.). The information to be entered is resumed in Table 4.5.

It is worth noting that the missing information appears in purple, and when the minimum data is specified the number of purple cells is reduced, and finally leaving all cells as shown in Fig. 4.25.

Once entered the operating conditions, the compositions for each of the defined streams must be entered. For this purpose, select the Property Data tab and then the *Hot Stream (1) Composition* option. Then, select the *Search Databank* button, and search for n-propanol adding it to the component list with the *Add* button. Finally, in *Composition* enter a value of *100*, because it represents a mass-based percentage composition.

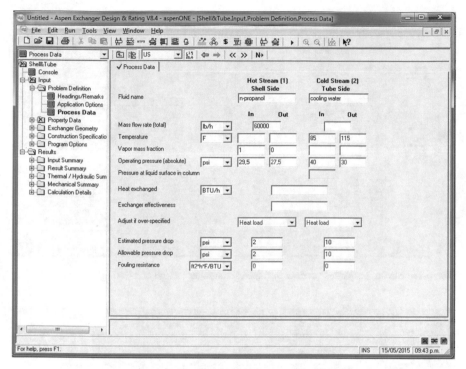

Fig. 4.25 Data entering screen in Aspen Exchanger Design and Rating®

The next step is to select the *Hot Stream (1) Properties* option in the selection tree, and click on *Get properties*. It worth noting that in this moment, the transport properties are estimated, these properties are required for thermodynamic and hydraulic calculation to evaluate the design. These calculations are performed in pressure and temperature ranges according to the operating conditions entered.

The same procedure is used to define cold stream, selecting the *Cold Stream (2) Composition* and *Cold Stream (2) Properties* options. The cold stream composition is 100 % water (Fig. 4.26).

All needed conditions for exchanger design were specified. Now, click the *Next* button and the simulator will start computing and generate a report as shown in Fig. 4.27. It is observed that the simulator performs exchanger calculations and seeks to reduce the equipment total cost varying geometric parameters. Results are reported in Table 4.6. The highlighted design is the simulation package suggestion.

Click on *Close* button. To observe this information again, select the *Results* tab, then *Results Summary* and *Optimization Path*. There specific geometric information can be obtained from the performed calculations to find the optimum design.

Now it is shown how to visualize all important aspects of selected optimum design. First, observe the specification sheet for the selected heat exchanger. Select the TEMA Sheet option, where it is reported in tables according to the TEMA standards. This specification sheet can be used for purchase purposes (Fig. 4.28).

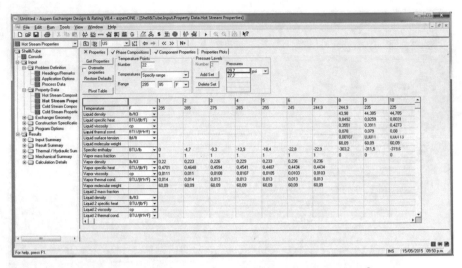

Fig. 4.26 Stream properties window in Aspen Exchanger Design and Rating®

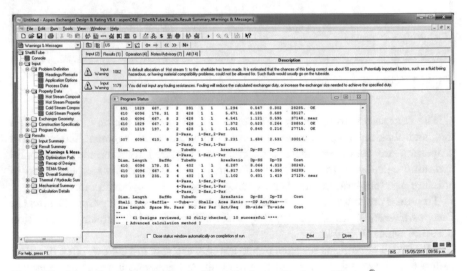

Fig. 4.27 Calculation report window in Aspen Exchanger Design and Rating®

Now, in the *Thermal/Hydraulic Summary* folder, all thermal and hydraulic parameters calculated for the optimum design can be obtained, as pressure drop, heat transfer coefficients, etc.

In the *Mechanical Summary* folder, detailed geometry information and drawings both general and tube arrangement as shown in Figs. 4.29 and 4.30 can be observed. With this information general equipment dimensions and some parameters, as tube length, arrangement, tolerance, etc. can be modified to correct the design performed by simulator.

Table 4.6 Design summary in Aspen HTFS + [®]

Design	Shell	Tubes	Pressure drop		Price (USD)	Status
	Size (in.)	Length (ft)	Shell	Tubes		
1	17.25	10	2.259	1.359	13,060	Near
2	17.25	8	1.666	7.312	12,367	OK
3	19.25	8	1.414	0.827	14,342	OK
4	19.25	8	1.581	4.444	14,213	OK
5	21.25	6	1.449	0.596	15,579	OK
6	21.25	6	1.3	2.751	15,380	OK
7	23.25	6	1.33	0.488	17,772	OK
8	24	6	1.267	0.47	17,874	OK

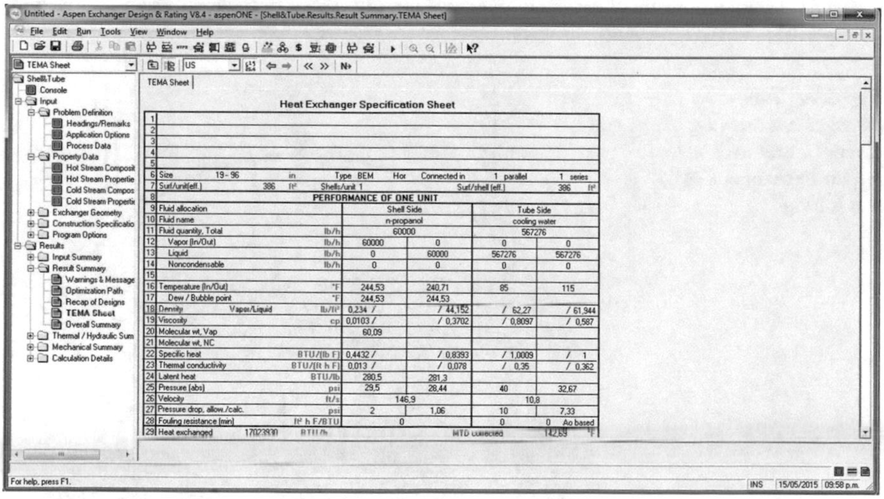

Fig. 4.28 TEMA specification sheet in Aspen Exchanger Design and Rating[®]

Other important window is located in *Results Summary*, called *Recap of Designs*, which shows what optimum designs were found with each modification in geometry, standards or construction material (*Rating mode*) or operating conditions (*Design mode*).

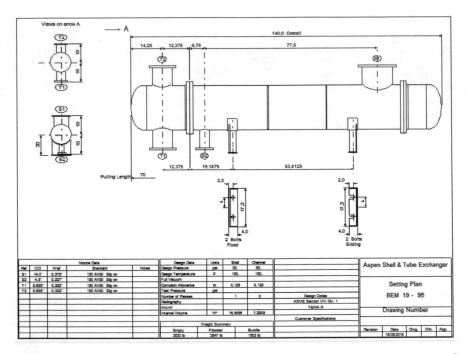

Fig. 4.29 General drawing of designed heat exchanger in Aspen Exchanger Design and Rating®

Fig. 4.30 Tube
arrangement drawing of
designed heat exchanger in
Aspen Exchanger Design
and Rating®

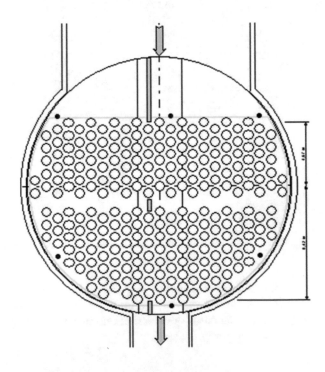

Once the design is completed, select the *Rating* mode to enter the available information according to example 12.1 and evaluate given geometry. For this purpose, open the Input folder, then select Problem Definition and then Application Options. Change the *Calculation Mode* option from *Design* to *Rating/Checking*.

When this option is activated, the *Exchanger Geometry* option also activates, where the geometric information is entered. For this example, the performed design used same length tube than those proposed in the example, but with a different tube arrangement. Regarding that tube arrangement is a parameter to be evaluated considering thermal and hydraulic behavior in each heat exchanger design.

To select the geometry data, select the *Exchanger Geometry* folder and then the *Geometry Summary* option where shell diameter, baffle spacing, tube length, and number of passes through the tubes are specified. For now, specify the inner diameter (ID) of 17.25 in. (optimum design diameter), with a tube length of 8 ft and two passes through the tubes (example information). Additionally specify a baffle spacing of 8 in.

Observe that when entering the shell diameter, the simulator by default define the outer diameter. In this window, it can be selected the header geometry and shell according to TEMA standards. The tube geometry and arrangement can be modified as well.

4.6.5 Results Analysis

Both simulators, Aspen Plus® and Aspen HYSYS® count with a large variety of models to calculate different heat exchange equipment. Some of these modules allow rigorously evaluating shell and tube heat exchangers. In the previous example was described how to evaluate equipment using both simulators, and with this information assess differences and similarities in the calculation.

In both programs, all geometry parameters without detailed mechanical information can be specified; this allows calculating pressure drop, film coefficients, among other important variables for chemical and process engineering. Despite that those modules cannot evaluate cost, vibration, and equipment real dimensions, the results can be exported to Aspen Exchanger Design and Rating®, where the calculation can be refined.

Another alternative available is the opposite process, i.e., enter the information directly into the Aspen Exchanger Design and Rating® interface and after the calculation, use generated file for inclusion in Aspen Plus® or Aspen HYSYS® to use information previously generated for a specialized software. This process is useful when a complete process is simulated with a medium or high detail level. Additionally, when there is a change in the main process simulation, the exchanger calculation is performed externally in Aspen Exchanger Design and Rating® and the corresponding simulation results are imported.

There are many differences when entering the data and how to calculate the equipment between Aspen Plus ® and Aspen HYSYS®. One of them is the TEMA specification; in Aspen HYSYS® you can specify the shell and heads, while in Aspen Plus® can only specify the TEMA type of the shell. Similarly, in Aspen Plus® the tubes material can be specified based on information of different material properties commonly used for this function. However, in Aspen HYSYS® is not allowed to specify the material such as, and instead one should provide the material properties.

In terms of results between simulators, there are some differences. In Aspen Plus®, a real area very close to the value obtained in the example exchanger was obtained. In Aspen HYSYS®, the results have small variations, especially in the water flow. This difference is due to the way both simulators calculate; Aspen HYSYS®, in order to include the n-propanol restriction, immediately calculates the necessary water flow, depending on the amount of transferable energy, and does not give the chance to define the final temperature. When looking at the results, there is a large decrease in water flow and a corresponding increase in temperature. However, both simulators allow achieving the required specification for the n-propanol outlet.

The simulators differ in the geometrical parameters reported. An example of this is the minimum required area; Aspen Plus® reports that value but Aspen HYSYS® omits this parameter. Overall, Aspen HYSYS® reports final results and very specific calculation, while Aspen Plus® provides values of other parameters that can lead to specific design adjustments.

It is impossible to carry out a detailed geometric design of the heat exchanger from Aspen Plus® or Aspen HYSYS® process streams; these programs are limited to evaluate given geometric configurations. Instead, Aspen Exchanger Design and Rating®, as it was observed, has tools to do so.

The calculation is developed in Aspen Exchanger Design and Rating® considers several variables (design, evaluation, and simulation modes) that other modules lack; between these cost analysis, vibration analysis, and several heuristic rules are implemented. This program allows rigorous calculation as detail exchanger mechanical design, something very close to real equipment, largely because the software was designed to make detailed design evaluation of existing heat exchange equipment.

Now a sensitivity analysis to observe the equipment performance can be done, considering that it is a distillation column condenser. Making a conceptual analysis can be considered that a change in top pressure can affect the energy consumption and therefore the amount of water required. Additionally, the scenario in which the n-propanol flow varies due to operational changes was studied.

4.6.5.1 Top Pressure Change

The top pressure is set at 15 psi g. Using Aspen Exchanger Design and Rating® is possible to vary the pressure in a range of 10 psi. This change may be due to vapors accumulation in the column that can lead to an increase in the pressure or, on the other hand, a leak in the system that makes it decrease.

The information shown below is obtained by making modifications to the base simulation explained above, and the following results were achieved (Fig. 4.31).

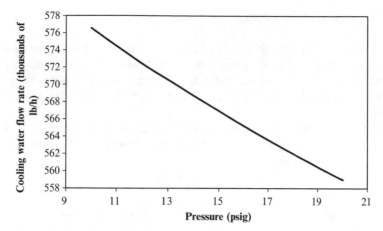

Fig. 4.31 Effect of the pressure change in the water flow

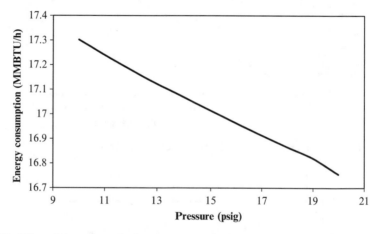

Fig. 4.32 Effect of the pressure in the energy consumption

Considering saturated steam from the distillation column, the top pressure has a remarkable effect on the water flow and the energy consumption. The observed trend is consistent with the theory because, gradually increasing the pressure, the latent heat of a pure substance decreases to be zero for the critical point (Fig. 4.32).

Considering energy consumption, it is not a completely linear behavior since the relationship between the pressure and the water flow is not so; remember that the PV behavior of a pure substance is shaped like a dome.

4.6.5.2   Propanol Flow Change

The *n*-propanol flow rate change is not so complex as to predict the change in pressure; however, it can cause problems in the heat exchanger because a greater flow should increase the equipment size, and an existing heat exchanger with known geometry will suffer vibration problems forcing the construction material (Fig. 4.33).

Here it can be seen that there is a direct relationship between the flow of *n*-propanol and water flow behavior can be inferred from the energy balance on the heat exchanger (Fig. 4.34).

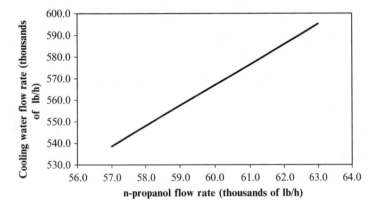

Fig. 4.33   Effect of *n*-propanol flow rate in the cooling water flow rate

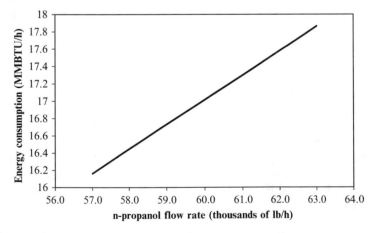

Fig. 4.34   Effect of the *n*-propanol flow rate in the energy consumption

In the case of energy consumption, it is possible to observe a direct relationship between the mass balance over the equipment and the ration between the flow of n-propanol and heat that must be removed to meet the requirements.

All these calculations can be performed manually; however, the simulator allows assessment of the operation of the equipment when in *Checking-Rating* mode or makes a comparison between the resulting geometries using it in *Design* mode.

4.7 Process Heat Integration

4.7.1 Introduction

The optimal use of energy is a big concern for industrial plants, it has been estimated that most companies are using 50 % more energy than it is needed (Alfa 2011). Several methodologies have been developed in order to have a more efficient use of energy in the process. These methodologies started to emerge during the 1970s, due to the industrial changes related to the oil crises. One of the responses to this problem was heat integration, based mostly on pinch analysis.

Since its creation during the 1970s, heat integration has been widely used among the different sectors of the processing and power generating industry over the last years. Its approach is to consider the different heat sources and sinks inside the process and propose alternatives to exchange heat inside the process, in order to make an optimal use of the energy.

Nowadays the approach of this technology has been broadened, in such a way that it does not referred to as heat integration but instead as process integration. In this sense, the International Energy Agency has defined process integration as follows: *Systematic and General Methods for Designing Integrated Production Systems ranging from Individual Processes to Total Sites, with special emphasis on the Efficient Use of Energy and reducing Environmental Effects* (Gundersen 2000).

New definitions have been proposed, as the aim of process integration has changed from just integrating the use of energy but to carry out similar integration for the mass streams in the process. In 2006, the Finnish process integration program, defined process integration as integrated and system-oriented planning, operation and the optimization and management of industrial processes (Timonen et al. 2006).

The following section is going to give an overview of the concepts and rules involved in heat integration, as well as an introductory example to use one of the software tools available, which is Aspen Energy Analyzer.

4.7.2   Theoretical principles

Before carrying out any heat exchanger network design it is necessary to analyze the process in order to have clearly defined guidelines to carry out the optimization of the system. According to (Gundersen 2013) the basic elements in process analysis are:

- Define performance targets ahead of design
- Calculate the composite curves, which are a representation for any "amount" of energy that has a "quality" (usually temperature)
- The pinch decomposition, which divides the process into a heat deficit region and a heat surplus region.

4.7.2.1   Targeting, Composite Curves, and Grand Composite Curves

In order to illustrate the basic process integration analysis steps, the following example will be used. A process has several streams which should be heat integrated as much as possible. The stream data is registered in the following table. Additionally cooling water and high-pressure steam should be used as cooling utility and heating utility respectively.

First, the energy targets have to be set. These refer to the heat targets and the unit targets. The targeting process defined by (Gundersen 2013) goes as follows:

Regarding the heat targets we have to know the minimum external cooling and heating for the system. As for the units target, it refers to the minimum number of units needed to integrate the process, this means just to exchange heat between the processes streams without achieving the heat targets, and the minimum number of units needed for maximum energy recovery, with this number of units needed to achieve the energy targets. This will give you the objectives when designing a heat exchanger network.

In order to define the targets it is necessary to introduce a methodology known as heat cascade. This methodology requires defining the minimum temperature difference for the streams to exchange heat, this value is usually known as ΔT_{min}. It is usually defined by a trade-off between the heat exchanger area (thus, the number of units) and the operational costs. The bigger the value for ΔT_{min} is, the bigger the driving force is, this leads to the use of less area for heat exchange; nonetheless the consumption of utilities will increase and this will mean that the energy costs will be higher. On the other hand when reducing the value for ΔT_{min}, the driving for will be less and the heat transfer will require more area but the amount of utilities needed will be reduced as well. The value for ΔT_{min} is set according to experience; however, it usually stays between 10 and 20 °C. For some industries, a lower ΔT_{min} value can be chosen. The implications and methods to select the appropriate

Table 4.7 Process streams information

Stream	Inlet temperature (°C)	Outlet temperature (°C)	$m\,C_p$ (kW/C)	ΔQ (MW)
H1	180	40	20	2.8
H2	100	70	90	2.7
C1	60	200	40	5.6
C2	50	170	30	3.6

value of ΔT_{min} are explained in detail in Sections 2.4 and 3.7 of Kemp (2006). Some example values registered by (Umbach 2014) are:

- Oil Refining—20–40 °C
- Petrochemical—10–20 °C
- Chemical—10–20 °C
- Low Temperature Processes—3–5 °C

For the example shown in Table 4.7 a minimum temperature difference of 10 °C will be selected.

Having set ΔT_{min} it is possible to carry out the heat cascade. The first step is going to define the temperature intervals. For this, the temperatures for the hot streams (those that can "give" energy) and the temperatures of the cold streams (those that "need" energy) have to be adjusted. This adjustment can be made by adding ΔT_{min} to the in- and outlet temperature of the cold streams and by subtracting this value for the in- and outlet temperatures of the hot streams or by adding half of it and subtracting half of it from the corresponding streams.

After adjusting the temperatures, the temperature intervals can be defined. Within each of these intervals, the total surplus of heat deficit of heat can be calculated as the difference between the different heat required by the hot and cold streams on each interval. The following equation shows the calculation for an interval i.

$$\Delta Q_i = \sum Q_{Hi} - \sum Q_{ci} = \Delta T_i \left(\sum mC_{pH} - \sum mC_{pC} \right) \quad (4.11)$$

Having the heat balance for each interval, the heat balances will add from the upper interval to the bottom one. With these "added values," it is possible to determine a value of heat that should be added from the top of the cascade, which will make the added energy balance zero at some point. This temperature is known as pinch point and it is a very important value for the design of heat exchanging networks. Table 4.8 contains the heat cascade carried out for the example.

From the heat cascade registered in Table 4.8 there are three important results. The first two of them correspond to the energy targets. In the last column the value on top corresponds to the minimum energy required to add from utilities, as well as the last value corresponds to the minimum "excess" of energy in the system, which has to be removed by utilities.

The final of these three values corresponds to the temperature where the value for the last column on Table 4.8 becomes zero. This point is known as pinch point.

Table 4.8 Heat cascade example

Hot streams T (°C)	Cold streams T (°C)	Streams in the interval	Delta T interval (°C)	Delta Q interval (kW)	Delta Q added (kW)	Delta Q adding heat (kW)
210	200					5200
Interval 1		C1	30	−1200		
180	170				−1200	4000
Interval 2		H1, C1, C2	80	−4000		
100	90				−5200	0
Interval 3		H1, H2, C1, C2	30	1200		
70	60				−4000	1200
Interval 4		H1, C2	10	−100		
60	50				−4100	1100
Interval 5		H1	20	400		
40	30				−3700	1500

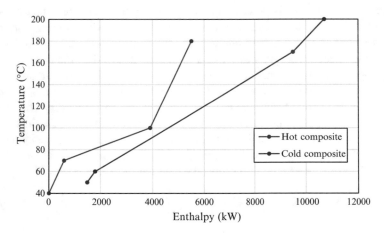

Fig. 4.35 Composite curves for the example

This point indicates where the system needs external heating and where it needs external cooling.

Another tool in order to determine these targets is the composite curve. This curve compares the energy of the cold and hot streams with their "quality" (temperature). The difference between the lower temperature end of both curves corresponds to the minimum cooling heat and the top differences corresponds to the minimum heating needed. Figure 4.35 shows the composite curves for the example.

It should be noted that the pinch point can be seen as the point where both curves get closer. This means that the driving force at the pinch temperature is minimum.

Another tool used for targeting corresponds to the Grand composite curve. This is a mix between both composite curves, it is very useful for cases where there are several utilities available. In order to generate this curve it is necessary to have a modified temperature for all the streams, this means add $\Delta T_{min}/2$ to the cold streams and subtract $\Delta T_{min}/2$ from the hot streams. The grand composite curve for the example can be seen on Fig. 4.36.

In Chapter 2 and 3 of Kemp (2006) and in Chapter 16 of Smith (2005), a more detailed explanation about the construction of these curves and the heat cascade can be found.

Having set the values for the minimum energy requirements, the only targets left to define correspond to the unit targets. The calculation of these values is straight-forward. For the minimum number of units, the calculation goes as follows (Table 4.9):

$$U_{min} = (N - 1) \tag{4.12}$$

$$N = n_c + n_h + n_{util} \tag{4.13}$$

For this example:

$$N = n_c + n_h + n_{util} = 2 + 2 + 2 = 6$$

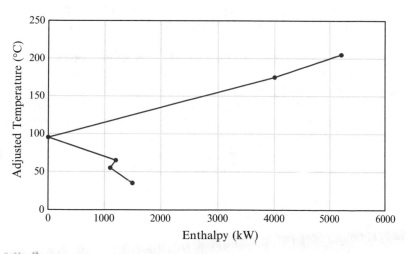

Fig. 4.36 Grand composite curves for the example

Table 4.9 Summary of targets for the heat integration example

$Q_{H,\ min}$	5200 kW
$Q_{C,\ min}$	1500 kW
Pinch	90/100 °C
U_{min}	5
$U_{min,\ MER}$	7

$$U_{min} = (N - 1) = 6 - 1 = 5$$

As for the minimum number of units for maximum energy recovery, the calculation goes as follows:

$$U_{min, MER} = (N - 1)_{above} + (N - 1)_{below}$$
$$N = n_c + n_h + n_{util}$$

For this example:

$$N_{above} = n_c + n_h + n_{util} = 2 + 2 + 1 = 5$$
$$N_{below} = n_c + n_h + n_{util} = 2 + 1 + 1 = 4$$
$$U_{min, MER} = (5 - 1)_{above} + (4 - 1)_{below}$$
$$U_{min, MER} = 7$$

4.7.2.2   General Rules for the Pinch Design Method

Once the targets have been defined and the process pinch point has been found, the next step is to start with the design of a heat exchanger network for the process. This method is explained in further detail in Chapter 4 of Kemp (2006) and Chapter 18 of Smith (2005).

The first step for the pinch design method is to separate two networks, one above and one below the pinch point. This means that it is not allowed to have heat exchange across the pinch point. Once these two regions have been defined, the design can be started. It should be noted that it does not matter which design is carried out first, below or above the pinch. However, the design of a heat exchanger network must be started at the pinch.

For the pinch heat exchangers, there are some rules set in order to place the heat exchangers. There are two cases: one for the case that is above the pinch and one for the case below the pinch.

For a pinch heat exchanger above the pinch the hot streams should be brought to pinch temperature without external cooling. When matching the streams the heat capacity of the hot stream must be less than or equal to the heat capacity of the cold stream (Umbach 2014).

As for the cold streams that are below the pinch, should be brought to pinch temperature without any external heating. In this case, when matching the streams the heat capacity of the hot stream must be greater than or equal to the cold stream. (Umbach 2014).

However when the matching conditions cannot be achieved, the streams can be split, such as they fulfill the matching conditions.

4.7.3   Aspen Energy Analyzer

Aspen Energy Analyzer (Aspen Technology Inc 2014) is the software which is included in the Aspen Engineering suite which focuses on energy integration. This software allows the user to design, optimize, and have a preliminary economic estimation for heat exchanger networks.

This software allows minimizing energy costs, offering an environment optimized to perform grassroot heat exchanger network design as well as retrofit cases. Its interface counts with tools to calculate the energy targets, capital costs and investment. Additionally it can be integrated with Aspen HYSYS® or Aspen Plus®, so that a heat exchanger network can be designed over a working steady state simulation.

4.7.3.1   Example: Manual Input Network

After opening Aspen Energy Analyzer the following options will appear.

In Fig. 4.37 it can be seen that right next to the save button, there are three options as follows: Open heat integration manager, create a new case, and create a new project (Fig. 4.38).

The heat integration manager allows seeing and managing all the cases and projects that are being used. A case is a small study that lets the user define the streams that will be integrated. A project is a combination of several cases (Fig. 4.39).

This data will be introduced in the project interface. After the data has been registered, the software will show additional information about the project such as the composite curve and the grand composite curve for the system (Fig. 4.40).

In the targets tab it is possible to see the minimum number of units and energy needed. Figure 4.41 shows the targets tab for this example.

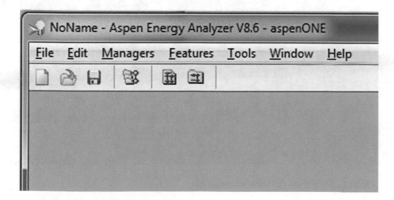

Fig. 4.37 Options for Aspen Energy analyzer

Fig. 4.38 Heat integration manager

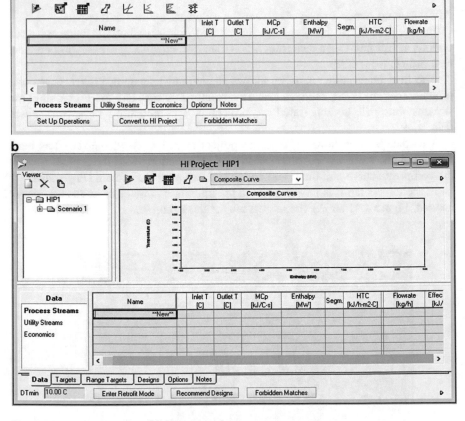

Fig. 4.39 (**a**) Case interface (**b**) project interface

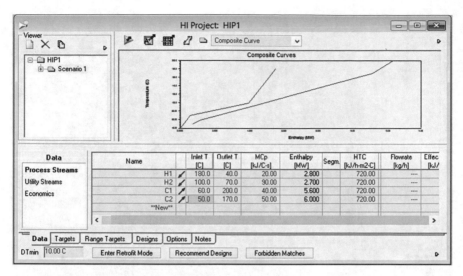

Fig. 4.40 Project interface after the data is registered

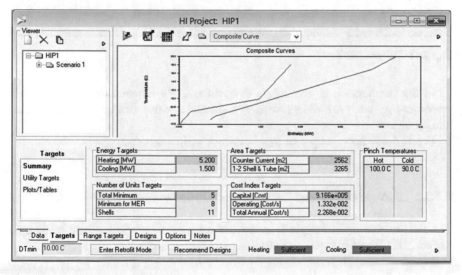

Fig. 4.41 Targets tab for the project

It is very important to consider these values as they will set the objectives for the design.

It is possible to click on the Scenario 1, and the available designs will be displayed. As shown in Fig. 4.42.

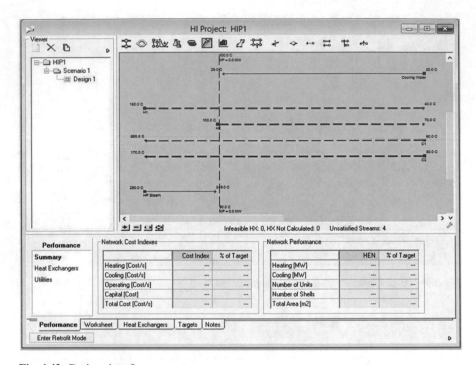

Fig. 4.42 Designs interface

In this interface, it is possible to design manually a heat exchanger network. Aspen energy analyzer allows the user to include heat exchangers, split streams, and optimize the network.

Figure 4.43 shows a manual input network and its targets.

4.7.3.2 Software Proposed Network and Optimized Network

The next step in this example is to use the software to propose a heat exchanger network and then improve this software proposed network.

In order to use this feature in Aspen Energy Analyzer, right click on the scenario folder and choose the option recommend designs (Fig. 4.44).

After choosing recommend designs, the following window will be displayed (Fig. 4.45).

Recommended designs have several options to customize. The first one is the allowed number of split branches, then the maximum number of designs (it can propose more than one). Additionally it counts with further options for the designs such as forbidden matches.

For this example, the proposed network is shown as follows (Fig. 4.46).

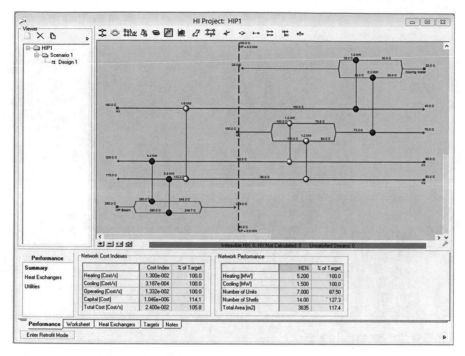

Fig. 4.43 Manual implemented heat transfer network

Fig. 4.44 Scenario folders

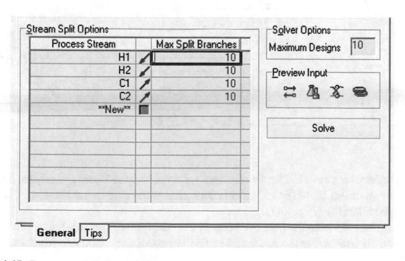

Fig. 4.45 Recommend designs window

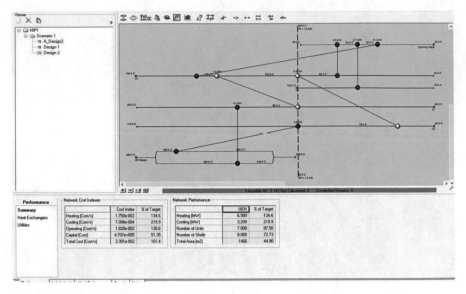

Fig. 4.46 Recommended design

No.	Hot utility	Heat exchangers	Cold utility
1	HP Steam	E-102 E-104 E-106	Cooling Water
2	HP Steam	E-102 E-104 E-103	Cooling Water

| Process Stream | Utility Stream | Loops | **Paths** | Degree of Freedom |

Fig. 4.47 Topology tool

In order to improve the heat exchanger network, Aspen Energy Analyzer counts with a tool called topology view. This tool allows the user to see the loops, paths, and controllability (Fig. 4.47).

It can be seen from the heat exchanger network, that the heat exchanger E-103 is transferring heat across the pinch and has a very small duty. Considering this, a first step to improve this proposed network is to remove the heat exchanger and compensating this duty through the path (Fig. 4.48).

Despite the recommended design does not offer many advantages over the manually proposed one, it can be used as a starting point for further improvements.

Apart from doing this manually Aspen Energy Analyzer counts with an integrated optimizer, which allows the user to select on which criteria should the network be optimized and the optimization variables (Fig. 4.49).

After running the optimizer the resulting network is shown next (Fig. 4.50).

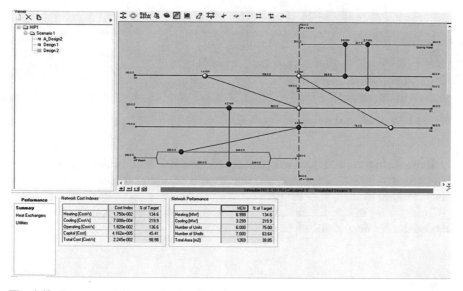

Fig. 4.48 Recommended network after slight improvement

Fig. 4.49 Optimizer window

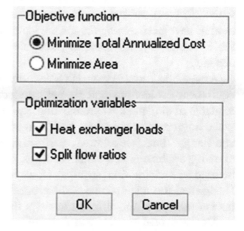

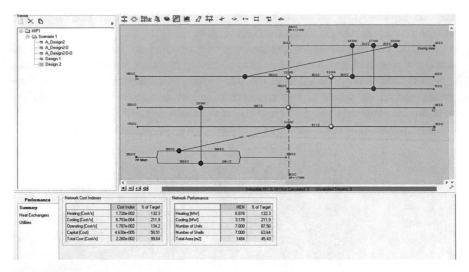

Fig. 4.50   Recommended network after optimizer

It can be appreciated that the optimizer reduces the duty for the heat exchanger E-102 to zero, which means that, that unit should be removed. This conclusion was also achieved by analyzing the paths.

This process of optimization, through paths and using the optimizer, can also be carried out for the manual input network as well.

4.8   Summary

The heat exchange modules allow calculation alternatives and are very versatile when different substances involve various processes types. Their calculation models have been among the most widely studied and still are used for calculating the different varieties, from thermodynamic exchangers to mechanically detailed designs.

Aspen Plus® and Aspen HYSYS® are very useful when making preliminary equipment evaluations and its performance in the process, which can be quite valuable in situations where existing equipment will be used. On the other hand, when a rigorous heat exchanger design is required, The Aspen Exchanger Design and Rating® interface allows to perform rigorous and reliable estimates, apart from allowing application-dependent settings, where the equipment results obtained can be used for construction purposes. However, integration between Aspen Exchanger Design and Rating® and one of the process simulators can be considered in order to take advantage of the different features they provide.

The presented models can become very accurate in their results, so that real equipment design from the data calculated by Aspen Exchanger Design and

Rating® can be made. However, it can also cause problems, as designers must have the sufficient criterion for assessing the validity of the simulator results and likewise the necessary modifications if a design and construction project is carried out using data obtained from the simulation.

Aspen Energy Analyzer counts with plenty of tools when it comes to the design of heat exchanger networks. However, they can lead to nonoptimal results. In order to make the best use of this tool it is necessary to have a broad knowledge of the principles involved in heat integration as well as the capabilities of the software. The engineering criteria should be above the suggestions from the software in order to propose networks that can be implemented in real cases.

4.9 Problems

P4.1 What is the physical meaning of the F_T factor included in the energy balance of the heat exchanger with multiple tube passes? What is its value for a tubes single-pass exchanger?

P4.2 What values can the F_T coefficient take?

P4.3 What variables should be taken into account when choosing which fluid goes through the tubes and what through the shell? Explain your answer.

P4.4 What is the fouling factor R_d? Why it should be included in the calculation? Does it change over time? If it changes, explain why it is important to study how the change happens.

P4.5 The objective of this exercise is to simulate a heat exchanger that uses water to cool a hydrocarbon mixture using three different methods: two Heaters connected with a stream of heat, a short calculation *HeatX* module and a calculations and detailed calculation *HeatX* module.

In Fig. P4.1 the simulation diagram is indicated; there have been used two *Dupl* operations that duplicate the streams conditions entering the exchanger in each of the three proposed methods.

Inlet Streams
- Hydrocarbon stream: 200 °C, 4 bar, 10,000 kg/h.
- Composition: 50 % w benzene, 20 % w styrene, 20 % w ethyl benzene, 10 % w water.
- Valid phases: vapor–liquid–liquid.
- Cooling water: 20 °C, 10 bar, 60,000 kg/h water 100 %.
- Choose the appropriate property package for hot and cold side of the system.

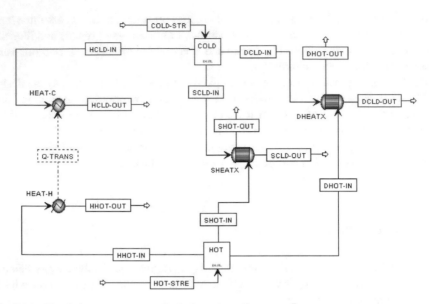

Fig. P4.1 Simulation arrangement of a hydrocarbon mixture cooler

For *Heater* and *HeatX* Short Calculation
- The hydrocarbon outlet stream must be a saturate liquid.
- No pressure drop was considered for both streams.

For the *HeatX* Detailed Calculation

1. Enter the geometric information:

   - Shell diameter 1 m, one tube passes. Hot fluid through the shell.
   - 300 plain tubes, 3 m long, 31 mm pitch, inner diameter (ID) of 21 mm, outer diameter (OD) of 25 mm.
   - All nozzles of 100 mm.
   - Five baffles with 15 % cut.

2. Calculate with *Rating* mode; exchanger specification requires that hydro-carbon stream leaves as saturated liquid.
   Required area:_____ m² Current area: _____ m²
   Hot outlet stream temperature: _____ °C
3. Change the calculation to *Simulation* mode and run again.
   Hot outlet stream temperature: _____ °C
4. Create heat curves with all required information for thermal design.

P4.6 A heat exchanger is required to cool 60,000 lb/h of acetone with a temper-ature of 250 °F and a pressure of 150 psia to 100 °F. To carry out this cooling, 185,000 lb/h of acetic acid are available, which is at 90 °F and 75 psia, and needs to be heated. There are four shell and tube exchangers, 1–2. Each exchanger has an internal shell diameter of 21.25 in and contains

270 plain tubes whose nominal diameter is ¾ in, 14 gauge BWG carbon steel, square tube arrangement with a pinch of 1 in., 16 ft length. Segmented baffles used 25 % cut and separate five in one another.

Determine if it is possible with one or more of these exchangers meet the cooling requirements of the acetone stream. Remember that if two, three, or four of these exchangers are arranged in series, such an arrangement would be equivalent to having an exchanger 2–4, 3–6, or 4–8, respectively. If the heat exchangers are not suitable to perform the required function, design a new system of heat exchange. Suppose combined fouling factor 0.004 h ft^2 °F/Btu.

Finally make the appropriate comments to the evaluation of heat exchange system, according to the results obtained in the simulation.

P4.7 70 kmol/h of saturate steam must be condensed at 1 atm and 101 °C, so it can be used later as a stripping fluid. Up to 700 kmol/h of ethylene at 40 °C and 1 atm are available. All steam needs to be condensed, and ethylene glycol is not recommended because the outlet temperature exceeds 90 °C. At the plant is suggested to use a vertical counter-current exchanger with steam as the fluid through the tubes. For this case pressure drop is not a problem.

(a) Perform a heat exchanger design in Aspen HTFS + ®.
(b) Take the results obtained in (a) and use the geometry in an Aspen Plus® simulation to evaluate the design.
(c) Take the results obtained in (a) and use the geometry in an Aspen HYSYS® simulation to evaluate the design.
(d) Compare the results.

P4.8 A Freon-12 stream has a flow of 100 kmol/h at 0 °C and 3 atm; it must be vaporized for use in a subsequent operation in plant. It is available for ethylene at 70 °C and 2 atm. At the plant is recommended to use 80 BWG tubes, and also that the pressure drop must be as minimum as possible.

(a) Perform a heat exchanger design in Aspen HTFS + ®.
(b) Take the results obtained in (a) and use the geometry in an Aspen Plus® simulation to evaluate the design.
(c) Take the results obtained in (a) and use the geometry in an Aspen HYSYS® simulation to evaluate the design.
(d) Compare the results.

P4.9 A heat exchanger is required to heat 52,000 kg/h of 50 % Ethane 25 % Propane, and 25 % Butane from 25 °C to 45 °C the stream will enter the exchanger at 3600 kPa and must not reach bubble point in the heat exchanger. The stream will be heated with isooctane at 120 °C and 700 kPa with a flow rate of 16,000 kg/h. Use a 1–2 shell and tube heat exchanger with 3/4-in., 16 BWG carbon steel tubes, 20 ft long, and 1-in. square pitch. Isooctane flows on the shell side. Assume a combineD fouling factor of 0.002 (h ft^2 °F)/Btu. Design a suitable heat exchanger considering a 30 % overdesign factor

P4.10 Glycerol at 100,000 lb/h enters the shell of a 1–4 shell and tube heat exchanger at 300 °F and is cooled to 140 °F with cooling water heated from 80 to 130 °F. Assume that the mean overall heat transfer coefficient is 100 Btu/(ft^2 h °F) and the tube-side velocity is 5 ft/s. Use ¾-in 16 BWG tubing (OD: 0.75-in. ID: 0.62-in arranged on 1-in. square pinch)

   a. Calculate the number of tubes, length of the tubes, and tube-side heat transfer coefficient
   b. Calculate the shell side heat transfer coefficient that allows an overall heat transfer coefficient of 100 Btu/(ft^2 h °F)

P4.11 Water is available as a cooling medium for cooling 30,000 kg/h of a gas stream with the composition shown in Table P4.1 from 450 to 80 °C. Propose preliminary 1–2 shell and tube heat exchanger design. The inlet temperature of the water is 25 °C and the maximum allowable outlet temperature is 50 °C

Table P4.1   Gas stream composition for problem P4.11

Gases	Mol%
H_2	0.36
H_2S	20.77
CO_2	0.03
N_2	1.34
Methane	26.30
Ethane	18.69
Propane	18.13
Butane	14.38

References

Alfa Laval (2011, 08 11) About us: Alfa Laval. http://local.alfalaval.com/en-gb/about-us/news/Pages/WasteHeatRecovery.aspx. Accessed 12 Jul 2014

Aspen Technology Inc (2014) Aspen Energy Analyzer. https://www.aspentech.com/products/aspen-hx-net.aspx. Accessed 25 June 2014

Butterworth D, Cousins L (1976) Use of computer programs in heat-exchanger design. Heat Exchanger Design and Specifications

Gundersen T (2000) A process integration primer—implementing agreement on process integration. International Energy Agency, Trondheim

Gundersen T (2013) Heat integration: targets and heat exchanger network design. In: Kleme's J (ed) Handbook of process integration (PI): minimisation of energy and water use, waste and emissions. Wood head Publishing, Cambridge, Chapter 4

Holman J (1999) Transferencia de calor. (10ª. reimpr.). Cecsa, México

Kemp I (2006) Pinch analysis and process integration: a user guide on process integration for the efficient use of energy. Elsevier, Amsterdam

Kern D (1999) Procesos de transferencia de calor (31ª. reimpr.). Cecsa, México

Mukherjee R (1998) Effectively design shell-and-tube heat exchangers. Chem Eng Prog 94:21–37

Smith R (2005) Chemical process design and integration, 2nd edn. John Wiley & Sons, Chichester

TEMA (1997) Standards of tubular exchangers manufacturers association, 9th edn. http://www.tema.org

Timonen J, Markku A, Huuhka P (2006) Competitiveness through integration in process industry communication. Helsinki: technology programme report 11/2006. National Technology Agency of Finland (TEKES)

Umbach JS (2014) Online pinch analysis tool. University of Ilinois at Chicago. http://www.uic-che.org/pinch. Accessed 20 June 2014

Chapter 5
Chemical Reactors

5.1 Introduction

Chemical reactors are the center of the all industrial chemical processes since they allow the transformation of the raw materials to products with high added value. This equipment defines the whole process, since the preparation process of raw materials depends on the reaction conditions as well as the effluents of the reactor determine the separation strategy and the difficulty to get the desired products with the adequate purity.

Several modules in the simulation packages allow simulations to model different types of reactors for different applications. Proper modeling of reactor is important for the correct functioning of the simulation and its proximity to the actual conditions. In this chapter the available modules on simulation packages are shown, and how the information is entered to simulate this important operation in the industry.

5.2 General Aspects

There are some terms to be introduced to take advantage more efficiently of the information that can be extracted from the simulators; and likewise ensure the quality of data to be entered to the calculation. Here are some fundamental concepts about chemical reactors, it should be noted that the simulators used in this text do NOT perform reactor design.

The design of a reactor refers to the calculation of its volume and dimensions to achieve a specific conversion. In these simulators this calculation cannot be performed, so it is always necessary to know beforehand the volume of the reactor; so that, the calculation of reactors in such simulators lastly corresponds to an evaluation of equipment already designed.

© Springer International Publishing Switzerland 2016
I.D.G. Chaves et al., *Process Analysis and Simulation in Chemical Engineering*,
DOI 10.1007/978-3-319-14812-0_5

5.2.1   Chemical Reaction

A chemical reaction is a process which one or more substances, called reagents, are transformed into one or more substances with different properties, those are known as products. In a chemical reaction, the bonds between the atoms that form the reagents are broken. Then, atoms are rearranged differently, forming new bonds and leading to one or more substances different from the initial substances (Harriot 2003; House 2007).

The general way to represent a chemical equation is shown below:

$$aA + bB \rightarrow cC + dD \qquad (5.1)$$

Where the substances which are to the left side of the arrow are the reagents, and the others located to the right side of the arrow, are the products.

5.2.2   Stoichiometry

The stoichiometry (from the Greek stoicheion, "element" and métrón, "measure") is the calculation of the quantities of reagents and products in the course of the chemical reaction.

These relationships can be deduced from the atomic theory, although historically it has stated without reference to the composition of matter, according to various laws and principles.

5.2.3   Conversion (Fogler 2008)

The conversion is related with efficiency of the reaction and having as a measure one of the substances present in the reaction, generally the limiting reagent. The definition of the conversion is shown below:

$$x \equiv \frac{\text{moles of A that react}}{\text{total moles of A}} \qquad (5.2)$$

Therefore, the conversion gives an idea of how efficient was the reaction from the amount of A that has been consumed to generate other substances. However, it only works to refer to the consumption of the reagent A, regardless of which product is being produced, in most cases, more than one reaction occurring simultaneously; in such cases it is also necessary to include the concept of Selectivity.

5.2.4 Selectivity

There are four (4) different types of reactions: reactions in parallel, series reactions, complex reactions, and independent reactions. Several reactions can occur in addition to the reaction of interest; and this reaction should be promoted over the others.

The selectivity indicates how favorable the production of one product is over another. That is because the rate of both reactions can be related. For example consider that the reaction A is the production reaction of interest (D) and the reaction (B) is a collateral reaction that produces (U). Then the selectivity can be expressed as:

$$r_{D/U} \equiv \frac{r_D}{r_U} = \frac{\text{formation rate of D}}{\text{formation rate of U}} \tag{5.3}$$

However, for calculation purposes, the total selectivity that related product streams instead of the reaction rates is generally employed.

$$r_{D/U} \equiv \frac{F_D}{F_U} \tag{5.4}$$

Experimental evidence shows that there is an inverse relationship between the conversion and selectivity. For a greater conversion, selectivity of the product of interest (D) is decreased as reducing conversion favors selectivity.

5.2.5 Reaction Kinetics

The reaction kinetics measures how fast the transformation of substance A to the substance B occurs. It mathematically means the change of the concentration of the substance "i" in time.

$$r_i \equiv \frac{dC_i}{dt} \tag{5.5}$$

When the reaction is homogeneous, that is, performed in the same phase, a power law model can be used as shown below:

$$r_A = -k[A]^{\alpha}[B]^{\beta} \tag{5.6}$$

Where the coefficients α and β are experimental. Also this kind of the reactions are named depending on the values taken by these constants, for example, if $\alpha = 1$ and $\beta = 1$. Then the reaction is order 1 for α, order 1 for β and global order 2.

5.2.6 Kinetic of Heterogeneous Reactions

In many cases the reactions cannot be carried out, therefore it is necessary to include a catalyst for the reaction to occur at a measurable extent. In some of these cases the catalyst is not in the same phase as the reactants, increasing the difficulty of modeling this type of phenomena.

When a heterogeneous catalyst is included, the overall reaction process has more steps. In this case, the reactants must reach the catalyst surface, react and then leave that surface. All this creates a concentration gradient, which considerably affects the global reaction kinetics.

For these cases, there is the LHHW model, which adequately represents some reactions where the diffusion of the reactants into the catalyst limits the reaction rate. However, for the heterogeneous reactions it can be applied also the power law model, but to apply it in this way, the transfer phenomena that happen in the catalyst are neglected.

$$r_{\text{LHHW}} = \frac{[\text{Kinetic Factor}][\text{Driving Force}]}{[\text{Adsorption}]} \tag{5.7}$$

The factor called Kinetic Factor is identical to that used previously for homogeneous power law kinetics. The Driving Force factor has the following form:

$$[\text{Adsorption}] = k_1 \prod_i^M C_i^\alpha - k_2 \prod_i^M C_i^\beta \tag{5.8}$$

The term referred as Adsorption must be entered in the following form:

$$[\text{Adsorption}] = \left[\sum_i^M K_i \left(\prod_j^N C_j^\gamma \right) \right]^m \tag{5.9}$$

Significantly, for the introduction of this type of kinetics in process simulators, it is necessary that the reaction rate will be in kmol/s m^3 if it is calculated on the volume of reactor, or in kmol/(s kg cat) if it is calculated related to the weight of catalyst.

In Sect. 5.8 it is shown in both simulators the use of these kinetics and reactions with heterogeneous catalyst.

5.3   Equations for Reactor Design

The design equations play an important role in the design of reactors, because they represent their material balance, and allow preliminary reactor design. For detailed information about deductions of these design equations refer to this chapter's references.

5.3.1   Continuous Stirred Tank Reactor

The Continuous Stirred Tank Reactor (CSTR) is one of the simplest reactors to model and design. It belongs to the category of kinetic reactors; meaning that, it is required to know the kinetic data of the system to design the reactor. It consists of a tank with a stirrer generally installed on top especially designed for the physical properties of the mixture (viscosity, density, compatibility of the reagents, etc.).

Sometimes an internal support can be installed for reactions that require a solid catalyst.

This reactor is generally used in liquid phase reactions and must ensure an adequate mixing of the components, so that the reaction is not restricted by mass or heat transfer limitations.

The design equation for this reactor is shown below:

$$V = \frac{F_{A0} - F_A}{-r_A} = \frac{x F_{A0}}{-r_A} \qquad (5.10)$$

With this equation it is possible to size a reactor that allows the reaction to reach certain conversion by knowing kinetic information. Also for a reactor of known dimensions it is possible to calculate the conversion that would be achieved using it at given conditions.

5.3.2   Plug Flow Reactor (PFR)

The plug reactors or PFR, consist of a cylindrical tube, sometimes tube bank, and is commonly used to carry out gas phase reactions.

In this type of reactor the reaction takes place through the reactor; so that the concentration varies along the length of the reactor and thus the reaction rate (unless zero order). Sometimes the reactor content can be heated or cooled by means of heat exchange devices in case of highly exothermic or endothermic reactions.

The design equation for this reactor is shown below:

$$\frac{dF_A}{dV} = r_A \tag{5.11}$$

In case of a packed bed reactor or PBR, there is an internal support containing the catalyst. The reaction rate changes along the length of the reactor. The material balance is performed around the support, so that the design equation is based on the weight of the catalyst contained in the reactor.

The equation for this reactor is shown below:

$$\frac{dF_A}{dW} = r_A' \tag{5.12}$$

where W represents the weight of the catalyst and r_A' the reaction rate relative to the weight of the catalyst.

5.3.3   Batch Reactor (Batch)

Batch reactors are widely used in small industries due to its versatility and its operating characteristics. Consisting of a stirred tank where predetermined amounts of reactants are fed and the operation is set as a function of the reaction time. Once this time is reached, the reactor is turned off, discharged, cleaned, and is enabled for another cycle of operation. This feature makes it important for small productions and different products.

The equation for this reactor is shown below:

$$\frac{dn_A}{dt} = r_A V \tag{5.13}$$

The corresponding material balance was carried out as a function of time because it can be considered a perfect mixing tank and thus the concentration does not change with the position.

5.4   Modules Available in Aspen Plus$^{®}$

There are three types of modules of reactors in the simulation packages: they can be based on balances, on thermodynamic equilibrium or on kinetics. The one based on balance, performs a material and energy balance to calculate the output of the reactor, the one based on the equilibrium calculates output conditions minimizing the Gibbs-free energy, and kinetic reactors are calculated taking into account the reaction kinetics. However, it should be noted that the Aspen Plus$^{®}$ and Aspen

HYSYS® simulators do NOT perform reactor design; they simply calculate the conversion using the data regarding flows, operating conditions, reactor volume, and reaction kinetics using the design equations corresponding to each reactor.

In Aspen Plus®, there are two modules based on the balance: *RYield* and *RStoic*. There are two modules based on the equilibrium: *RGibbs* and *Requil*, and three modules based on kinetic: *RCSTR*, *RPlug*, and *RBatch* (Table 5.1).

The input for the reactions must be done on the *Reactions > Reactions section* where the type of reaction and the parameters required are selected depending on its application. In *Reactions > Chemistry* route it can be entered as special chemical reactions such as dissociation or salt precipitations.

Table 5.1 Summary of modules available in Aspen Plus® for chemical reactors

Icon	Name	Description
Ideal reactors		
RYield	RYield	Allows to perform the simulation based on individual yield of component, in kg or kmol of product per kg of feed, it has two features: the inclusion of reactions is not necessary and it does not ensure an atomic balance in the reactor although yields are normalized to allow the mass balance
RStoic	RStoic	Allows to perform the simulation of reactions by providing a molar extension (product flow) or any fractional molar conversion of one of the reactants, it provides the mass balance in the equipment and makes an estimate of the heat of reaction
REquil	REquil	This module performs the calculation of the condition for chemical and thermodynamic equilibrium for reactions supplied by an approach stoichiometry, evaluates the thermal load, and the equilibrium constant for the reactions involved. It cannot solve three-phase vaporization
RGibbs	RGibbs	It calculates chemical and phase equilibrium by minimizing Gibbs-free energy, this module does not require the provision of reactions and therefore does not calculate the chemical equilibrium constants
Kinetic reactors		
RCSTR	RCSTR	The module RCSTR models rigorously stirred tank reactors, using the assumption of perfectly mixed, that is, the reactor has the same composition of the product stream, it can handle kinetic and equilibrium reactions
RPlug	RPlug	The module Rplug models rigorously tubular reactors with and without packaging, the module assumes perfect mixing in the radial direction and that there is not mixing in axial direction, it enables the inclusion of coolant which flows counter-current or in parallel, and it requires knowledge of the kinetics of the reaction
RBbatch	RBatch	The model RBatch allows to simulate batch reactions where kinetic information must been known. The feeding streams can be continuous or discontinuous. Batch specific programming is required

5.5 Available Modules in ASPEN HYSYS®

For the case of Aspen HYSYS® there are two modules based on the balance: Yield Swift Reactor and Conversion Reactor. There are also two modules based on the equilibrium: Equilibrium Reactor and Gibbs Reactor. And two modules based on the kinetics: Plug Flow Reactor and CSTR (Table 5.2). It must be noted that HYSYS® does not have any module for batch reactions, although, the module CSTR can be executed in dynamic mode for that purpose.

As in Aspen Plus ®, it is necessary to set up the reactions at a specific site of the interface and to organize them into groups. For the case of Aspen HYSYS® the reaction data must be entered at the beginning of the simulation when the model components and properties are set up in the Reactions tab. However, it must be

Table 5.2 Summary of modules available in Aspen HYSYS® for chemical reactors

Icon	Module	Characteristics
Y	Yield swift reactor	Yield Shift Reactor is supported by the use of tables of data for the modeling of reactors and makes jump calculations. The operation can be used in complex reactors where there is a model. There are two methods of setting the reaction in a reactor Yield Shift: Only yield or conversion percentage. Depending on what information it provides, the reactor automatically uses the appropriate equation to solve the reaction
C	Conversion reactor	Conversion reactor is a vessel in which is carried out conversion reactions, only is allowed linking reaction groups which include conversion reactions. Each reaction in the set occurs until the specified conversion is reached or until a limit reactant is completely consumed
E	Equilibrium reactor	The reactor of equilibrium is a module appropriate for equilibrium reactions. The reactor exit streams are in chemical and physical equilibrium. The group of reactions linking the module can contain an unlimited number of equilibrium reactions, which are solved simultaneously or sequentially. Neither, the components or the mixing process required to be ideal because Aspen HYSYS® can calculate the activity of each component in the mixture based on the fugacity of the mixture and the pure components
G	Gibbs reactor	Gibbs reactor is unique, because it can work with or without a reaction set. It calculates the output compositions as the chemical and phase equilibrium, it does not require specifying the stoichiometry to calculate de product composition. The condition of minimum Gibbs-free energy is used to calculate the composition of the product mixture
	CSTR	Equivalent to the module RCSTR of the Aspen Plus®. It can simulate a batch reactor using the dynamic mode of the module
	Plug flow reactor	Equivalent to the module RPlug of the Aspen Plus®. It simulates chemical reactions occurring into tubular reactors making the calculation along its length

arranged in sets manually and must be associated in group reactions to model properties so that the simulator can do the appropriate calculations.

Once this step is made, the reactor module can select which group of reactions to use. It should be noted that the groups must contain the same type of reactions (Equilibrium, kinetic, conversion or equilibrium).

5.6 Introductory Example of Reactors

5.6.1 Problem Description

To illustrate the use of the modules associated to the chemical reactors, in the following example the production of ethyl acetate is studied through the esterification of acetic acid with ethanol; this reaction is known, because it is limited by the chemical equilibrium.

$$CH_3COOCH_2CH_3 + H_2O \tag{5.14}$$

Usually, the reaction is carried out in batch reactors at atmospheric pressure and reflux temperature.

The existing kinetic for this reaction is of the power law type, with:

$$k_1 = 1.9 \times 10^8 \; E_{a1} = 5.95 \times 10^7 \frac{J}{kmol} \tag{5.15}$$

$$k_2 = 5.0 \times 10^7 \; E_{a2} = 5.95 \times 10^7 \frac{J}{kmol} \tag{5.16}$$

This is a first order reaction with respect to each one reactant of both directions (second global order) and the base of the concentration is molarity.

5.6.1.1 Aspen Plus® Simulation

In this example, shown in Fig. 5.1 will be generated to illustrate the behavior of the modules of chemical reactors, including: RStoic, RGibbs, RPlug y RCSTR.

The compounds involved are included in the reaction (ethanol, water, ethyl acetate, and acetic acid) and the model NRTL-HOC is selected as a thermodynamic package.

The conditions of the FEED flow are shown in Table 5.3:

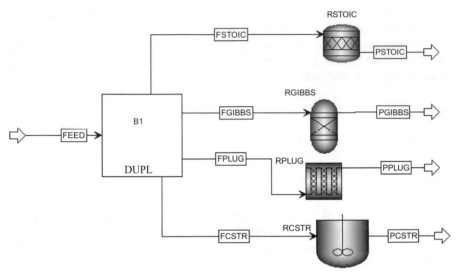

Fig. 5.1 Flow diagram of the example proposed in Aspen Plus®

Table 5.3 Feed flow
conditions of the introductory
example

Variable	Value
Temperature	75 °C
Pressure	1 atm
Mass flow [kg/h]	
Water	160
Ethyl alcohol	8600
Acetic acid	11,570

5.6.1.2 Stoichiometric Reactor (RSTOIC)

Once feeding flow is specified, then each one of the reactors is specified; the stoichiometric reactor RStoic presents the configuration window shown in the Fig. 5.2.

In this window the reactor thermodynamic condition is introduced by selecting two of the displayed options (pressure, temperature, heat load, or vapor fraction), in this particular step a temperature of 75 °C and 1 atm. of pressure is introduced.

Next, click on the Reactions tab shown in Fig. 5.3.

In this window, the new reaction is created by clicking on the *New* button, after carrying out this action, the following window appears, where the specifications of the reaction are entered. Here the involved compounds, differentiated as products and reactants, are introduced in the reaction; although the stoichiometric coefficients are introduced in the reaction, providing information about reaction conversion, either molar conversion of the molar conversion of some reactants or like

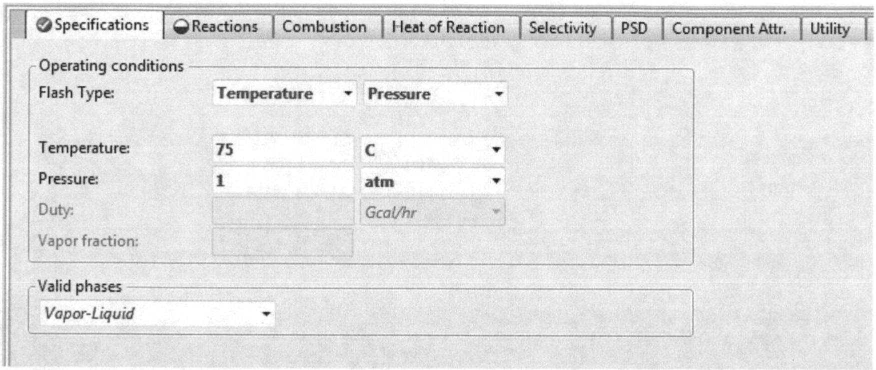

Fig. 5.2 Module RStoic configuration in Aspen Plus®

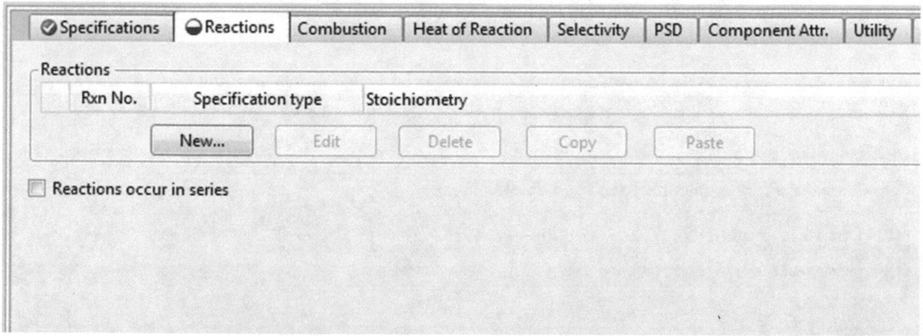

Fig. 5.3 Reactions input in the module RStoic in Aspen Plus®

molar extension (product flow). In this case, a 0.65 fractional conversion of ethanol is specified. Thus, this reaction is configured and the window is reached as shown in the Fig. 5.4.

With the above information, stoichiometric reactor configuration is completed; then, the Gibbs reactor must be performed.

5.6.1.3   Gibbs Reactor (RGIBBS)

The Gibbs reactor presents the configuration window shown in Fig. 5.5. It is possible to specify the thermodynamic condition at which the calculations are performed. In this particular case, pressure is 1 atm and temperature is 75 °C.

In the configuration window the calculation options can be set as well. Figure 5.6 shows the list of allowed options. In this case both the phase and chemical equilibrium calculated.

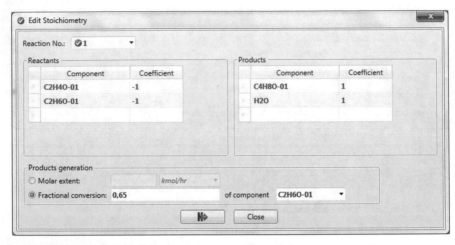

Fig. 5.4 Specification window of the reaction in Aspen Plus®

| Specifications | Products | Assign Streams | Inerts | Restricted Equilibrium | PSD | Utility |

Calculation option:

Calculate phase equilibrium and chemical equilibrium

Operating conditions

Pressure:	1	atm
◉ Temperature:	75	C
◯ Heat Duty:		Gcal/hr

Phases

Maximum number of fluid phases:

Maximum number of solid solution phases: 0

☐ Include vapor phase

☐ Merge all CISOLID species into the first CISOLID substream

Fig. 5.5 RGibbs module configuration in Aspen Plus®

As a next step, click on the products tab (see Fig. 5.7), where is possible to input its specifications; by default, the option *RGibbs considers all components as products*.

5.6.1.4 Kinetic Reactors

There is a particularity for rigorous reactor calculations. The involved reactions must be specified outside of the reactor module. So, the first step in this simulation

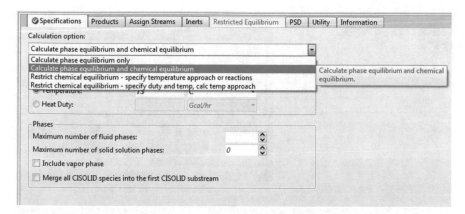

Fig. 5.6 Options window for RGibbs calculating module in Aspen Plus®

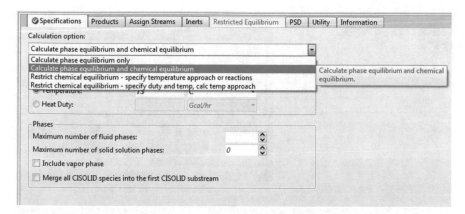

Fig. 5.7 Product specifications for RGibbs module

is to install the reaction. In the navigation tree Data Browser, the item Reactions is selected, next, in the Reactions submenu a new set of reactions is created, in this step a menu appears which allows to choose the type of kinetic expression to use. In this case power law or POWERLAW is selected (see Fig. 5.8).

After creating the R-1 group, reactions information is completed. For this purpose, a new reaction is introduced once again by clicking on *New* button, where a new window as shown below appears (Figs. 5.9 and 5.10).

Available fields must be completed, including the stoichiometric coefficients and the exponents associated with kinetic expression. Once these data are supplied, click on Next button to proceed to the kinetic data specification.

This window displays information associated with the kinetic expression including the reaction phase and the concentration base provided above. The procedure for the reverse reaction is repeated by introducing the parameters specified at the beginning of the example. After the kinetic reaction is set, proceed to define the desired kinetic reactor.

Fig. 5.8 Reaction type selection menu in Aspen Plus®

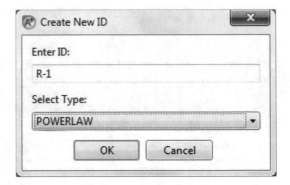

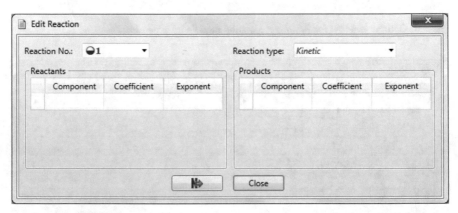

Fig. 5.9 Window to specify reactions for several reactors in Aspen Plus®

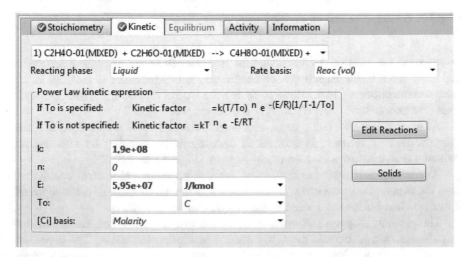

Fig. 5.10 Window to specify kinetic data in Aspen Plus®

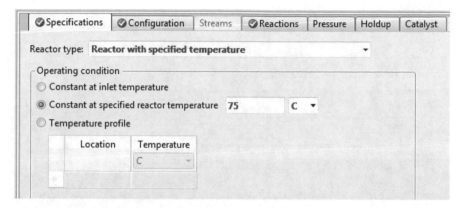

Fig. 5.11   RPlug Module configuration in Aspen Plus®

5.6.1.5   Plug Reactor (RPLUG)

For a PFR reactor configuration, select the Setup window of the *RPlug* block (see Fig. 5.11).

First, the thermal condition of the reactor must be specified, taking into account the process requirements; these options are given below:

- Reactor with specified temperature.
- Adiabatic reactor.
- Reactor with coolant at constant temperature.
- Reactor with coolant in co-current.
- Reactor with coolant counter-current.
- Reactor with temperature-coolant profile.
- Reactor with external heat flow profile.

In this case, due to the lack of information, option 1 is selected. Reactor with specified temperature and its value is set to 75 °C. Then, select tab Configuration where the length and diameter of the equipment is specified; the values are shown in Fig. 5.12.

The next step is to select the Reactions tab, where the reaction set created earlier is entered. In this way the configuration of tubular reactor is completed.

5.6.1.6   Stirred Tank Reactor (RCSTR)

As in the case of the plug reactor (PFR), before setting up the stirred tank reactor (CSTR), it is necessary to specify the set of reactions that will be used outside the module. Then, the Setup window of RCSTR block, where the following table (see Fig. 5.13) appears, must be selected. At this point 2 thermodynamic conditions under which the equipment operates must be defined. Temperature, pressure, or

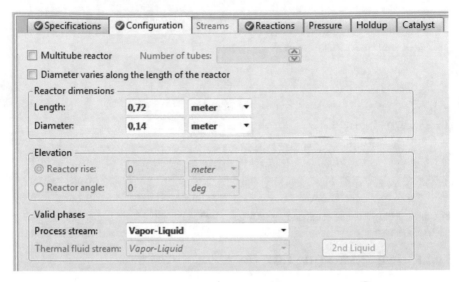

Fig. 5.12 Geometric configuration of the reactor module PFR in Aspen Plus®

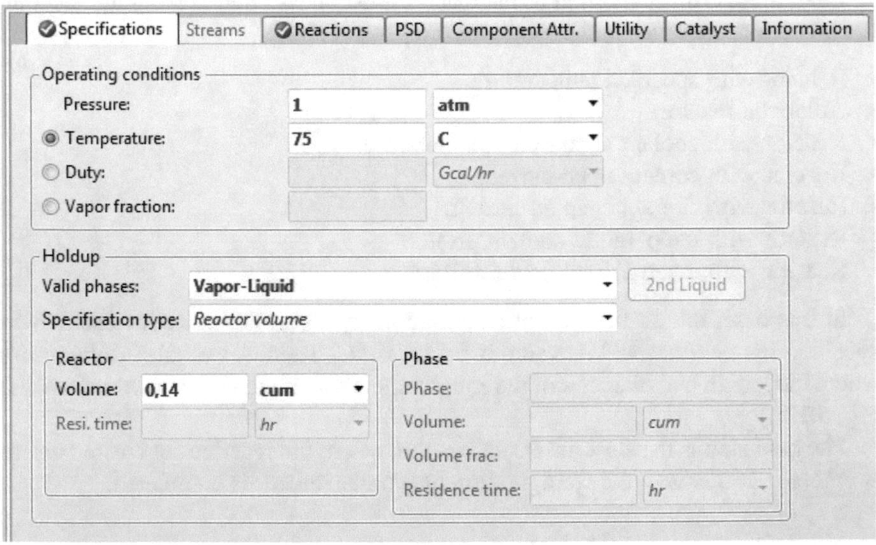

Fig. 5.13 RCSTR module configuration in Aspen Plus®

heat load can be defined. In this case, pressure (1 atm.) and temperature (75 °C) are defined.

Similarly, in the specifications window other aspects of the reactor may be defined, such as: the phases which the calculation is valid and the volume or residence time at which the calculations are made. In this particular case, the valid phases are Vapor–Liquid and a volume of 0.14 m^3 is defined. Next, in the Reactions tab, the reaction set previously created is loaded.

5.6.2 Simulation in Aspen Hysys®

To compare the results, the flowsheet shown in Fig. 5.1 must be completed, by introducing the components ethanol, water, acetic acid, and ethyl acetate, and using the NRTL thermodynamic model.

Once the diagram is built, note that the single reactor is gray with yellow rim, (partially defined), it is the Gibbs reactor, because in this reactor model the definition of the present reactions is not required.

Once the flow chart is built, it is necessary to define the feed flows. In the case developed in Aspen Plus® a module called Duplicator was used to define all reactant streams from a single stream; Aspen HYSYS® procedure is different and it is described once the process stream "GIBBS Feed" is defined.

Using the data shown in Table 5.3, the stream FSTOIC is defined. Once this stream is specified the other three streams are defined from it (Fig. 5.14).

First, the properties window of the current FGIBBS is opened by double clicking on its icon (Fig. 5.15). Then, click on the Define from other Stream button and the sub-window shown in Fig. 5.15 appears. This allows selecting the stream from which the data is imported. For this example is the stream FSTOIC. Click OK.

Once the current FGIBBS is defined, the two remaining feed streams are defined in an analogous way.

Now, it should be noted that despite of defining the reactor feed streams, these reactors remain undefined. The first step to specify these reactors is to create the set of reactions that will be considered in each reactor.

5.6.2.1 Reaction Specifications and Reaction Sets

To input the corresponding chemical reactions, go to the *Properties* button by clicking on it. In this section, go to the Reactions tab and click on *Add Reaction* (Fig. 5.16), the window shown in Fig. 5.17 is displayed, where the type of reaction is selected. Initially the option kinetic is selected; then, the required data in the stoichiometry tabs (Stoichiometry), base (Basis), and parameters (Parameters) must be entered.

In stoichiometry, the stoichiometric coefficients of the reaction are provided, positive for products and negative for reactants, the values for the reaction orders of each component in forward and reverse directions in the kinetic expression are also introduced. Data is available in Eqns. (5.14), (5.15), and (5.16). Then, in the Basis tab the concentration expression, the reaction phase, and units of the kinetic constant are specified. Finally, the parameters of the pre-exponential factor and

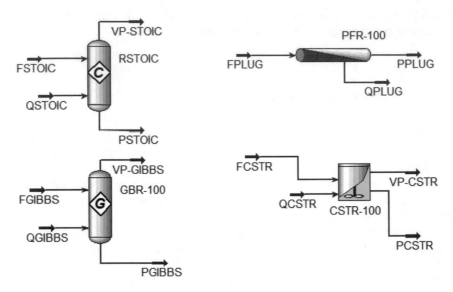

Fig. 5.14 Flow diagram of the proposed example in Aspen HYSYS®

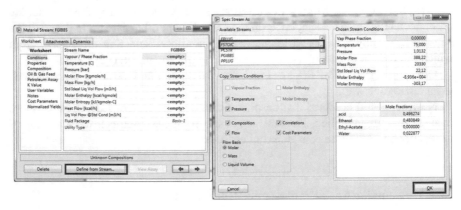

Fig. 5.15 Definition of one stream from another in Aspen HYSYS®

the activation energy for the reaction are introduced in the respective tab, in forward and reverse directions (Eqns. 5.15 and 5.16) (Figs. 5.18, 5.19, and 5.20).

In this way the reaction is completely defined, now it is necessary to configure the reaction set. In this case, it consists of a single reaction. Also it is necessary to associate the reaction set to the property package. Aspen HYSYS® has this tool to create these reaction sets to group reactions with similar characteristics (kinetics, conversion, equilibrium) in a single set that is added to the calculation module.

On the Reactions tab, click the reactions set that was created previously. This is shown in the window from Fig. 5.21. The set of reactions can be named as well as to select the active reactions in it.

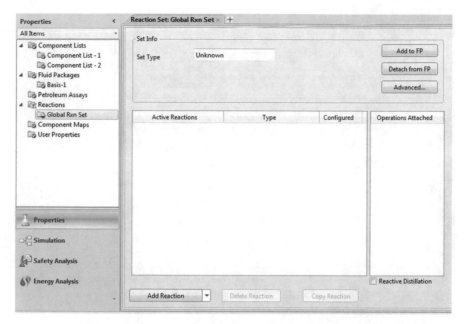

Fig. 5.16 Entry window for Aspen HYSYS® reactions

Fig. 5.17 Selection
window of the type of
reaction in Aspen HYSYS®

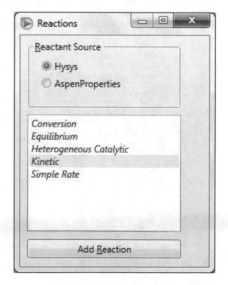

The only remaining step to define the set of reactions is to associate the thermodynamic properties package, this is done by clicking the add fluids package Add to FP button. Then a sub-window appears where the property package is selected by clicking on Add Set to Fluid package. This allows the simulator to have sufficient information to perform their respective calculations. Remember to

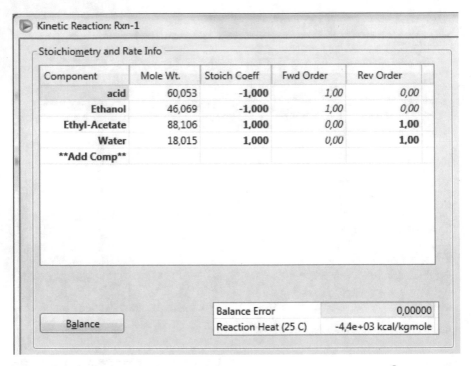

Fig. 5.18 Definition window for kinetic reaction stoichiometry in Aspen HYSYS®

Basis	
Basis	**Molar Concn**
Base Component	**Ethanol**
Rxn Phase	*LiquidPhase*
Min. Temperature	*-273,1 C*
Max Temperature	*3000 C*

Basis Units	*kgmole/m3*
Rate Units	*kgmole/m3-s*

Fig. 5.19 Kinetic reaction base in Aspen HYSYS®

always associate the property package; otherwise, the simulator cannot perform the calculation for the reactor (Fig. 5.22).

Since the reaction that was just configured is of the kinetic type, it only works for the kinetic reactors, PFR and CSTR. It is necessary to create a new conversion type

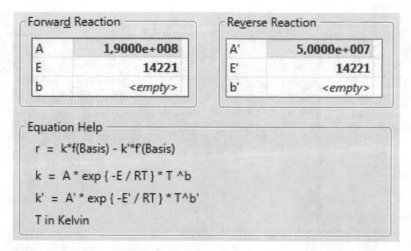

Fig. 5.20 Parameters definition of kinetic reaction in Aspen HYSYS®

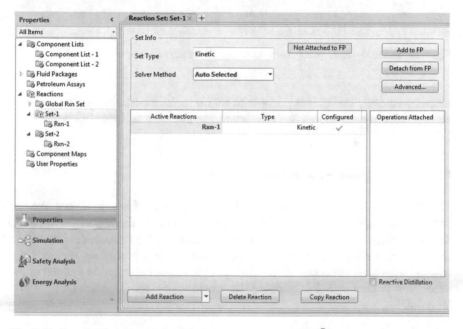

Fig. 5.21 Specification sets window of reactions in Aspen HYSYS®

reaction where the stoichiometry of the reaction and the conversion from one reactant (limiting reactant) are introduced. In addition, it is necessary to create a new set of reactions involving the conversion reaction, to operate in the conversion reactor, this procedure is analogous to the one developed for the reactions set for

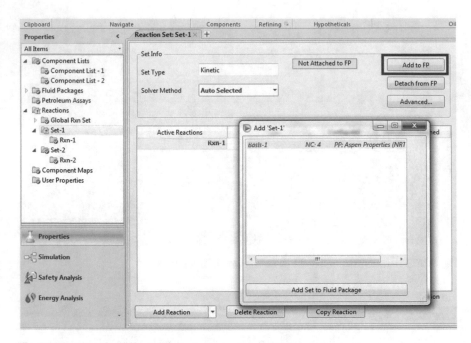

Fig. 5.22 Thermodynamic package selection window for the reactions set

kinetic reactors. Finally, it is possible to return to the simulation environment by closing the windows and click on Return to Simulation Environment button.

5.6.2.2   Conversion Reactor Configuration

The first step in the conversion reactor configuration is to associate the set of reactions with the equipment; this is done by opening the properties window. Select the Reactions tab and assign the set of reactions. The reaction set should be the one containing the conversion type reaction. This procedure specifies the reactor and results are obtained. However, the fractional conversion defined in the base environment can be modified by clicking on the *Conversion%* button (Fig. 5.23).

5.6.2.3   Kinetic Reactor Configuration

In the definition of kinetic reactors it is necessary, at least to specify the kinetics of the reactions occurring in the equipment, as well as its size, meaning that it runs in a rating mode.

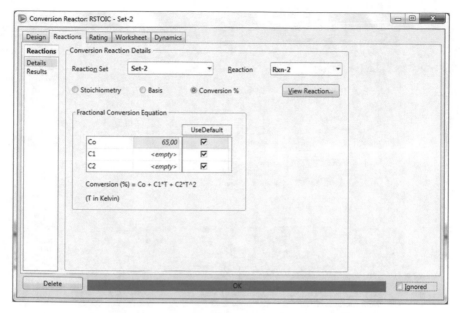

Fig. 5.23 Reactions tab of conversion reactor module in Aspen HYSYS®

5.6.2.4   CSTR reactor

As mentioned above, the first step in the specification of kinetic reactors is to assign the reactions with their respective kinetic expressions, this is done by selecting the set of reactions where kinetic expressions were included (Fig. 5.24).

Now, it is necessary to specify the size of the equipment. In this case, a volume of 0.14 m³ is selected, which is introduced into the Rating tab of the properties window of CSTR reactor (Fig. 5.25). Also, it is important to change the liquid volume percentage to 100 %. Thus the specification of CSTR reactor is concluded. Remember that Aspen Plus® and Aspen HYSYS® simulations do NOT perform reactor design, so it is necessary to size the equipment beforehand to perform the calculation successfully.

For the tubular reactor (PFR), a similar procedure is performed to make the specification of CSTR. The first step is to assign the set of reactions. The information about the dimensions of the reactor is provided in the Rating tab, for this case a tube reactor of 0.14 m diameter and 0.72 m length is specified (Fig. 5.26).

5.6.3   Results Analysis

Table 5.4 presents a summary of the results of the product streams of the reactors simulated on Aspen HYSYS®, additionally Table 5.5 presents the values achieved

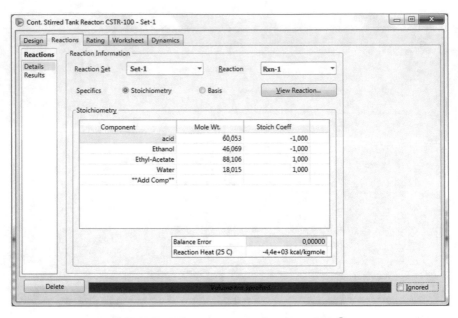

Fig. 5.24 Reactions tab of the CSTR reactor module in Aspen HYSYS®

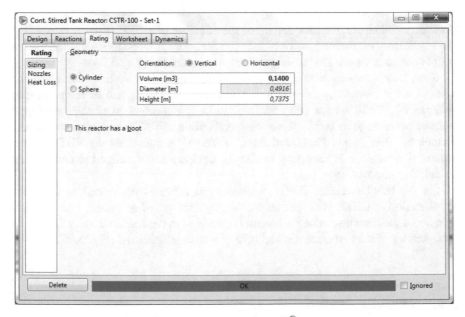

Fig. 5.25 Rating tab of the CSTR reactor in Aspen HYSYS®

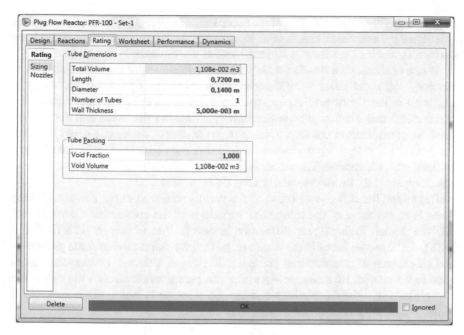

Fig. 5.26 Rating tab of the PFR reactor in Aspen HYSYS®

Table 5.4 Aspen HYSYS® results for the different reactors achieved

Properties		Gibbs	Conversion	CSTR	PFR
Molar composition of liquid phase	Ethanol	0.152	0.168	0.176	0.173
	Acetic acid	0.167	0.184	0.192	0.188
	Ethyl acetate	0.329	0.312	0.305	0.307
	Water	0.352	0.335	0.327	0.330
T outlet [°C]		75	75	75	75
Vapor fraction		0	0	0	0
Ethanol conversion		68.4 %	65 %	64.78 %	63.9 %

Table 5.5 Results achieved for the different reactors in Aspen Plus®

Properties		Gibbs	Conversion	CSTR	PFR
Molar composition of liquid phase	Ethanol	0.063	0.168	0.169	0.171
	Acetic acid	0.079	0.184	0.185	0.186
	Ethyl acetate	0.418	0.313	0.311	0.310
	Water	0.441	0.335	0.334	0.333
T outlet [°C]		75	75	75	75
Vapor fraction		1	0	0	0
Ethanol conversion		67.9 %	65.0 %	64.7 %	64.5 %

from Aspen Plus®. In order to make a comparison between the various parameters calculated in both simulators; it is important that the composition, thermodynamic condition, and reactant flow in all feed streams are the same.

When analyzing the results achieved in a single simulator for different types of reactors, the most striking differences occur in terms of conversion. As it is expected in the Gibbs reactor, because the calculation is based on the condition of chemical and thermodynamic equilibrium, it allows the maximum conversion possible of reactants in one step. Moreover, in the case of the kinetic based reactors, the conversion results are similar; indicating that the volumes of each one allow to achieve such conversion. The reader is invited to calculate the PFR reactor volume and compare it to the volume used in the CSTR reactor.

Regarding the differences between the results achieved in the two simulations there is an evidence of the substantial variations of the conversion calculation of Gibbs reactors. Perhaps, this difference is due to that in Aspen HYSYS ® the NRTL-HOC model is not in the database, so that the estimation of some properties and calculation of chemical and phase equilibrium is different, causing the difference in the results. However, another important consideration refers to neglect the vapor phase into the calculations because the reaction is taking place only in the liquid phase and this could modify the results.

5.7   Propylene Glycol Reactor Example

5.7.1   General Aspects

Industrial Propylene glycol (1,2-propanediol) is a high purity material which is produced by hydrolysis of propylene oxide at high temperature and pressure with an excess of water. This is a liquid distillate product with a purity specification of 99.5 %. Propylene glycol is a clear, viscous, nontoxic, water soluble, and hygroscopic liquid. This process is constrained by the current supply offer of propylene since this is derived from petroleum.

Propylene glycol is used in different products such as engine coolants, polyester resins, latex paints, and heat transfer fluids, among others. It is also used as a solvent and heat transfer medium or as a chemical intermediate product.

This example is proposed to build the model in steady state propylene production reactor, as it is shown in Fig. 5.27. Initially it is fed with 3600 kg/h of water (WATER) into the reactor mixed with a second stream (OXIDE) of 1085 kg/h of propylene oxide and 1050 kg/h of methanol. In real operation of the reactor, the sulfuric acid is used as catalyst to be dissolved in the water stream with a concentration of 0.1 % by weight. The methanol is placed into the mixture in order to improve the solubility between the propylene oxide and water. This mixture enters a CSTR reactor (REACTOR) where the hydrolysis reaction of propylene oxide is

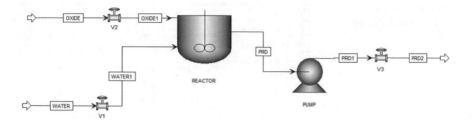

Fig. 5.27 Process scheme of the reaction to produce propylene glycol

carried out with more than 90 % of conversion. From the reactor the reaction products are removed in liquid phase (PRD).

The hydrolysis reaction of propylene oxide has been widely studied and various kinetic expressions have been published. Most of them expressed in terms of the oxide concentration. This example is proposed to use a second-order kinetic expression with respect to propylene oxide concentration.

$$C_3H_6O + H_2O \rightarrow C_3H_8O_2 \qquad (5.17)$$

$$-r_{C_3H_6O} = 9.15 \times 10^{22} \times \exp\left(\frac{-1.556 \times 10^8}{RT}\right) \times C_{C_3H_6O}^2 \qquad (5.18)$$

with:

$r_{C_3H_6O}$: Reaction rate in kmol/m^3 s.
$C_{C_3H_6O}$: Molar concentration of the propylene oxide in kmol/m^3.
Activation energy must be J/kmol in consistence with the kinetic equation.

5.7.2 Process Simulation in Aspen Plus®

Inlet process conditions and the configuration of some equipment are shown in Table 5.6 and Table 5.7.

Initially the process flow diagram according to the information is specified according to Fig. 5.27. The components that are part of the simulation are: propylene oxide, water, methanol, and propylene glycol (1,2-propanediol). The thermodynamic model used is NRTL.

The hydrolysis reaction of propylene oxide is highly exothermic. To control the temperature, this reaction is carried out in liquid phase and the reactor operates at high pressure (3 bar), thereby the evaporation of the mixture decrease, and further temperature increasing is controlled by placing an excess of water. In Fig. 5.28 the setup screen of the CSTR reactor appears. Note that the valid phases have been defined as Liquid Only and the reactor operation is adiabatic (duty is 0).

Table 5.6 Inlet flow conditions entering the reactor

Stream	Water	Oxide
Temperature (°C)	24	24
Pressure (bar)	6	6
Component flow (kg/h)		
H_2O	3600	–
C_3H_6O	–	1085
CH_4O	–	1050

Table 5.7 Equipment configuration from process diagram

Equipment	Reactor	Pump	Valve
Temperature (°C)	Adiabatic	–	–
Volume (m³)	1.14	–	–
Pressure (bar)	3	6	3
Type	CSTR	–	Equal percent globe

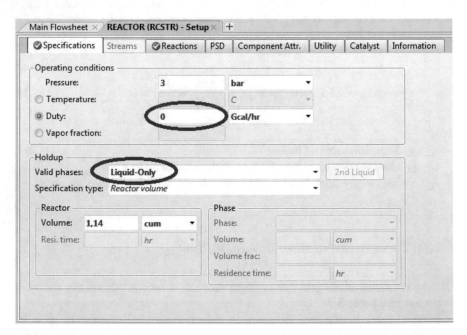

Fig. 5.28 Configuration of CSTR reactor of propylene glycol in Aspen Plus®

The objective of the installation of the valves within the flowsheet is to illustrate in a later chapter its use as control valves. Now, the maximum pressure drop in each one (3 bar on all valves) is specified and those are calculated in design mode to estimate the flow coefficient (Cv) and the percentage of opening. The simulator has within its database information corresponding to curves of control valves commercially available. This information, used in the calculation and design of the valve, allows adequately to define the required diameter.

It is important to make a correct choice of the valve size at this stage of the simulation so that later the process has a good controllability when in dynamic state. In this case the valves V1 and V2 are 1 in. diameter, globe type, manufactured by Neles-Jamesbury series V810_equal_percent_flow. Valve V3 has the same configuration except for the diameter which is 1.5 in. In Fig. 5.29 is shown how to configure the control valves.

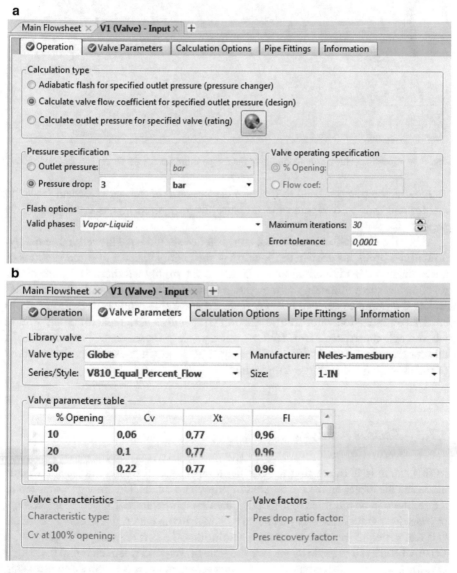

Fig. 5.29 Control valve specification. (**a**) Definition of the calculation type and pressure drop, (**b**) valve characteristics selection

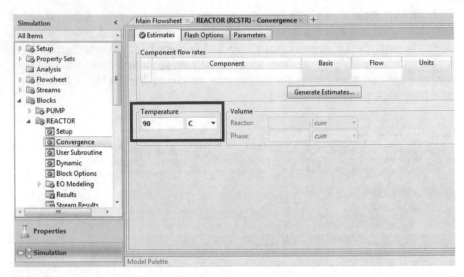

Fig. 5.30 Specifying the estimated value of the temperature for the CSTR reactor in Aspen Plus®

Finally, the pump is used to increase the pressure of the outlet stream from the reactor and thus to overcome the pressure drop required across the valve V3 to regulate the flow.

When the simulation is executed, convergence should be achieved immediately. Sometimes problems arise in the solution of the energy balance in some versions of the simulator. This is normal and it is due to the highly exothermic nature of the reaction and the nonlinearity of the system represented in higher-order kinetics that makes it difficult to achieve convergence. To solve this problem it is necessary to enter an initialization value for the reaction temperature in the reactor configuration tab through the route: *REACTOR > Convergence > Estimates > Temperature*. In this case it can be started with 90 °C. In Fig. 5.30 is shown the specification window of the estimated value of the temperature.

5.7.3   Results Analysis

In this example is interesting to note the temperature at which the reaction takes place and the effect on the conversion of propylene oxide. The flow of water fed to the reactor can be modified to affect the reaction temperature. Likewise, the change in water flow simultaneously generates a change in the material and energy balance that can generate multiple steady state solutions. To further study this process, a sensitivity analysis is made by varying the flow of water between 2500 and 9300 kg/h in order to see the effect it has on the reaction temperature. It is proposed to use a step size of 100 kg/h. After performing the analysis in forward direction and plotting it, a second analysis scouring the water flow in the reverse direction is done, that is from 9300 to 2500 kg/h with a step of −100 kg/h. The objective is to detect

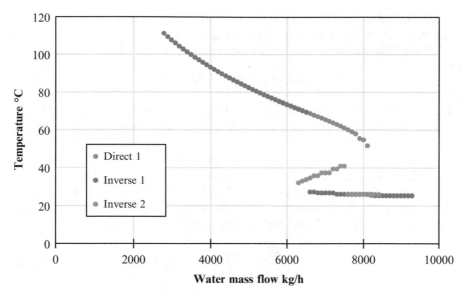

Fig. 5.31 Sensitivity analysis of the water flow on reaction temperature

different solutions of steady state for the same input conditions. The results are shown in Fig. 5.31.

It can be easily seen that the area of multiple solutions is between 6500 and 8700 kg/h of water. If the respective conversion were calculated, it would be found also one multiplicity zone which matches with the zone determined for temperature. In general, it can be said that the reaction should be operated under water flows at 6000–3000 kg/h, to ensure high conversion and appropriate temperature control. It is also noted that higher values of flow of water decrease the reaction temperature, reducing the conversion of propylene oxide and leading the reaction to the zone of instability and multiple stable states.

This problem is more interesting from a dynamic point of view trying to determine the control strategy for the system which ensures stable operation with high conversions. This problem is addressed in Chap. 8.

5.8   Methanol Reforming Reactor

5.8.1   Problem Description

Methanol plays an important role as a feedstock for the production of various chemicals, such as formaldehyde, acetic acid, biodiesel, and gasoline. Additionally, methanol has been considered as a reservoir of hydrogen which overcomes many of the problems associated with transportation and storage. The hydrogen production can be conducted from the reforming reaction of methanol in vapor phase. To this it has been found that the most active catalysts are based on ZnO–CuO–Al$_2$O$_3$ catalysts. This reaction achieves high conversion using the following reaction:

$$CH_3OH + H_2O \leftrightarrow CO_2 + 3H_2 \tag{5.19}$$

The simulation performed corresponds to the experimental setup and kinetic models reported in Tesser's work. In this work a single PFR reactor tube with a length of 12 cm and a diameter of 4 cm was used. In this case cylindrical catalyst pellets were used with a height and a diameter of 0.5 cm with a void fraction of 15 %. The system uses thermal oil as the refrigerant and operates between 125 and 325 °C. A feed flow between 1.621×10^{-6} and 2.341×10^{-6} and kmol/s was used. After these tests four kinetic models for the reaction were achieved, two of them correspond to the LHHW form and the other two to a power law.

$$r = \frac{\eta k_M b_M p_M}{1 + b_M p_M + b_W p_W} \quad \text{(Model 1)} \tag{5.20}$$

$$r = \frac{\eta k_M b_M p_M}{1 + b_M p_M + b_W p_W + b_H p_H} \quad \text{(Model 2)} \tag{5.21}$$

where the constants are defined as follows:

$$k_M = k_M^0 e^{\frac{-\Delta E}{RT}}; b_i = b_i^0 e^{\frac{-\Delta H_i}{RT}}; i = M, W, H \tag{5.22}$$

where, M is methanol, W water, and H hydrogen.

The parameters for both models are represented in Table 5.8.

For the power law model the kinetics equations are defined as follows:

$$r = k \, p_M^a \, p_W^b \, p_{CO_2}^c \, p_{H_2}^d \quad \text{(Models 3 and 4)} \tag{5.23}$$

where

$$k = k^0 e^{\frac{-\Delta E}{RT}} \tag{5.24}$$

The values of the parameters are shown in Table 5.9

To simulate all those models, the following conditions are required (Table 5.10).

5.8.2   Simulation in Aspen Plus®

To start the simulation in Aspen Plus®, the four components involved in the reforming reaction should be set up (Eqn 5.19). The thermodynamic package to

Table 5.8 Parameters for models 1 and 2 for LHHW kinetic

Parameter	Units	Model 1	Model 2
k^0_M	mol/(hgcat)	3.063×10^{10}	6.142×10^9
ΔE	cal/mol	25,799	24,331
b^0_M	atm^{-1}	2.365×10^{-2}	2.122×10^{-1}
ΔH_M	cal/mol	-8211	-7906
b^0_W	atm^{-1}	1.605×10^{-1}	1.845×10^{-2}
ΔH_W	cal/mol	-4639	-4334
b^0_{H2}	atm^{-1}	$-$	4.531×10^{-5}
ΔH_{H2}	cal/mol	$-$	-7509

Table 5.9 Parameters for models 3 and 4 in power law model

Parameter	Model 3	Model 4
a	0.389	0.235
b	-0.151	0.216
c	0	0
d	0	0.436
k_0 (mol/(gh))	5.609×10^9	2.948×10^6
E_a (cal/mol)	24,163	19,028

Table 5.10 Operating conditions for the reforming reactor example

Parameter	Value
Pressure	5 atm
Feed temperature	200 °C
Heating temperature	400 °C
Catalyst density	1.115 g/cm^3
ΔH_R	$-13,900$ cal/mol
U	0.018 kJ/(m^2 s K)
Molar ratio W/M	1.8
Feed flow	2.341×10^{-6} kmol/s

be included is UNIQUAC. Once these two steps are completed, the flowsheet presented in Fig. 5.32 is installed.

Four reactors, each one with different kinetic model are included in Fig. 5.32; however, their settings are exactly the same and corresponds to the reactor used by Tesser (Rawlings & Ekerdt 2009). The data for each reactor are shown in Fig. 5.33.

Once these parameters are specified, it is necessary to include the data of the catalyst in the Catalyst tab. There, the density data of the catalyst particles and the empty bed space are included. These fields are located in the process description and Fig. 5.34 shows how these should be introduced in the simulator.

In order to run the simulation, include the reactions in each one of the respective reactors. In Fig. 5.32, each reactor is shown with the name of the kinetic model which performs the calculation. Thus, the Reactions tab in the navigation tree includes four sets of reactions, two for power law and two for LHHW kinetics. The change in the types of kinetics is done by selecting the type of reaction to

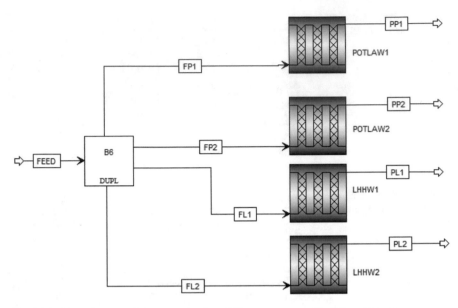

Fig. 5.32 Flow diagram of the example, four different reactors with different kinetics

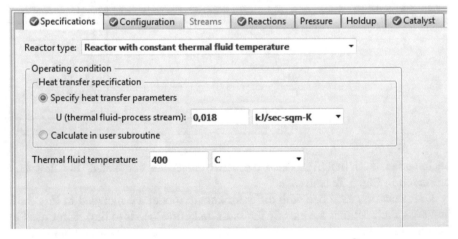

Fig. 5.33 Tab specifications for methanol reforming reactor in Aspen Plus®

include in the reaction set (Fig. 5.35). It is important to note that the kinetic constants must be in international system units so they can be included in Aspen Plus®. The reader is invited to calculate the values of the different constants in the International System of Units.

For power law kinetic, the data is input as in the above example. On the other hand, for the reaction kinetics of the type LHHW the data input is different. Now, the denominator of the kinetic expression is included corresponding to transport

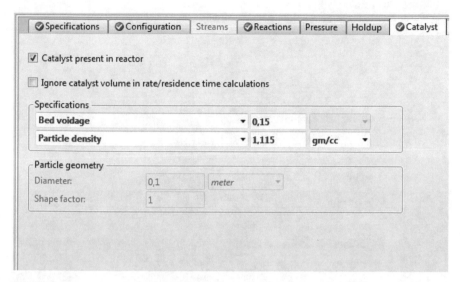

Fig. 5.34 Data of the catalyst reforming reaction of methanol in Aspen plus®

Fig. 5.35 Specification of
the type of reaction to be
included in Aspen Plus®

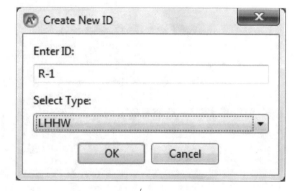

phenomena occurring in heterogeneous catalysis. In the simulator it is located in the
Adsorption button. The data input for model LHHW-2 is illustrated below
(Figs. 5.36 and 5.37).

   To include the numerator of the kinetic equation, you select the Driving Force
button. Another window appears where the information for the numerator can be
provided term by term. This is illustrated in Fig. 5.38. In the two models Term 2 is
not included because it is not present in the driving force equation.

   For the LHHW-1 reaction the setup is carried out in a similar way.

   Once the kinetics of the type LHHW and power laws are specified, associate
each reaction set to the corresponding reactor. After this the simulation can be run.

   After convergence, go to the Profiles tab for each reactor and use the Plot Wizard
tool (at the end of the Plot menu) to retrieve the composition profiles for each
reactor. In Fig. 5.39 the profiles achieved with each kinetic model are shown.

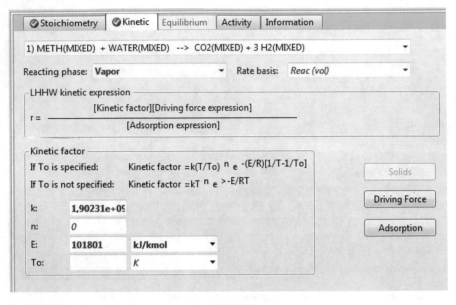

Fig. 5.36 Kinetic tab for the reaction called LHHW-2

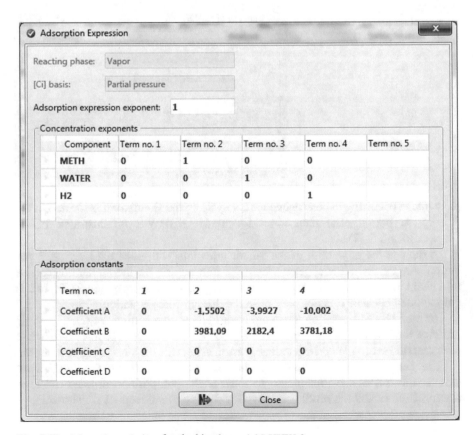

Fig. 5.37 Adsorption window for the kinetic model LHHW-2

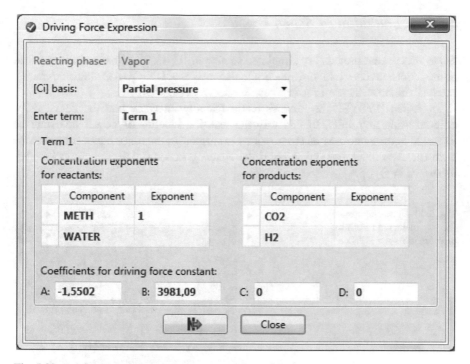

Fig. 5.38 Driving Force window for the kinetic model LHHW-2

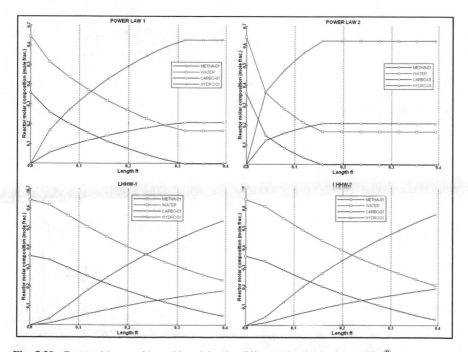

Fig. 5.39 Composition profiles achieved for the different kinetics in Aspen Plus®

5.8.3   Simulation in Aspen Hysys®

Now, these same models are simulated in Aspen HYSYS®. First include the four components involved in the reaction; also use the UNIQUAC thermodynamic model. The flow diagram for this case is shown in Fig. 5.40.

In Aspen HYSYS® the kinetics has a friendly interface and the data can be entered using any units. In this way the reaction kinetics of power law may be included in the same way as the previous example. On the other hand, for reactions with LHHW model must be selected the Heterogeneous Catalytic reaction model, as shown in Fig. 5.41.

Fig. 5.40 Flow diagram for the four methanol reforming reactors in Aspen HYSYS®

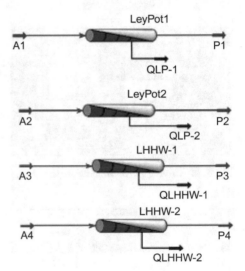

Fig. 5.41 Selection of the kinetic type model in Aspen HYSYS®

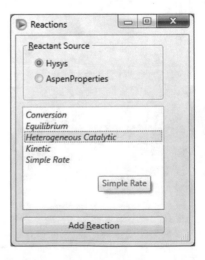

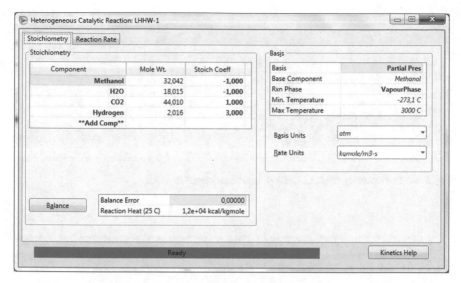

Fig. 5.42   Basis and Stoichiometry tabs for the LHHW-1 kinetic information input

Fig. 5.43   Numerator tab for LHHW-1 kinetic information input

After that, another window appears in which you can specify the different parameters of the reaction. The first two tabs allows to include the stoichiometry of the reaction and their units, here, the information data is included as it was made in the power law reaction, the other two corresponds to the numerator and denominator for a kinetic model of LHHW type (Fig. 5.42).

To include the numerator and denominator of the kinetic model data it should be done as shown in Figs. 5.43 and 5.44.

The kinetic model LHHW-2 is also included this way. After four reactions are included, four sets are generated, one for each reaction, and each of these is associated with the property package. Now the reactors should be configured, as in the case of Aspen Plus®, all four reactors have the same geometric specifications and for the catalyst. Fig. 5.45 shows how these data are included.

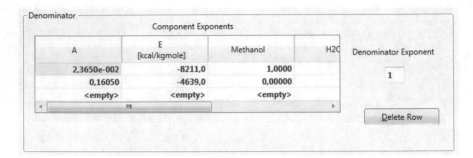

Fig. 5.44 Denominator tab for LHHW-1 kinetic information input

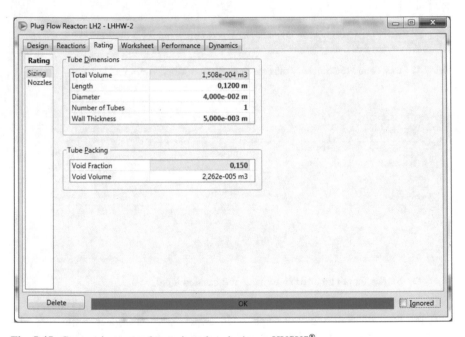

Fig. 5.45 Geometric reactor data to introduce in Aspen HYSYS®

Including the reaction, is also specified the data for the reaction catalyst, such as particle density and size. Figure 5.46 shows the data for the LHHW-1 reaction.

By repeating this procedure on all reactors, the simulator should have produced results in each one. Then the composition profiles for each reactor can be plotted, they are in the Performance tab. The profiles achieved are shown in Fig. 5.47.

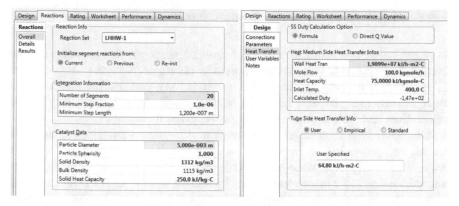

Fig. 5.46   Heat transfer and Reaction data in Aspen HYSYS® reactors simulation

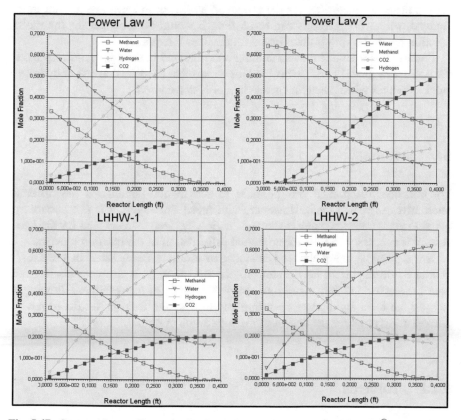

Fig. 5.47   Composition profile results for the different kinetics in Aspen HYSYS®

5.8.4   Analysis and Results Comparison

In general, after comparing the results of simulations, is possible to realize that the trend is very similar to that the one reported in the paper. However, for power law 2, the reaction on both simulators gets a result for a composition profile whose change is faster than for the other cases, because the model includes hydrogen in its calculations and their generation is still faster, given its exponent.

Moreover, it is possible to appreciate the difference between heterogeneous models and power law, as in the first one, the reaction is slower than the power laws. This is expected since the heterogeneous models include the effects of transport taking place in the catalyst, and by adding these there is a slower reaction because the reactants must not only react but must also get to the point of reaction.

The main difference between the two models is in the diffusion term where the model 2 includes the hydrogen produced, while model 1 discards it. This difference affects the rate which takes for the produced hydrogen to leave free a point in the catalyst, where the reactants can react, thus it would be expected for the model reaction two to be slower than for the model 1 as it happens.

Finally, closer kinetic models and the one obtained in the article, correspond to the power law model 1 and LHHW 1. This means that the diffusion rate of hydrogen can be omitted when a heterogeneous model is considered.

5.9   Summary

For chemical reactors, it was possible to see in both simulators the different alternatives to model them. However, with these simulators, it is not possible to perform reactor design, but a review of previously sized reactors, and for the latter case, it is necessary to have experimental information of the reaction kinetics.

It was possible to establish the existence of multiple steady states in a reactor, a characteristic important for the start-up and operation. It is also possible to study how the reactor and feed conditions the performance of the reaction. Allowing to suggest the most appropriate operating conditions, and from its results to have a clear starting point for the design of the separation system.

5.10   Problems

P5.1 The configuration of a CSTR reactor for the production of propylene glycol emphasizing the multiplicity of steady states and how to determine it through a sensitivity analysis was illustrated in Sect. 5.7. Furusawa et al. (1969) studied the stability of reporting CSTR kinetics reactor of first order respect to propylene oxide. Develop for this exercise a similar analysis varying the data of the reaction kinetics according to the following expression:

$$-r_{C_3H_6O} = 4.71 \times 10^9 \times \exp\left(-\frac{18}{RT}\right) \times C_{C_3H_6O} \tag{P5.1}$$

with:

$r_{C_3H_6O}$: Reaction rate in kmol/m^3 s.

E: Activation energy in kcal/gmol.

$C_{C_3H_6O}$: Molar concentration in kmol/m^3.

How many steady states can be identified? The area of multiplicity corresponds to the same area found using the second-order kinetics?

P5.2 Is desired to evaluate the design of a reactor for the production of allyl chloride according to the main reaction.

$$Cl_2 + C_3H_6 \rightarrow CH_2 = CH\text{-}CH_2Cl + HCl$$

And the secondary reaction produces 1,2-dichloropropane.

$$Cl_2 + C_3H_6 \rightarrow CH_2Cl\text{-}CHCl\text{-}CH_3$$

The reaction mixture containing 4 mol of propylene per mole of chlorine and get into the reactor at 392 °F at a rate of 0.85 lbmol/h. The operating pressure can be assumed constant and equal to the feed with a value of 29.4 psia. Initially it is expected to use a tubular reactor with 2-in. of inside diameter, jacketed and working with boiling ethylene glycol as a coolant, so that the temperature inside is getting constant at 392 °F. The heat transfer coefficient inside is 5 Btu/h ft^2 °F.

The speed equations of the two reactions, expressed in lbmol/h ft^3, are:

$$r_1 = 206000e^{\frac{-27200}{RT}} p_{C_3H_6} p_{Cl_2}, T(°R) \text{ Main Reaction}$$

$$r_2 = 11.7e^{\frac{6860}{RT}} p_{C_3H_6} p_{Cl_2}, T(°R) \text{ Secondary reaction}$$

The activation energy is in Btu/lbmol and the partial pressure in atm.

(a) Simulate the reactor operation using ethylene glycol as a coolant in the conversion and determine how the conversion varies according to the length of the reactor.

(b) Simulate the adiabatic operation of the reactor and determine how the conversion varies according to the length of the reactor.

(c) Compare the results previously obtained with the results that will be obtained in a CSTR reactor of 0.83 ft^3.

P5.3 Adapted from (Froment and Bischoff 1990). The oxidation reaction of o-xylene to produce phthalic anhydride is highly exothermic and is carried out in a PFR reactor using a melt salt as a refrigerant (NaNO3). The o-xylene is mixed with air before entering the reactor trying to have very small amounts of oil into the mixture. Under these conditions there is a large excess of oxygen which causes the reaction to be pseudo-first order with respect to the concentration of o-xylene.

$$A + 3B \rightarrow C + 3D$$

where A, B, C, and D represent o-xylene, oxygen, phthalic anhydride, and water, respectively.

A mixture of 6900 kg/h having 1.4 % mol of o-xylene is fed, 20.8 % molar of oxygen, and 77.8 % molar of nitrogen. The feed pressure is 1 atm and a temperature of 625 K. The reaction is carried out in the vapor phase, in a reactor with 3000 tubes of 1 in. of diameter and 2 m of length. In the reactor there is no pressure drop and heat transfer coefficient is $U = 77.37$ kcal/h m^2 K. Sodium nitrate (NaNO$_3$) is used as refrigerant which is fed in counter-current at 30 kg/s, with a pressure of 1 atm and a temperature of 620 K. In order to accelerate the reaction is used vanadium pentoxide V$_2$O$_5$ as a catalyst with a particle diameter of 3 mm and a density of 1300 kg/m^3.

The kinetic reaction of pseudo-first order reported by Froment obeys to the expression:

$$r_A = k p_A p_B$$

Where,

$$\ln k = 19.837 - \frac{13636}{T}$$

expressed in kmol/kg of catalyst atm^2 h.

Perform the simulation in Aspen Plus® and Aspen HYSYS®. Then perform a sensitivity analysis in Aspen Plus® varying coolant flow between 10 and 60 kg/s. Finally, compare the results operating the reactor with the coolant fed in concurrent and counter-current.

P5.4 Adapted from (Fogler 2008). The styrene can be produced from ethyl benzene by the following reaction:

$$\text{Ethylbenzene} \rightarrow \text{Styrene} + H_2$$

However, several irreversible secondary reactions also occur:

$$\text{Ethylbenzene} \rightarrow \text{Benzene} + \text{Ethylene}$$
$$\text{Ethylbenzene} + H_2 \rightarrow \text{Toluene} + \text{Methane}$$

Ethyl benzene is fed at the rate of 0.00344 kmol/s in a 10.0 m³ PFR reactor with inert water vapor at a total pressure of 2.4 atm. The molar ratio of steam/ethyl benzene is at the beginning [so that, parts (a) and (b)] 14.5: 1, but it might vary. Given the following data, determine the molar output flow rates of styrene, benzene, and toluene at the following inlet temperatures when the reactor is operated adiabatically.

(a) 800 K
(b) 930 K
(c) 1100 K
(d) Find the inlet ideal temperature for the production of styrene with one ratio of water vapor/ethyl benzene of 58:1.
(e) Determine the water vapor/ethyl benzene ideal radio to produce styrene at 900 K.

Additional information:

$$\rho = 2137 \, \frac{kg}{m^3} \, \text{catalyst pellet}$$

$$\phi = 0.4$$

$$K_{pl} = \exp\left\{ b_1 + \frac{b_2}{T} + b_3 \ln T + [(b_4 T + b_5)T + b_6]T \right\} \text{atm}$$

$$b_1 = -17.34$$

$$b_2 = -1.302 \times 10^4$$

$$b_3 = 5.051$$

$$b_4 = -2.314 \times 10^{-10}$$

$$b_5 = 1.302 \times 10^{-6}$$

$$b_6 = -4.931 \times 10^{-3}$$

The rate law for the formation of styrene (Es), benzene (B), and toluene (T) respectively are as follows. (Ethyl benzene = Eb)

$$r_{1Es} = \rho(1 - \phi)\exp\left(0.08539 - \frac{10\,925}{T} \right) \left(P_{EB} - \frac{P_{Es}P_{H_2}}{K_{pl}} \right) (kmol/m^3s)$$

$$r_{2B} = \rho(1 - \phi)\exp\left(13.2392 - \frac{25\,000}{T} \right) (P_{EB}) (kmol/m^3s)$$

$$r_{3T} = \rho(1 - \phi)\exp\left(0.2961 - \frac{1100}{T} \right) (P_{EB}P_{H_2}) (kmol/m^3s)$$

References

Furusawa T, Nishimura H, and Miyauchi T (1969) Experimental study of a bistable continuous stirred tank reactor, J Chem Eng Japan, 2, pp. 95

Fogler HS (2008) Elementos de Ingeniería de las Reacciones Químicas, 4th edn. Pearson Prentice Hall, Naucalpan

Froment G, Bischoff K (1990) Chemical reactor analysis and design, 2nd edn. John Wiley & Sons, New York

Harriot P (2003) Chemical reactor design. Marcel Dekker, New York

House JE (2007) Principles of chemical kinetics, 2nd edn. Elsevier, San Diego

Rawlings JB, Ekerdt JG (2009) Chemical reactor analysis and design fundamentals, 2nd edn. Nob Hill Publishing, Madison

Chapter 6
Gas–Liquid Separation Operations

6.1 Introduction

From early twentieth century, there was awareness of shortcut and rigorous methods to calculate distillation columns. By that time, shortcut methods constituted the primary design tool, since it was necessary to carry out calculations by hand and rigorous methods' implementation was a complex task that could demand several days and sometimes, even weeks. Error derived from applying shortcut methods was compensated with overdesign (Henley and Seader 1998).

The arrival of the computers allowed a leap in using rigorous methods and their rapid development in order to apply them for the design of complex columns. The rigorous methods demonstrated greater accuracy, and therefore are currently the mostly applied tools for columns design, restricting the use of shortcut methods to preliminary calculations that allow us to determine approximate values of main designing variables (reflux ratio, number of theoretical states, phase feeding, etc.).

The final design of a multicomponent separating equipment through balance stage models makes it necessary for a rigorous determination of the temperatures, pressures, flows, compositions, and heat transfer speeds in every stage (Henley and Seader 1998). This determination requires further adjustments with the stage efficiency specification and accurate knowledge of each component physical properties, this can be made through the solution of mass and energy balance equations, as well as from ratios that describe the phase equilibrium in each stage; generally, these constitute an algebraic equation highly nonlinear system that involves complex solution procedures (Henley and Seader 1998).

The equilibrium stage concept has been used in the rigorous calculation of distillation columns, and during over 100 years applied in the modeling of liquid–liquid distillation operations, absorption, stripping, and extraction. This calculation was developed in its beginning through approximate methods that involve calculating stage by stage and equation by equation, which demonstrated numerical instability when implemented in digital computers. Holland, in 1963, developed

© Springer International Publishing Switzerland 2016
I.D.G. Chaves et al., *Process Analysis and Simulation in Chemical Engineering*,
DOI 10.1007/978-3-319-14812-0_6

modifications to the mentioned methods which demonstrated a significant success, calling it the Method theta-θ. Later the so-called bubble point methods (BPM), Flow sum rates method (SRM), simultaneous correction methods (SCM), and inside-out algorithms (IOM), among others (Henley and Seader 1998).

Distillation columns constitute a large field in the process industry, since it is one of the most widespread separation operations. One of the main uses for distillation is for oil refining, for this reason the literature is extensive for modeling, analysis, and suggestions for the operation. Furthermore, the fuel alcohol and many others industries require more demanding distillation operations, as in the case of the extractive distillation, azeotropic distillation and reactive distillation.

In this vast topic we can also include stripping columns, absorption columns, liquid–liquid extraction and any operation that requires the use of this type of equipment with industrial purposes.

Throughout this chapter it is indicated how to insert and obtain the necessary information for the dynamic dimensioning and analysis for these separation systems. Likewise the review of literature related to basic concepts and the physical principles which explain the separation processes is recommended.

6.2   Available Modules in Aspen Plus®

Column simulation comprises liquid–liquid distillation, absorption, and extraction operations. Simulators have different modules to develop these estimates, which similarly to heat transfer equipment, use shortcut calculation strategies and detailed separation equipment. In this way, thus, with this tool it is possible to design equipment as from the conceptual analysis to the design of internal and detailed evaluation of already existing equipment. In conclusion, these modules may simulate such complex operations in the assisted distillations, absorbing, stripping, and crude distillation, among others.

Aspen Plus® has a large range of models to carry out column calculations. There are nine models, which allow you to carry out from shortcut calculations to complex distillations. Each has different convergence options and calculation methods for the different operations to simulate.

6.2.1   Shortcut Methods

Shortcut methods are simple calculation procedures developed to relate incoming and outgoing streams with the number of the system equilibrium stages. They are called like this because they involve a global treatment of stages without considering in detail the temperature and composition change in each one, in order words, internal profiles are not calculated.

Shortcut methods allow us to have an approximate idea of the necessary number of stages, reflux ratio, feeding stage position, and components distribution.

Generally, constant flows in every stage or constant relative volatility (α_{ij}) of components are assumed. Due to different assumptions, these methods should be applied with caution. For example, both the Fenske equations for minimum number of stages, and the Underwood equation for minimum reflux ratio, are based on the separation of key components. Both methods assume constant relative volatilities at the average conditions. Similarly, Gilliland correlation is used. The traditional application of these equations in design generally consists on determining the minimum number of stages (through Fenske equation), the minimum reflux ratio (with Underwood equation). Then these results are used in Gilliland equation to calculate optimal values.

6.2.1.1   Fenske Method

The method developed by Fenske allows us to calculate the minimum number of stages required to produce the desired separation under total reflux conditions. Given the mass balance and the relative volatilities of the feed, distillate and bottom conditions, the equation is solved, whose difficulty lies on finding the corresponding compositions. The method applies for the cases where relative volatility does not vary appreciably in the column; when this occurs, its use allows us to determine initial conditions that then are incorporated to calculation rigorous methods.

6.2.1.2   Underwood Method (Eckert and Vanek 2001; Thomas 1991)

The method is mostly used to calculate the minimum reflux ratio along with Brown-Martin, Colburn and K. Venkateswara Rao—A. Raviprasad methods. The Brown-Martin Method and Colburn method provides satisfactory results but are too complex to be considered within shortcut methods group (Pham and Doherty 1990).

The method works with constant relative volatilities, and as input data uses distillate stream composition and the liquid–vapor ratio of the feeding mixture.

6.2.1.3   Gilliland Method (Glasser et al. 2000)

This method allows us to calculate the number of ideal stages necessary to perform the separation, after specifying the number of minimum stages and the minimum reflux ratio. For the calculation uses correlations found as from experimental data that relate the number of stages with the reflux ratio.

Aspen Plus® has three modules implemented for shortcut calculations; these are: DSTWU, DISTL, and SCFrac. Below is a listing of the characteristics of each module (Table 6.1).

Table 6.1 Shortcut calculation module in Aspen Plus® for distillation columns

Icon	Name	Application
DSTWU	DSTWU	This module uses the Winn-Underwood-Gilliland method for simple columns. In it is specified the desired recovery of defined components as light key and heavy key, and with these defined parameters the method calculates: the number of necessary theoretical stages for the separation, minimum reflux ratio, feeding stage, and heat duties required. Likewise, it is possible to obtain a curve of the reflux ratio versus the number of theoretical stages, to be able to propose a design with reduced total costs
Distl	DISTL	This module enables simulation of multistage columns with a feeding stream and two products (*Rating*). We can also specify if the condenser is total or partial. The calculation of output composition is made through Edmister approach; additionally equimolar flows and constant relative volatilities are assumed
SCFrac	SCFRAC	This module is useful to simulate complex distillation columns that have only one feed; optionally it may have stripping vapor streams and has any amount of products. This model is mainly used to model atmospheric and vacuum crude units. The calculation supposes constant relative volatilities for each section and that the section to section liquid flow is negligible. Furthermore, this model can make free-water calculations in the condenser

6.2.2 Rigorous Methods

A rigorous method models a column as a group of equations to calculate operating conditions of the column. Minimum specifications for the calculation through a rigorous method are:

- Composition, flow, and thermodynamic condition of feeding streams.
- Number of stages of the column.
- Separation requirements.
- Feeding stages and corresponding heat exchange equipment, side streams, etc.
- Column pressure profile.

The group of equations to be solved is known as MESH (*Material, Equilibrium, Summation, and Heat*) equations, and refers to global and component balance equations, energy balances, and the phase equilibrium equations. MESH equations describe the behavior of a distillation column in steady state and are obtained as from the concept of equilibrium stage. The set of equations required in a stage is made up by mass balances for each component (C equations), the equilibrium relations among phases for each component (C equations), summation restrictions of molar fractions (one for the vapor phase and one for the liquid phase) and the energy balance equation. What means that there are $2C + 3$ equations to solve, where C represents the number of components, and $2C + 3$ the unknown variables represented in the molar fractions of components in the liquid and vapor phases, molar flows of liquid and vapor outgoing from the stage and the stage temperature.

An equilibrium stage is similar to a flash tank in steady state that contains liquid or vapor in the feed stream, whose composition, temperature, and pressure are defined. The separator produces two streams: one of vapor, constituted by the most volatile components and one liquid, made up by heavier components. According to the equilibrium stage, the mentioned streams are in phase equilibrium and inside the separator there is a perfect mixing condition; in this manner are mass and head transfer phenomena caused by diffusional effects are neglected.

The MESH variables, known as the column state variables, are (Henley and Seader 1998): stage temperatures, internal liquid and vapor flows, and compositions of the liquid and vapor streams that exit from each stage. This set of variables is the one obtained as from the solution of equations that describe the column.

6.2.2.1   Stage by Stage Methods (Henley and Seader 1998)

With these algorithms we can obtain a solution that includes all the conditions in every stage, as well as the products' properties. Initially conditions are set in one of the stages and the rest is calculated by an appropriate iterative procedure. Generally, the top and bottom stage conditions are known or can be estimated easier. Consequently, the calculations made stage by stage from both ends of the column to the feeding plate or from one end to the other. In both cases the compositions and conditions calculated shall coincide with the known or estimated ones. Therefore, the problem convergence is reached when a criterion is satisfied, which generally consists of verifying that the global and component balances are met.

Methods stage by stage are useful when one wants to learn the number of stages under known conditions of the feed for a given separation. Classical examples of the stage by stage procedures are the ones from Lewis and Matheson (1932) and Thiele and Geddes (1933), widely used ever since their conception, for manual calculation, which lost their effect after their implementation in digital computers, where numerical instability problems were detected which made them not applicable to complex column designing.

The Lewis-Matheson method is an iterative procedure where a number of stages is taken as freedom grades, the desired separation is specified for two key components, reflux ratio, pressure in column, and the location of the feeding plate. Initially top and bottom compositions are assumed, stage by stage calculation is made from both ends of the column to the feeding stage, where the ratios of key components must coincide. The mass balance in the feeding stages acts as convergence criterion. If the balance is not satisfied, estimates for top and bottom compositions have to be adjusted to repeat the calculation (Henley and Seader 1998).

The Thiele and Geddes (1933) method, on the other hand, requires specifying the number of equilibrium stages over and under the feeding stage, the reflux ratio, distillate flow, column pressure and the component and conditions of food. The temperature and vapor or liquid flow profiles in stages are the iteration variables that shall be assumed initially. Mass balances are solved for each stage starting from an end of the column to the other, or from both ends to the feeding stage.

Temperature profile is corrected then solving the molar fraction summation equations and, finally, stage flows are adjusted by the energy balance. The procedure is repeated until all the equations are satisfied.

6.2.2.2   Bubble Point Methods

The BPM derives its name from using equilibrium equations, and specifically the bubble temperature as iteration variable, and the summation equations, of compositions to calculate the column temperature profile (Henley and Seader 1998).

The Wang and Henke method is the first of this type of methods; it is used to calculate distribution of components in complex columns, and allows for a fast calculation, and numerical stability. This method uses a tridiagonal matrix to calculate flows of components or compositions, which are then used to calculate temperatures through the resolution of the bubble point equation (Treybal 1996).

The BPM generally work best for tight boiling point systems or for ideal systems or near to the ideal behavior, where the composition has a more relevant effect upon the temperature than the latent evaporation heat. Among the most significant methods are the Theta-θ method, K_b method, and the constant composition method (Henley and Seader 1998).

6.2.2.3   Sum Rates Method

The SRM is used to model absorbers and strippers. This method works appropriately with wide boiling intervals, specifically in those that contain noncondensable components.

The energy balances in each stage are used to find, through a solution algorithm of Newton–Raphson, the column temperature profile. Compositions do not have a considerable influence on the temperature calculation; on the contrary, the calorific effects and latent heats of vaporization do have them. The flows are calculated by the tridiagonal matrix method.

The SRM can be applied to distillation columns, but the algorithm equation does not allow to model the condenser and the reboiler along with the other stages of the column, because the condenser and reboiler duties shall be specified along with the reflux ratio, what converts energy balances into independent functions. The SRM is available at the ABSBR option from the RadFrac module calculation of the simulator Aspen Plus (Aspen Technology, Inc. 2001).

6.2.2.4   2 N Newton Methods

The 2 N Newton methods are called like this because they manage two equations per stage, for a total of $2 \times N$ functions and variables for the column with the Newton–Raphson method. A difference from the BPM and SRM methods, which calculate temperatures and flow rate separately, in the methods 2 N Newton these variables are calculated in simultaneous manner.

These methods have demonstrated good results for broad boiling point mixtures including refinery fractionators, absorbers, and strippers. Among the methods that outstand the most within this group are the Tomich method and the 2 N Newton–Raphson Method.

6.2.2.5   Simultaneous Correction Methods

Among the rigorous methods, the SCM are the most popular, specifically the Naphtali and Sandholm method. In these methods, all the equations are solved using a Newton–Raphson technique through the appropriate selection of the MESH variables and equations. Additionally several modifications have been made in order to include additional equations and variables to solve distillation columns that involve the formation of three phases (Henley and Seader 1998).

One of the main advantages of these methods in respect to the methods BPM, SRM, and 2 N is their capacity to work with highly nonideal systems where the equilibrium constants and the enthalpy depend strongly on the composition, while the methods BPM, SRM, and 2 N use the tridiagonal matrix to calculate compositions independently to the calculation of equilibrium constants and enthalpies, which are calculated with the previous iteration compositions.

One of the main advantages of the SCM lies on its high sensitivity, which requires appropriate initialization values that are generally obtained through BPM or SRM methods.

The Naphtali and Sandholm Method is part of commercial simulator solution algorithms such as CHEMCAD, PRO II, TSWEET, and Aspen Plus®, among others (Aspen Technology, Inc. 2001; Henley and Seader 1981).

6.2.2.6   Double Iteration Methods (*Inside-Out Methods*)

The inside-out method was developed by Boston and Sullivan (1974). This technique uses two models to calculate the required thermodynamic properties: a model supported on parameters approximate to the volatility and enthalpy calculation, and a second model strictly rigorous where parameters used in calculation said properties are assessed. The MESH equations are solved always with the approximate model in a cycle called "inside," where parameters defined for the approximate model are updated with the strictly rigorous model calculation, in a cycle called "outside." For this reason, this method and its versions are known as *Inside-Out Methods*, IOM.

IOMs have been positioned as the most recommendable for distillation operations' rigorous calculations for most of the standard, azeotropic and highly nonideal mixtures with heterogeneity problems for the liquid phase, at the same time have demonstrated to be considerably stable and rapid to converge into a solution (Krishnamurthy and Taylor 1985a, b).

The main advantage of these methods is that they enable flexibility in the problem specifications.

A Boston method with firm support, provided with a wide range of characteristics and options to solve different types of columns is available in the calculation modules RadFrac and MultiFrac in Aspen Plus® (Aspen Technology, Inc. 2001).

6.2.2.7   Relaxation Methods

Relaxation method finds the solution in steady state of a column as from the MESH equations dynamic expressions, which is solved by successive approximations. The column initialization is made with starting data from the column such as the liquid retained in every stage along with the food composition at the bubble point (Henley and Seader 1998).

These methods are not very widespread since they require greater numerical effort and calculation cycles; however, they were the first to describe the dynamic behavior of distillation operations, and their versions are generally used for this purpose (Krishnamurthy and Taylor 1985a, b).

6.2.2.8   Homotopy-Continuation Methods

The algorithm for the Homotopy-Continuation Methods (HCM) is based on the awareness of a solution resulting from an approximate equation set $(a(x))$, which is followed by another set of rigorous equations $(b(x))$, which one wants to solve. Both sets are connected through mathematical homotopy expressions (Henley and Seader 1998; Krishnamurthy and Taylor 1985a, b) 23]: $h(x,p) = p(b(x)) + (1-p)$ $(a(x)) = 0$, where p is a homotopy parameter. The function $a(x)$ can be defined as $b(x) - b(x_0)$, with x_0 as any initialization value.

A great advantage of these methods is that they allow to explore aspects on the simultaneous solution evaluation, characteristics of the multiple steady states that can be reached by an azeotropic distillation operation or specific reactive, and which can be studied through the different ways that can be chosen for the homotopy p parameter integration. Additionally these methods are the recommended ones for problems where the rest have failed.

As conclusion, we can state that currently there is a wide range of possibilities, in calculation algorithms and solution procedures, for the distillation column problems. In Fig. 6.1 there is a brief methodology with selection criteria for calculation algorithm for a specific problem.

Rigorous models enable a complete calculation of the composition, temperature, flow, and pressure profiles in columns. They enable to establish design specifications to guarantee an appropriate operation. Furthermore, they allow both the design and the evaluation of plate or packing sections. Below are the rigorous calculation models available at Aspen Plus® (Fig. 6.2).

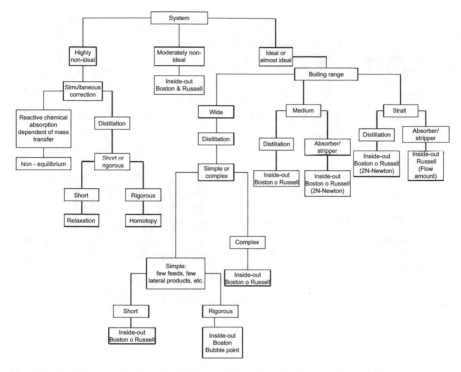

Fig. 6.1 Decision tree for the calculation methods of gas–liquid separation equipment

Fig. 6.2 Icon of *RadFrac* module in Aspen Plus®

6.2.2.9   Radfrac

This is a model that enables to simulate any type of vapor–liquid fractioning operation. Among these operations are included:

- Simple distillation
- Absorption
- Absorption with reboiler
- Splitting
- Splitting with reboiler
- Extractive and azeotropic distillation
- Reactive distillation (equilibrium, conversion, electrolytic, etc.)

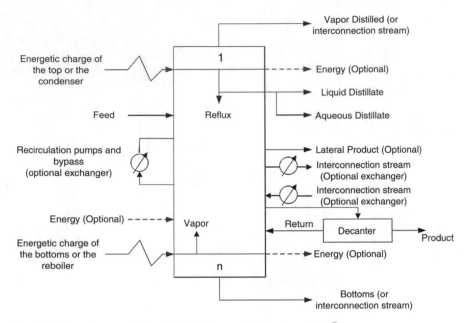

Fig. 6.3 General connectivity of the *RadFrac* module in Aspen Plus®

Likewise, RadFrac can calculate the following systems:

- Two-phase systems
- Three-phase systems
- Systems with boiling points very near or very far apart
- Systems that have strong nonideal behavior

RadFrac also can manage a free-water phase or a second liquid or solid phase in any stage of the column. It also enables *Pumparounds* in the column, among many other functions (Fig. 6.3; Table 6.2).

Additionally, the model enables the following calculations:

- Design specifications on column, such as components recovery or their fractions in a given stage.
- The plate and pack design and evaluation can be carried out in the column; for this there is a database with the different parameters of packs and plates according to several manufacturers.
- Specification of the Murphree efficiencies for column calculations.
- It is likely to make thermal and hydraulic analysis for a column, and in this manner modify the operation conditions and improve its performance.
- It has convergence algorithms:

  - Inside-out
  - Newton
  - Sum rates

Table 6.2 Rigorous calculation modules for distillation columns in Aspen Plus®

Icon	Name	Description
MultiFrac	MULTIFRAC	This is a rigorous model that is useful to simulate multiple separating units connected to each other. Used to simulate air separating columns, absorbers and splitters combinations, among others. Its calculation assumes equilibrium stages; however is it possible to specify the efficiency of Murphree
PetroFrac	PETROFRAC	This is a module that has any necessary item to simulate any type of complex separating operation carried out in the oil refining industry. This module can detect an aqueous phase and decant it at any stage of the column. This module assumes equilibrium state calculation, but allows specifying the evaporation efficiencies or Murphree efficiencies. This module could also be used to dimension and evaluate plate or packed columns, and in this last case, PetroFrac enables simulation of random packs and structured
RadFrac	RATEFRAC (From version 7 it is joined to module RadFrac)	This rigorous model is nonequilibrium, that means it is based on the calculation of mass and heat transfer rates for the different calculations of the column, so that the separation degree is given by these rates. The calculation does not involve empirical factors and is used to design and evaluate plate and packed columns
Extract	EXTRACT	This is a rigorous module to simulate liquid–liquid extractors; used in evaluating extractors. In it multiple food, heaters/freezers, and lateral streams can be handled. In order to develop the simulation, this module calculates the different distribution coefficients

6.3   Modules Available in Aspen Hysys®

Aspen HYSYS® has a powerful module that enables the calculation of different types of columns; this includes multistage fractioning, crude units (atmospheric and vacuum), demethanizing columns, extractive distillations among others, where each can be calculated through equilibrium stages or using nonideal stages. Each stage has one or more feeding streams, as well as liquid or vapor outlets, and lateral heaters or coolers.

In this module, columns with *pumparounds* or lateral splitters can be specified. Although very few columns involve this type of additional equipment, it can virtually calculate any type of column.

In order to carry out the column simulation there are predetermined modules that correspond to different operation configurations and in analogue manner has a shortcut method to estimate them. However, contrary to Aspen Plus®, Aspen HYSYS® has an additional internal flowsheet especially for designing specific columns with their respective accessories.

6.3.1   Predefined Columns

Aspen HYSYS® has different predefined columns that correspond to several of the most frequent configurations for separating operations (Table 6.3).

All of these columns start from the same models which have the inner interface for columns; however, they are adjusted according to the application required by the column along with its accessories and convergence models.

6.3.2   Shortcut Calculation Model

Aspen HYSYS® also has a shortcut method for distillation columns that use the Fenske–Underwood method to calculate simple columns with reflux. In this mode, the minimum reflux of Underwood is established, and the minimum stage number of Fenske. Through a specified reflux rate it is possible to calculate vapor and liquid flow in enrichment and splitting sections, the condenser and reboiler duty, the number of ideal stages required and the optimal feeding stage (Fig. 6.4).

Table 6.3 Predefined modules available in Aspen HYSYS® for gas–liquid separating equipment

Icon	Type of column
	Column with reboiler and condenser (partial or total)
	Absorber with condenser (partial or total)
	Absorber
	Absorber with reboiler
	Three-phase distillation
	Liquid–liquid extractor

Fig. 6.4 Shortcut method
module in Aspen HYSYS®

Fig. 6.5 Icon blank column sub-flowsheet interface in Aspen HYSYS®

Fig. 6.6 Icon access to column environment from a column already designed in Aspen HYSYS®

The shortcut method is only an estimate of the column performance, and for this case, is only limited to columns with simple reflux. In order to obtain more actual results a rigorous model of column shall be used; however, the shortcut method calculates very well previous estimates to use rigorous models.

6.3.3   Column Interface

As mentioned above, Aspen HYSYS® has a special internal interface to calculate and design columns. There, all the equipment sets that could eventually involve the designing of one of these separating units are introduced. This interface is accessed through the button shown in Fig. 6.5.

This interface can also be accessed through the column information diagram using the button of Fig. 6.6.

Column environment corresponds to a *flowsheet* where only relevant modules are provided for the column designing. Below is an example of the use of this interface:

As shown in Fig. 6.7, the object palette in this environment only has equipment relevant for the column designing, as well as balance operations and Dynamics options. There are displayed several known equipment sets such as pumps, separators, exchangers, valves, and mixers, among others. However, new models are added for the reboiler, the condenser, and column sections (Table 6.4).

In this manner, a column can be designed with additional equipment of a freer manner, inclusive allowing the use of logical operations and control equipment.

6.4   Distillation Introductory Example

6.4.1   Problem Description

In order to illustrate the use of the corresponding modules both in Aspen Plus® and Aspen HYSYS®, distillation of styrene and ethylbenzene is studied. This separation

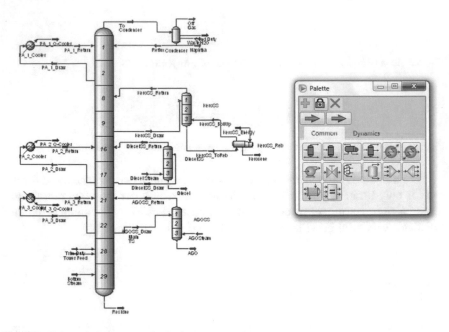

Fig. 6.7 Column interface to calculate a *topping* column

Table 6.4 Internal modules to calculate columns in Aspen HYSYS®

Icon	Unit
	Condensers:
	Total
	Partial
	Three phase
	Column sections (can be main section and side splitters)
	Reboiler

system is of utmost importance, since the styrene production is made using ethyl-benzene as raw material and later is it necessary to separate them in order to purify the product and recirculate the reactant.

Let's say we want to separate 27,550 lb/h from a mixture of styrene and ethyl-benzene that is at 110 °F and atmospheric pressure. The mass composition of the stream is shown in Table 6.5. The required distillation column operation conditions shall be determined (distillate flow rate, reflux ratio, number of stages, etc.).

The goal is to recover 99.2 % of the ethylbenzene by the column top using a molar reflux rate of 6. The styrene recovery percentage is of 2.5 %, taking a pressure at the top of 45 Torr and in the bottom of 105 Torr. Consider the use of a total condenser.

Table 6.5 Feeding composition to the distillation column

Component	Mass fraction
Ethylbenzene	0.5843
Styrene	0.415
Heptadecane	0.0007

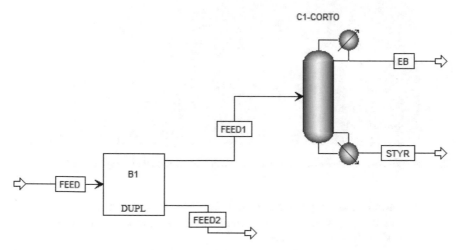

Fig. 6.8 Configuration for shortcut calculation in Aspen Plus®

6.4.2   Simulation in Aspen Plus®

Initially a simulation is opened with English units. The corresponding components were input and NRTL is used as a thermodynamic model. Now in the *flowsheet* a *DSTWU* model is input which will allow you to determine basic conditions with which later the shortcut calculation is started. In order to implement the modules in a more comfortable manner, the use of the tool called *Stream Duplicator* is established, which is in the tab *Manipulators* under the name *Dupl*. This tool enables to duplicate streams, that is to say, generate many streams with the same specifications from the one entered to the said module (Fig. 6.8).

Once the configuration in *flowsheet* is completed, the data from the entry stream (*FEED*) is entered, which corresponds to the inlet of the module *Dupl*. The resulting stream from the said module is called *FEED1*. The shortcut calculation module is called *C1-CORTO*.

In order to enter the corresponding data in the shortcut calculation module, keep in mind the information provided from the problem description. In the box *Reflux ratio* positive values indicating the reflux ratio or negative values indicating the factor that shall multiple the minimum reflux calculated by the simulator can be entered (Fig. 6.9).

With this information the problem is completely defined and then click on the button *Next* for the simulator to carry out the corresponding calculation. As soon as

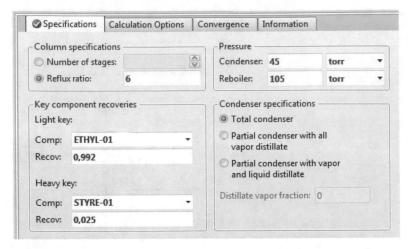

Fig. 6.9 Window of data input into the shortcut calculation module in Aspen Plus®

the calculation motor reports the calculation has been made satisfactorily, the shortcut calculation results can been seen clicking on the module icon and selecting the option *Results* on the left tree.

Here you can see the results provided by the shortcut calculation: minimum reflux ratio, actual reflux ratio; minimum number of stages, number of actual stages, optimal feed stage, heat duties and temperatures in both the condenser and the reboiler, and finally the distillate to feed ratio.

This information is used to specify appropriately the rigorous calculation added below (Fig. 6.10).

Then add a *RadFrac* (rigorous calculation) module from the object palette, name it *C2-RIG*. Connect to this module a new stream called *FEED2* coming from the module *Dupl*. Also connect the top stream, *ETHYLB*, and the bottom stream, *STYRENE*. Make sure the top stream is connected in the corresponding one to the liquid distillate, since the condenser of the distillation column is total.

Observing the results provided by the shortcut calculation, we can see that the molar reflux ratio of 6 corresponds to 1.23 times the minimum reflux ($R_{min} = 4.887$), this value stays within the interval recommended in the literature, generally between 1,2 and 1,5 times the minimum reflux (Taylor and Krishna 1993). This is because at higher reflux ratio, the column has a higher liquid load, and thus the column diameter increases. The energetic duty both from the reboiler and the condenser increases while the number of stages diminishes, increasing the operating costs and reducing initial investment costs.

On the other hand, when using a low reflux ratio many more stages are required to accomplish the desired separating grade; however, both the diameter as the energy duties diminish. Operating costs are low but initial investment costs are higher. For this reason an optimization shall be carried out to determine the ratio that minimizes costs from both operating and initial investment costs. The reflux

Summary	Balance	Reflux Ratio Profile	⊘ Status	

Minimum reflux ratio:	4,88488	
Actual reflux ratio:	6	
Minimum number of stages:	29,7544	
Number of actual stages:	54,6352	
Feed stage:	25,3221	
Number of actual stages above feed	24,3221	
Reboiler heating required:	1,92418e+07	Btu/hr
Condenser cooling required:	1,87058e+07	Btu/hr
Distillate temperature:	131,693	F
Bottom temperature:	182,099	F
Distillate to feed fraction:	0,585726	
HETP:		

Fig. 6.10 Results provided by the shortcut calculation module in Aspen Plus®

ratio interval introduced above is the region where 90 % of the times the optimal point is found (Fig. 6.11).

Now we have to select the shortcut calculation information that is used to appropriately specify the rigorous calculation and that enables to obtain composition, flow, and temperature and pressure profiles throughout the column.

Performing a degrees of freedom analysis on the system, the conclusion is that at this point two values have to be specified, which can have top/feed or bottom/feed ratios, top or bottom flows, reflux flow or reflux ratio, provided these values are independent (Fig. 6.12).

After this information is determined, in the tab _Streams_ specify the stage on which the feed is input, in this case stage 25. Finally, in the tab _Pressure_ are specified the pressures of the condenser (stage 1) and the pressure drop through the column for the pressure at the bottom to be 105 Torr, as required by the statement (Figs. 6.13 and 6.14).

With this information, click on the button _Next_ in order for the simulator to carry out the corresponding calculation. As soon as the calculation is completed, in the option _Profiles_ you can see the way tabulate composition, temperature, pressure and liquid and vapor flows profiles throughout the column (Fig. 6.15).

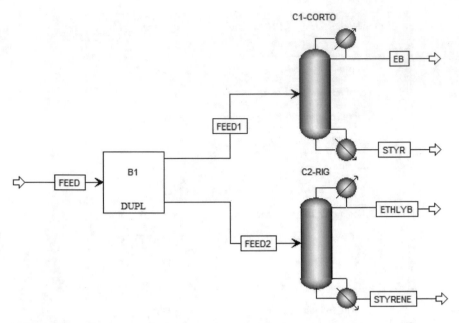

Fig. 6.11   Final flow diagram including the rigorous calculation module in Aspen Plus®

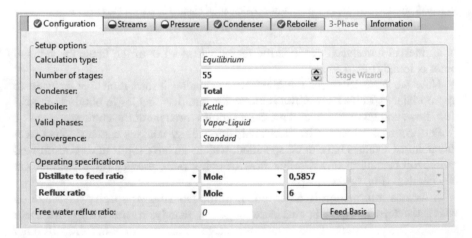

Fig. 6.12   Information input window in the rigorous calculation module in Aspen Plus®

In the taskbar of *Home* option appears the *Plot* tab that allows to plot the data registered in the *Profiles* tab. It should be noticed that the *Profiles* tab must be open in order to access the option *Plot* in the menu.

Figure 6.16 shows the main window of the mentioned tool, where there is a brief explanation on its uses. Clicking the down button you can see all the types of

Fig. 6.13 Information input window on streams for the rigorous calculation module in Aspen Plus®

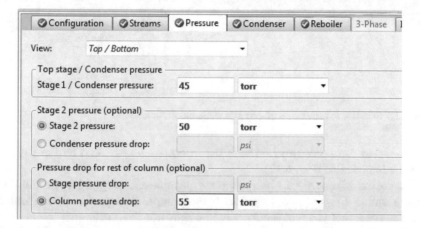

Fig. 6.14 Window of pressure profile input for the rigorous calculation module in Aspen Plus®

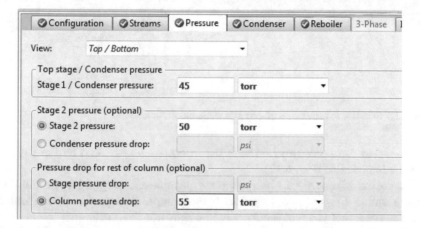

Fig. 6.15 Profiles generated by the rigorous calculation module in Aspen Plus®

Fig. 6.16 Main window
of the option *Plot* available
in Aspen Plus®

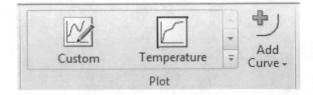

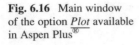

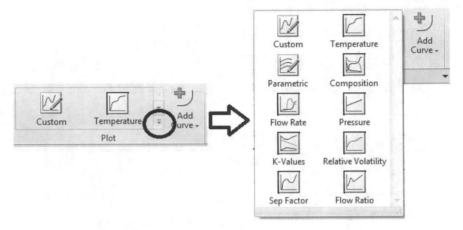

Fig. 6.17 Window to select type of graphics available in Aspen Plus®

graphics that can be constructed with the information generated by the Rigorous
Calculation Module (Fig. 6.17).

For the exercise purpose, observe the composition and temperature profile. Click
on the option *Composition* to generate the composition profile throughout the
column. After that you will see the window shown in Fig. 6.18.

Here select all the components and specify in *Select phase* the option *Liquid*.
Click *OK*. A new tab is opened and in the upper menu the option *Format* is
displayed (Fig. 6.19). There are several format options regarding to the title as
well as the name of each axis, the composition legend and others, are defined.

After specifying appropriately the graphic that displays the corresponding composition profile, the process previously described can be repeated in similar manner
to generate the temperature profile throughout the column (Figs. 6.20 and 6.21).

6.4.3   Simulation in Aspen Hysys®

For the distillation column simulation in Aspen HYSYS®, a new simulation shall be
started. Add the necessary components, input the model *NRTL* as property package
and, finally, enter in the simulation environment.

Fig. 6.18 Window of
composition profile
configuration in Aspen
Plus®

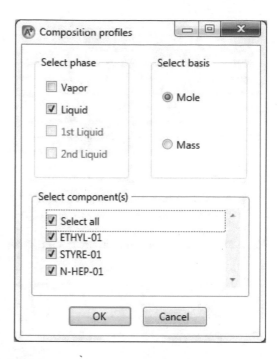

Fig. 6.19 Window of graphic configuration for composition profile in Aspen Plus®

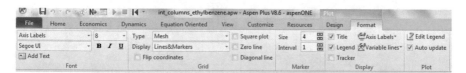

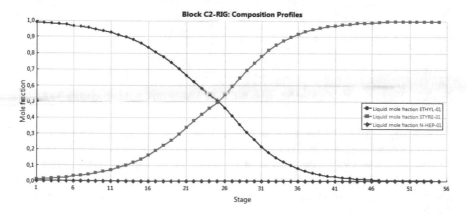

Fig. 6.20 Composition profile generated by the rigorous calculation module in Aspen Plus®

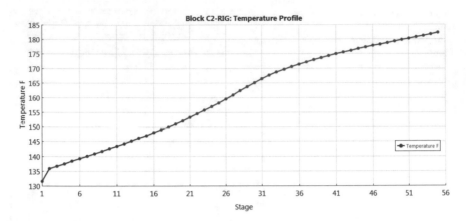

Fig. 6.21 Temperature profile generated by the rigorous calculation module in Aspen Plus®

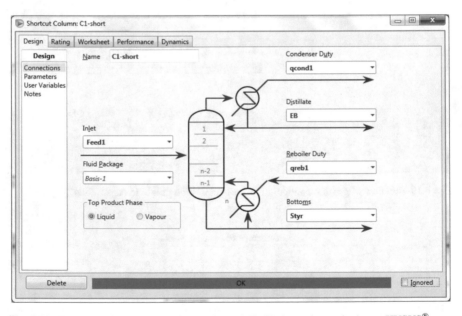

Fig. 6.22 Shortcut calculation module interface of distillation columns in Aspen HYSYS®

Similarly to what has been performed in Aspen Plus®, Aspen HYSYS® also has a flash calculation module. To select it, go to the operations pallet and select the module *Shortcut Distillation*. Name it *C1-short*. Said module has as entrance a stream called *Feed1*, a top stream, *EB*, and a bottom stream, *Styr*. Additionally, the name of the energetic streams both for the condenser (*qcond1*) and the reboiler (*qreb1*) must be specified (Fig. 6.22).

Once the values corresponding to the connection are specified, input the specifications into the tab *Parameters*. Here introduce the pressure in the condenser as in the reboiler and the corresponding mole fractions of the key components in the two product streams. This is one important difference with respect to Aspen Plus where mole recoveries are defined in the shortcut calculations. However, keep in mind that in Aspen HYSYS® key component mole fractions are expressed based on the stream where the component is not wanted.

After doing the mole balance for the desired recovery, in the option *Light Key in Bottoms* you have to enter 0.010 for the ethylbenzene mole fraction. In the option *Heavy Key in Distillate* enter 0.0012 for styrene mole fraction. These two values correspond to the mole recovery specified in Aspen Plus® calculations. Then pressure is included both in the condenser and the reboiler.

As soon as this is specified, in the box *Minimum Reflux Ratio* there is a value of 4.336 which is calculated with the entered data. Finally, in the box *External Reflux Ratio*, enter a value of 6. With this, the module is completely specified (Fig. 6.23).

As performed in Aspen Plus®, the information corresponding to the shortcut calculation is taken to initialize appropriately the rigorous calculation that requires data that we do not have with the initial information. For this reason, in Fig. 6.24 is the tab *Performance*, which shows the results obtained in the shortcut calculation.

In Fig. 6.25 the current flow diagram up to this point is shown.

First, a stream called *Feed2* is created with the same conditions as stream *Feed1*. In Aspen HYSYS® there is not any tool such as the module *Duplicator* in Aspen Plus®, reason for which a stream is created using the operations pallet and double clicking on it the stream properties window is displayed. In the lower part is the

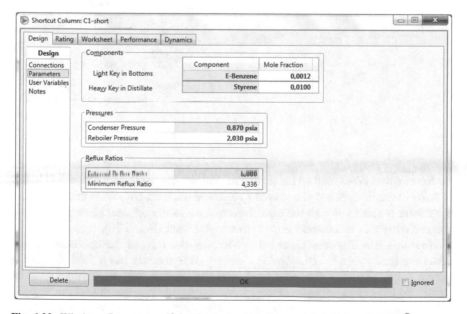

Fig. 6.23 Windows *Parameters* of the shortcut calculation module in Aspen HYSYS®

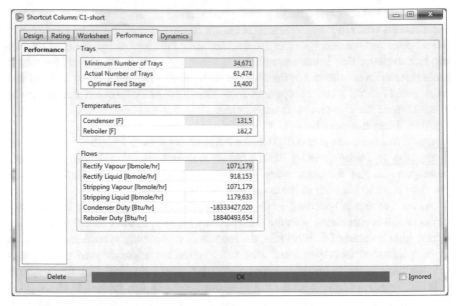

Fig. 6.24 Window _Performance_ of the shortcut calculation module in Aspen HYSYS®

Fig. 6.25 Flow diagram
with the shortcut calculation
module in Aspen HYSYS®

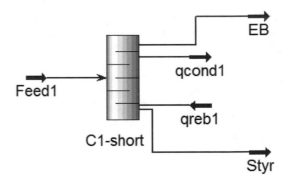

button _Define from Other Stream_ where you can select from what stream you want
the information taken, and on the right part you can see the values that were chosen
in order to confirm. Select the stream _Feed1_ and click on _OK_. With this procedure
the stream is specified with the same information as the original stream. Likewise
you can copy the information in any stream you want (Fig. 6.26).

Now install a Rigorous Calculation Module. For this, in the operations pallet
select the module called _Distillation Column_. This module has a feed stream the
stream _Feed2_. The top stream, _EthylB_, and the bottom stream, _Styrene_, shall be
connected in the same manner. Energy streams for the condenser and the reboiler
are on _qcond2_ and _qreb2_, respectively. Here keep in mind that the shortcut

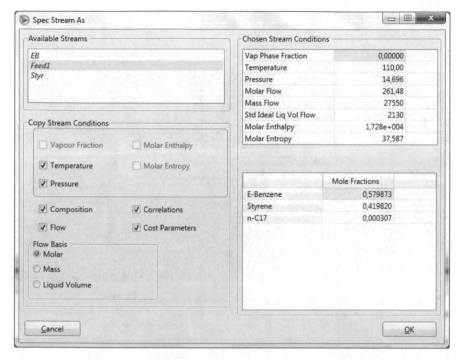

Fig. 6.26 Window to import information from a stream to another in Aspen HYSYS®

calculation gave a result of 61 stages and an optimal feeding stage in stage 16. Here, for comparison purpose, the same results obtained in Aspen Plus® will be specified. In Aspen HYSYS®, the condenser is not taken into account for the stage count, and when entering the stage value in the rigorous calculation window, the reboiler is not taken into account, reason for which the number of resulting stages shall be entered in the shortcut calculation less one (the reboiler). When the corresponding data has been entered, click on the button *Next* (Fig. 6.27).

The following window asks for the pressures in the column ends. Enter the data reported in the problem description. As soon as entering the corresponding values, click on the *Next* button (Fig. 6.28).

In the following window, enter the top and bottom estimated temperatures to improve the module convergence; however, for this exercise, it is not necessary. Click on the button *Next* (Fig. 6.29).

Finally, in the window shown in Fig. 6.30, the reflux ratio and distillate flow can be provided. Keep in mind that the specified reflux ratio was 6. In the *Worksheet* tab of the shortcut calculation module the distillate flow rate that shall be entered in the rigorous calculation module can be observed. Once the corresponding data is introduced, click on the button *Done*.

Now, when returning to the module main window (Fig. 6.31), click on the option *Specs* to check that there are zero degrees of freedom. This determines if the column was specified correctly and if it is likely to get an answer.

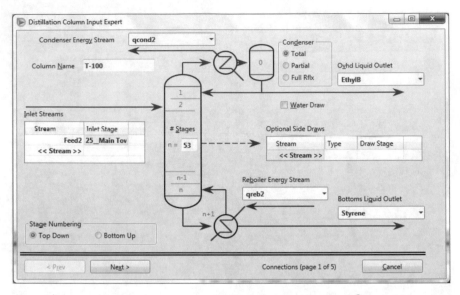

Fig. 6.27 Main window for rigorous calculation module in Aspen HYSYS®

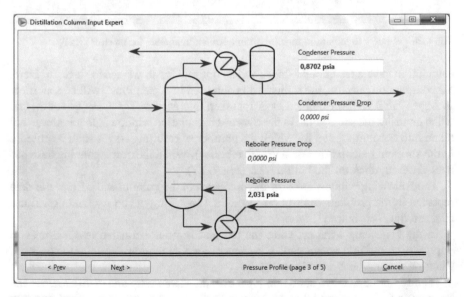

Fig. 6.28 Entry window for the pressure profile in the rigorous calculation module in Aspen HYSYS®

In this window you can see all the specifications Aspen HYSYS® has by default, and in the upper right part if it is active or not. On the left bottom part, in the box *Degrees of Freedom,* is reported if the module has specified or not completely specified. You can add specifications by clicking on the button *Add* in the section

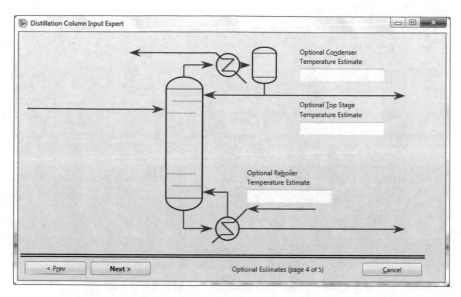

Fig. 6.29 Temperature estimate entry window in the rigorous calculation module

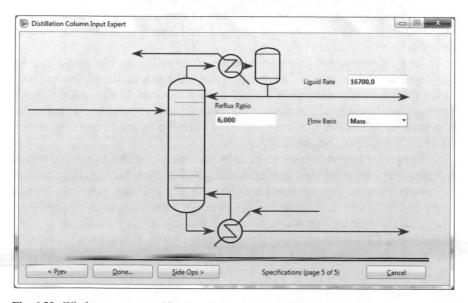

Fig. 6.30 Window to enter specifications in the rigorous calculation module

Column Specifications. Additionally, on the right lower part, the parameters used to calculate error are listed. These parameters can be modified to prevent convergence problems in case of greater complexity.

Fig. 6.31 Window *Specs* of the rigorous calculation module in Aspen HYSYS®

Finally, click the button *Run* on the lower part for the simulator to perform the module calculation. When the lower bar changes from red to green, indicates that the module has accomplished convergence and that the results are already available.

In the *Performance* tab all the information of the calculated distillation column behavior is shown. As soon as the corresponding window is displayed, flows as well as the entry and outlet streams of the column are reported. In the option *Column Profiles* you can see in tabular manner the temperature, pressure, liquid, and vapor flow profiles. Additionally, on the upper part, the reflux ratio and evaporation rate on which the column operates is displayed.

In the option *Plots* one can see graphically the interest Profiles. For effects of this exercise composition profile and temperature profile are reported to compare with the results obtained from Aspen Plus® (Fig. 6.32). For this, select the option *Compositions* in the section *Tray by Tray Properties* and click on the button *View Graph*. In this manner a profile is displayed that can be very similar to the one shown in Fig. 6.33. In the same way the temperature profile that appears in Fig. 6.34 can be constructed (Fig. 6.35).

6.4.4   Results Analysis and Comparison

The distillation simulation for the ethylbenzene-styrene system is performed in the two simulation packages, Aspen Plus® and Aspen HYSYS®, finding considerable differences in the results. Taking into account that the same property package (NRTL) was used in both simulations, and that the shortcut calculation is basically

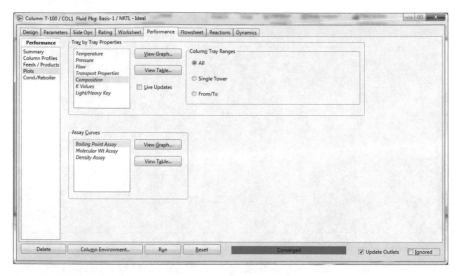

Fig. 6.32 Window performance $\geq$ *Plots* of the rigorous calculation module

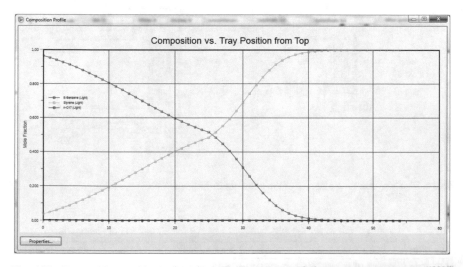

Fig. 6.33 Composition profile obtained from the rigorous calculation module in Aspen HYSYS[R]

the same, the only likely response is that the binary interaction parameters are different. Below is the report of the shortcut calculation results in each simulator (Table 6.6).

With this information we can see that the two simulators have slightly different values as to the minimum reflux ratio and in the global efficiency of the plate; however, this is reflected in an increase of five stages and the difference in the feed stage for the two calculations. This involves differences in the temperature and

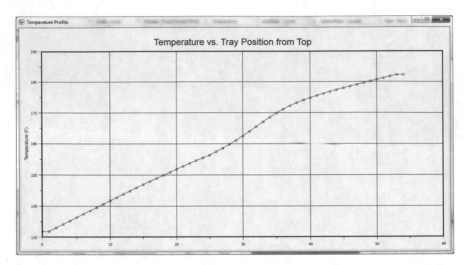

Fig. 6.34 Temperature profile obtained from rigorous calculation module in Aspen HYSYS®

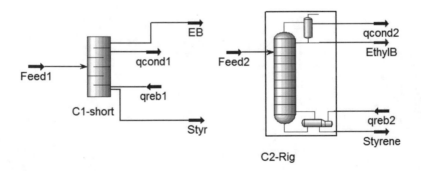

Fig. 6.35 Complete flow diagram of the columns in Aspen HYSYS®

Table 6.6 Results obtained for the ethylbenzene–styrene distillation in both simulators

Variable	Aspen Plus®	Aspen HYSYS®
Minimum number of stages	29.754	34.491
Number of actual stages	54.659	61.194
Global efficiency	54.4 %	58.5 %
Feeding stage	25.333	16.646
Minimum reflux ratio	4.888	4.336
R/R_{min}	1.23	1.38
Distillate/feed ratio	0.5857	0.5850
Condenser temperature (°F)	131.5	131.5
Reboiler temperature (°F)	182.2	182.2

composition profiles, specifically in the composition which is affected in greater proportion.

Profiles comparison is shown in Figs. 6.36 and 6.37.

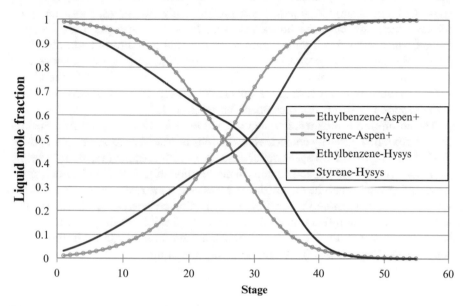

Fig. 6.36 Comparison of composition profiles with the information obtained

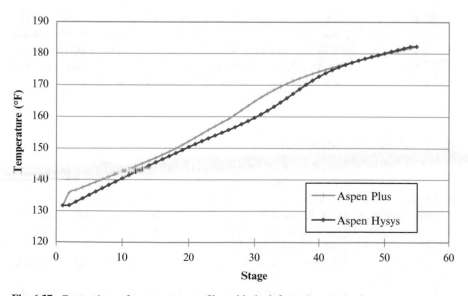

Fig. 6.37 Comparison of temperature profiles with the information obtained

6.5   Absorption Introductory Example

Absorption is a unit operation where one or more components are diluted in a large
gas stream, removed by the action of a nonvolatile liquid solvent. Placing the gas in
contact with an absorption liquid, where preferentially the interest components are
soluble, these condense the liquid, releasing heat. The absorption efficiency is
favored when the solvent temperature decreases, the pressure is high, absorbent
flow is high and when the molecular weight of the absorbent is low. However, the
solvent molecular weight is limited by the liquid–vapor equilibrium in the higher
state of the absorption column. If the solvent is too light, there may be losses due to
high evaporations that restrict the operation feasibility.

The absorption is the most frequent method used for CO_2 removal from natural
gas and from fuel gases by contact with solutions of NaOH or mono-ethanol-amine
(MEA), in which case we speak about absorption with chemical reaction since the
gas reacts with the solvent and remains in the solution. If the reaction is irreversible,
the resulting liquid shall be disposed. For cases where the reaction is reversible, the
solvent can be regenerated with stripping or distillation operations. There are also
cases where the absorption is carried out by purely physical mechanisms and where
basically the separation occurs as consequence from high solubility in one of the
components, from the gas stream in the absorption liquid.

6.5.1   Problem Description

Acetone that is found in an air stream will be absorbed using water as solvent. The
air stream is entering the absorption column with a flow rate of 100 kmol/h, and has
a 3 % molar of acetone at 1 atm and 30 °C. The water molar flow is 200 kmol/h at
20 °C and 1 atm. The column has six theoretical stages (Adapted from: *Separation
Process Engineering*), Appendix of Chapter 12 (Urdaneta et al. 2002).

6.5.2   Process Simulation

To draw the absorption equipment some of the icons available in module
RADFRAC are used, preferably several of the ones not including condenser or
reboiler (although any of them can be used). Enter the feeding such as: gas feed in
the bottom (*AIR-ACE*), bottom liquid product (*ACE-SVTE*), liquid feed on top
(*SVTE*) and top gas product (*AIR-LIM*) (see Fig. 6.38).

In the window configuration of block *RADFRAC* the following information is
entered (Table 6.7).

In this manner, since the condenser and the reboiler are not specified, the
operation to be simulated corresponds to the absorption column.

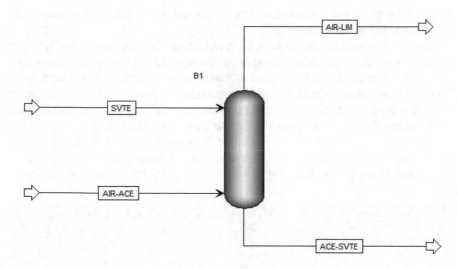

Fig. 6.38 Flow diagram for absorption column in Aspen Plus®

Table 6.7 Information to enter in the absorption column module

Cell	Value
Condenser	None
Reboiler	None
Convergence	Petroleum/wide boiling

Fig. 6.39 Window specification of stream location in the block at Aspen Plus®

Entering to the *Streams* tab of the block, as convention for the liquid stream is entered the option *Above* in stage 1 and the gas as *On Stage* in the last stage (Fig. 6.39). Below, in the convergence option of the absorption column block, in the

Basic tab, shall appear in the Sum Rates algorithm (*Sum Rates*). The maximum iterations value is set as 50.

Components (water, acetone, air) are introduced and the property method NRTL is selected. Streams are entered at atmospheric pressure, fed gas has an acetone concentration of 3 % molar, at 30 °C and the feed flow rate is 100 kmol/h. Water flow rate is 200 kmol/h, which enters at 20 °C and the column has six stages. Initially the objective is to obtain a top gas stream with a molar fraction of acetone of 0.003 maximum. Click on the button *Next* in order to execute the corresponding calculations.

Check the temperature profile entering in the option *Profiles* (Fig. 6.40). Locate maximum temperature and register it, check if the acetone composition specification is met in the gas product (Fig. 6.41). This is found by clicking on the option

TPFQ	Compositions	K-Values	Hydraulics	Reactions	Efficiencies	Properties	Key Components	Thermal Analysis	Hydraulic Analysis	Bubble Dew Points

View: All   Basis: Mole

Stage	Temperature	Pressure	Heat duty	Liquid from (Mole)	Liquid from (Mass)	Vapor from (Mole)	Vapor from (Mass)	Liquid feed (Mole)	Liquid feed (Mass)
	C	bar	Gcal/hr	kmol/hr	kg/hr	kmol/hr	kg/hr	kmol/hr	kg/hr
1	22,2267	1,01325	0	200,851	3631,99	99,568	2859,67	200	3603,06
2	24,0358	1,01325	0	201,369	3653,39	100,419	2888,59	0	0
3	25,1746	1,01325	0	201,743	3672,69	100,937	2910	0	0
4	25,4644	1,01325	0	201,927	3689,98	101,311	2929,3	0	0
5	24,5987	1,01325	0	201,9	3709,29	101,495	2946,59	0	0
6	21,6037	1,01325	0	200,432	3725,87	101,468	2965,9	0	0

Fig. 6.40 Temperature profile along the column generated in Aspen Plus®

	Material	Heat	Load	Work	Vol.% Curves	Wt. % Curves	Petroleum	Polymers	Solids

Display: All streams   Format: GEN_M   Stream Table   Copy All

	ACE-SVTE	AIR-ACE	AIR-LIM	SVTE
Mass Flow kg/hr	3725,87	2982,48	2859,67	3603,06
Volume Flow cum/hr	3,782	2487,53	2413,27	3,608
Enthalpy Gcal/hr	-13,65	-0,151	-0,164	-13,663
Mole Flow kmol/hr				
C3H6O-01	2,8	3	0,2	
H2O	197,371		2,629	200
AIR	0,261	97	96,739	
Mole Frac				
C3H6O-01	0,014	0,03	0,002	
H2O	0,985		0,026	1
AIR	0,001	0,97	0,972	

Fig. 6.41 Stream results, for a six stage column with a solvent flow rate of 200 kmol/h

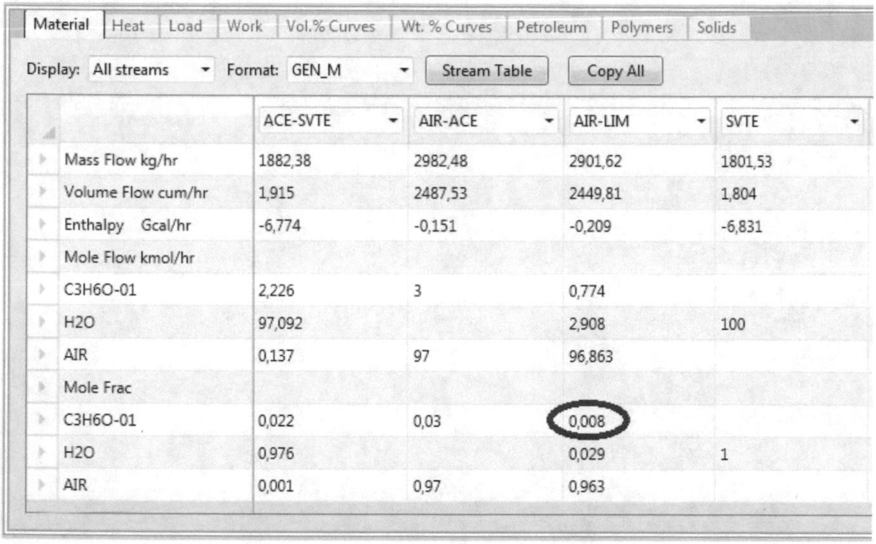

	ACE-SVTE	AIR-ACE	AIR-LIM	SVTE
Mass Flow kg/hr	1882,38	2982,48	2901,62	1801,53
Volume Flow cum/hr	1,915	2487,53	2449,81	1,804
Enthalpy Gcal/hr	-6,774	-0,151	-0,209	-6,831
Mole Flow kmol/hr				
C3H6O-01	2,226	3	0,774	
H2O	97,092		2,908	100
AIR	0,137	97	96,863	
Mole Frac				
C3H6O-01	0,022	0,03	0,008	
H2O	0,976		0,029	1
AIR	0,001	0,97	0,963	

Material Heat Load Work Vol.% Curves Wt. % Curves Petroleum Polymers Solids

Display: All streams ▾ Format: GEN_M ▾ Stream Table Copy All

Fig. 6.42 Stream results, for a six stage column with a solvent flow rate of 100 kmol/h

Stream Results. When observing the liquid product composition, it is found that the acetone is too diluted (Fig. 6.41), situation that surely is a problem for the purification of this stream. What can be done to increase this molar fraction?

An option is to decrease water flow to half (100 kmol/h) and watch if the specification is met and if it increases the concentration in the liquid product. We observe that although the liquid stream comes out of the equipment at a greater concentration (Fig. 6.42), it does not meet the specification in the gas product.

As the specification is not met, could we increase duplicating the number of stages in the column ($N = 12$). Does the gas stream concentration considerably improve? Likely not (Fig. 6.43). This is because, although the number of contact stages between phases is increased, the driving force is not increased.

Now, modify the temperature from the two input streams to the equipment at a value of 10 °C. Once again execute the simulation with 12 equilibrium stages and a solvent flow rate of 100 kmol/h. The results obtained correspond to the ones shown in Fig. 6.44.

As it can be noted, in this manner the design specification is met. It is evident that the stream temperature within the column has a considerable effect on the absorption efficiency.

6.6 Enhanced Distillation

Enhanced distillation is a very useful tool in chemical processes when the system has partial azeotropy or immiscibility among its components. Those difficulties make it impossible to perform a conventional distillation and require the

| | Material | Heat | Load | Work | Vol.% Curves | Wt. % Curves | Petroleum | Polymers | Solids |

Display: All streams ▾ Format: GEN_M ▾   Stream Table   Copy All

	ACE-SVTE ▾	AIR-ACE ▾	AIR-LIM ▾	SVTE ▾
Mass Flow kg/hr	1886,53	2982,48	2897,48	1801,53
Volume Flow cum/hr	1,92	2487,53	2455,67	1,804
Enthalpy Gcal/hr	-6,773	-0,151	-0,21	-6,831
Mole Flow kmol/hr				
C3H6O-01	2,335	3	0,665	
H2O	96,97		3,03	100
AIR	0,138	97	96,862	
Mole Frac				
C3H6O-01	0,023	0,03	0,007	
H2O	0,975		0,03	1
AIR	0,001	0,97	0,963	

Fig. 6.43 Stream results, for a 12 stage column with a solvent flow rate of 100 kmol/h

| | Material | Heat | Load | Work | Vol.% Curves | Wt. % Curves | Petroleum | Polymers | Solids |

Display: All streams ▾ Format: GEN_M ▾   Stream Table   Copy All

	ACE-SVTE ▾	AIR-ACE ▾	AIR-LIM ▾	SVTE ▾
Mass Flow kg/hr	1928,93	2982,48	2855,07	1801,53
Volume Flow cum/hr	1,959	2323,42	2348,09	1,787
Enthalpy Gcal/hr	-6,887	-0,165	-0,126	-6,848
Mole Flow kmol/hr				
C3H6O-01	2,669	3	0,331	
H2O	98,229		1,771	100
AIR	0,148	97	96,852	
Mole Frac				
C3H6O-01	0,026	0,03	0,003	
H2O	0,972		0,018	1
AIR	0,001	0,97	0,979	

Fig. 6.44 Stream results, for a 12 stage column with a solvent flow rate of 100 kmol/h, streams entering at 10 °C

introduction of new analysis tools to make decisions on the operation design. Below the basic concepts to understand that operation are listed. For further information, look up the references at the end of the chapter.

6.6.1 Residue Curves Map

The structure and properties of phase equilibrium diagrams for the azeotropic multicomponent mixtures are based on the definition of *azeotropy*. For this reason, as starting point it is important to clearly understand this concept. There are different definitions reported in the literature, but many of them cannot be generalized, in other words, are not applicable to all the situations.[1] *Azeotrope* is a word from ancient Greece that translates "bubbling without any change," what means that the vapor generated in a distillation has the same composition of the liquid with which it is in equilibrium (Rodríguez et al. 2001). However, a broader definition establishes that an azeotropic state is the one where the composition of each component is the same one in all the coexisting phases (Tao et al. 2003), providing the possibility of the existence of more than two phases. In fact, not any of both definitions is correct for the vapor–liquid–liquid heterogeneous systems, because the boiling temperature derivative with respect to the composition is not defined in the azeotropic point (Rooks et al. 1998), in such a way that the composition of each component is different in each of the phases.

In a more general way, an azeotropic state is defined as the state in which the mass transfer occurs while the composition of each phase remains constant, but not necessarily identical (Diamond et al. 2004). In this way it is likely to generate the necessary conditions for specific situations, as the heterogeneous azeotropic distillation and reactive distillation. A more useful way to classify azeotropic mixtures is through its deviations from the Raoult law. For a homogenous multicomponent mixture in vapor–liquid equilibrium, the equilibrium constant for each of the species i, K_i, is defined as:

$$K_i = \frac{y_i}{x_i} = \frac{\gamma_i^L f_i^L}{\Phi_i^V P} \tag{6.1}$$

Where x_i and y_i are the molar fractions of the species i in liquid and vapor phases, respectively, in the equilibrium. The nonideality is expressed within terms of deviations from the activity coefficient unit for the liquid phase γ_i^L, and the fugacity coefficient for the vapor phase ϕ_i^V; f_i^L if the fugacity of the component is pure in the liquid phase. At low pressures, the fugacity coefficient from the vapor phase is equivalent to one and the fugacity of the component i pure in the liquid phase is made equivalent approximately to the saturation pressure P_i^S at the temperature T; in this manner, (6.1) becomes:

---

[1] Circumstances on which the formation of two liquid phases occur or chemical reason are cases in which the azeotropy definition is not valid.

$$K_i = \frac{y_i}{x_i} = \gamma_i^L \frac{P_i^S}{P} \qquad (6.2)$$

An Azeotrope of maximum boiling point exists when there is a negative deviation from Raoult Law ($\gamma_i^L < 1.0$). In this point, dew point and bubble point curves reach a minimum in the P–x–y diagram for a determined temperature, and for this temperature and pressure a maximum of these two curves in the diagram T–x–y is reported. This situation is illustrated in Fig. 6.45 for the acetone–chloroform system, in which we can see the Azeotrope of maximum boiling point in $x_1 = y_1 = 0.35$ and a temperature of 64.5 °C. For this system, a conventional distillation column with a feed that has an acetone composition under the azeotropic point allows us to obtain by top pure chloroform and as bottom stream the azeotropic mixture. In the same manner for a feed stream with acetone composition over the azeotropic point, the distillation column has as top product the pure acetone and as bottom product the azeotropic mixture.

An Azeotrope with minimum boiling point exists when there is a positive deviation from the Raoult Law ($\gamma_i^L > 1.0$). In this point, the dew point and bubble point curves reached a maximum in the P–x–y diagram for a determined temperature, and for this temperature and pressure, a minimum of these two curves is reported in the T–x–y diagram. In this type of systems the top product that is obtained in a conventional distillation column always is the azeotropic mixture, and according to the feeding mixture composition, one of the pure components is obtained as a bottom product. An example of this type of systems is given in Fig. 2.2 for the methanol–methyl acetate mixture, which reports a minimum boiling point Azeotrope in 53.6 °C and for a molar methanol composition of 0.33 (Fig. 6.46).

When the positive deviations from Raoult Law are sufficiently large ($\gamma_i^L \gg 1.0$), a phase separation can occur and a minimum boiling point heterogeneous Azeotrope is generated, in which the vapor phase is in equilibrium with both liquid phases. In cases where there are heterogeneous azeotropes, the two-liquid phase region is superposed to the vapor–liquid equilibrium region, as illustrated in Fig. 6.47. Additionally, the mixture boils the liquid global composition which equalizes the vapor phase composition, but the three coexistent phases have different compositions.

At an azeotropic point, equilibrium constants from all the species are one, and through simple distillation it is not likely to accomplish any separation. For this reason, in nonlinear dynamic an Azeotrope is known as *fixed or stationary point*, and through a stability analysis, its occurrence possibility can be determined in a distillation operation. Likewise, through the nonlinear dynamic has been possible to develop design and operability methodologies, stability analysis and occurrence of multiple steady states with a highly mathematical work and which has gained considerable acceptance during the last decade.

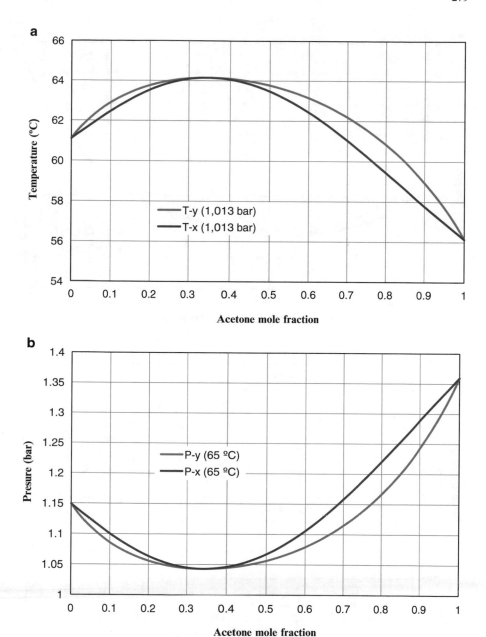

Fig. 6.45 Vapor–liquid equilibrium diagram for the acetone–chloroform mixture calculated with Aspen Plus®: (**a**) Diagram T–x–y; (**b**) Diagram P–x–y

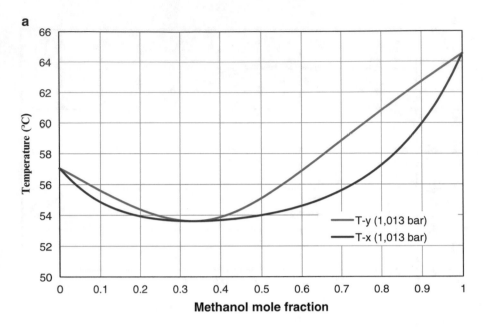

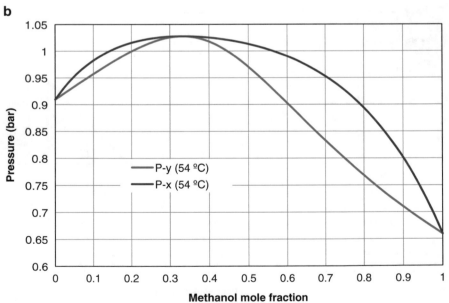

Fig. 6.46 Vapor–Liquid equilibrium diagrams for the methanol–methyl acetate mixture calculated with Aspen Plus®: (**a**) Diagram *T–x–y*; (**b**) Diagram *P–x–y*

Fig. 6.47 Schematic diagram of a binary mixture with an Azeotrope, a constant pressure

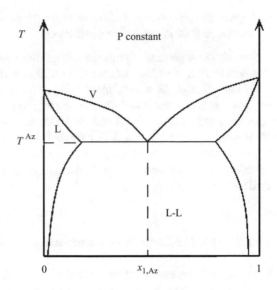

6.6.1.1   Residue Curve Maps for Ternary Mixtures

Distillation systems are generally designed by simulation or through trial and error processes where it is necessary to assume an initial configuration for the column and estimate all the operation parameters. Nowadays, several commercial simulators provide the chance to design column sequences with graphic utilities such as residue curve maps (RCM). These diagrams are an indispensable tool for interpreting the behavior and feasibility of the separation of a homogeneous azeotropic mixture. Because the separation from a binary azeotropic mixture is impossible with the conventional distillation, improvements have been introduced that try to displace the Azeotrope in an economic manner. The modification in pressure is the first alternative to be considered; however, not always the azeotropic composition is affected significantly with changes in pressure (Black 1980). It is then when the decision to add a third component known as separating agent is made.

Separation agents used in distillation operations are divided into four categories (Diamond et al. 2004), according to how the separation is produced:

1. *Liquid separating agents that do not induce liquid phase formation in the ternary mixture;* this type of separation is called homogeneous azeotropic distillation and the extractive distillation is a special case of it.
2. *Liquid separating agents that induce the formation of two or more liquid phases in the ternary mixture:* this type of separation is known as heterogeneous azeotropic distillation.
3. *Separating agents that react with* one of the components from the binary azeotropic mixture (reactive distillation).

4. *Separating agents that dissociate ionically* in the binary mixture displacing the azeotrope (saline extractive distillation).

Not all the separating agents are within these four categories and for this reason it is necessary to develop methods that enable the determination of the feasibility to use them. The MCR allows identifying quickly separation alternatives within a process, and is constituted in the starting point to select the alternative that optimizes the process global economy (Dennis and Megan 2001a, b, c; Dyk and Nieuwoudt 2000; Manan and Bañares-Alcántara 2001). Below we have the most important features that accompany this type of diagram, along with several basic criteria to select separation agents that are used in both azeotropic and extractive distillation operations.

6.6.1.2   Residue Curve Map Construction

RCM are constructed in ternary diagrams, which provide information about the possible separation means in columns that operate at total reflux (Dennis and Megan 2001c; Henley and Seader 1981). In a column operating at total reflux, the mass balance in a section is met when the vapor rising from a n stage has the same composition than the liquid falling from a superior stage $n-1$. Liquid and vapor streams that abandon a stage are in phase equilibrium. Concentration profiles in the liquid phase are represented in approximate manner through series of lines that connect the composition values in every stage and that constitute a residue curve (Tolsma 1999).

A simple manner to interpret the construction of the RCM consists on the description of a simple distillation, in which a liquid mixture is subject to boiling within a container. At any instant, the vapor (rich in the most volatile component) that is being generated is removed from the container, and assumed that it is in equilibrium with the remaining liquid mixture under the assumption of a perfect mixture. As the most volatile is constantly removed, the remaining liquid phase composition and temperature continuously change throughout time and are moved to composition regions of the heavier components and higher temperatures. The trajectory of changes in the composition in liquid phase, starting from the initial point, is called a *residue curve of simple distillation or residue curve,* and the set of all the possible curves for a specific mixture is called *residue curve map* (Manan and Bañares-Alcántara 2001).

If the simple distillation of Fig. 6.48 is considered, where the vapor abandoning the container is in phase equilibrium with the liquid perfectly mixed, in order to meet the mass balance we require that the speed in which the liquid disappears to be exactly the same as the speed in which the vapor escapes:

Fig. 6.48 Schematic
diagram of a simple
distillation

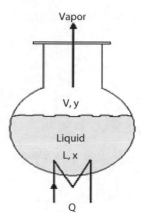

$$\frac{dL}{dt} = -V \tag{6.3}$$

Where L represents the total number of moles from the remaining liquid and V the vapor flow in moles per unit of time. Likewise, a balance of the component leads to:

$$\frac{d(Lx_i)}{dt} = -Vy_i \tag{6.4}$$

For $i = 1,2,3,\ldots,C-1$ components. Expanding (6.4) we obtain:

$$L\frac{d(x_i)}{dt} + x_i\frac{d(L)}{dt} = -Vy_i \tag{6.5}$$

Replacing (6.3) in (6.5) and reorganizing, we obtain:

$$\frac{dx_i}{dt} = \frac{V}{L}(x_i - y_i) \tag{6.6}$$

Where L and V change with time.

Taking into account that the interest is centered in the residue curve map knowledge and, specifically, the relative change of the composition, calculations can be simplified by the introduction of a new modified time variable, in the following manner:

$$d\xi = \frac{V}{L}dt \tag{6.7}$$

Here, ξ is a dimensional variable that varies within 0 and $+\infty$. When $t=0$, $\xi=0$, and when $t=t_f$ (the period of time in which the container is emptied), $\xi=\infty$.

Reorganizing (6.6), we have:

$$\frac{dX}{d\xi} = X - Y \tag{6.8}$$

Depropanizer Column

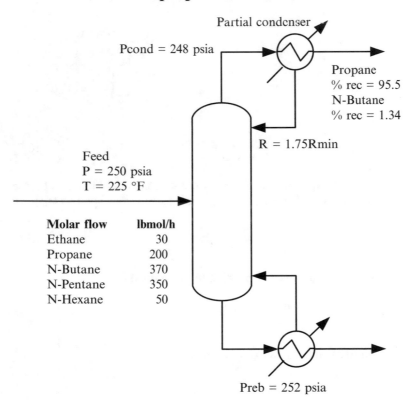

Where X represents the state vector of $(C-1)$ molar fractions in the independent liquid phase and Y the corresponding equilibrium vector of the molar fractions in the vapor phase (Manan and Bañares-Alcántara 2001).

The algorithm followed to solve the differential equation system described by (6.8) is rather simple and can be implemented through a programming language or in a worksheet as Microsoft Excel. The steps contemplated in the calculation are:

- Set a starting composition for the liquid mixture: x_1, x_2, and x_3.
- Set a value for the pressure P.
- Calculate the values of y_i and T through a vapor–liquid equilibrium model (activity coefficient, state equation, etc.).
- Carry out the numerical integration of (6.8) through Euler or fourth order Runge Kutta numerical methods. In every step of the integration it is necessary to repeat

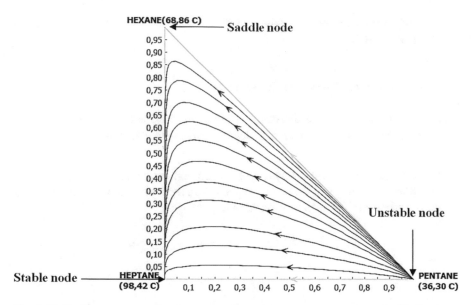

Fig. 6.49 Residue curve map for the pentane–hexane–heptane system generated with the simulator Aspen Plus®, $P = 1$ atm

vapor–liquid equilibrium calculations for the new compositions of the liquid being obtained.
- Repeat the integration until you find stability in T and x_i values.

Figure 6.49 shows the RCM for the pentane–hexane–heptane system generated in the simulator Aspen Split from Aspen Technology. There, all the trajectories indicated by the arrows were originated in the pentane vertex (the lightest component) and end in the heptane vertex (the heaviest component). The arrows point toward the temperature increase direction and decrease of volatility; in this manner, the liquid mixture initially enriches in hexane and then the arrows progressively go further from this vertex indicating heptane enrichment.

Because all the curves were originated in the pentane vertex, this point is the curve source and it is called *unstable node* or of low boiling point. Likewise, the heptane vertex behaves like a curve destination and is called *stable node* or of high boiling point. The vertex corresponding to the hexane is called *saddle node* or of intermediate boiling point and behaves as origin and destination of residue curves. Note that the lightest component is the unstable node, the heaviest component is the stable node, and the intermediate boiling point component is the saddle node (Malagón 2010; Diamond et al. 2004; Wahnschafft et al. 1994).

In the case of extractive distillation, where a separation agent is added to an azeotropic mixture that does not form new azeotropes with any of the initial mixture components, the residue curve map does not have any distillation regions (Diamond et al. 2004), that is to say, it is delimited by a stable node, an unstable one and at least a saddle node.

6.6.1.3   Properties of Residue Curve Maps

The most outstanding properties of the residue curve maps (Diamond et al. 2004; Henley and Seader 1981; Manan and Bañares-Alcántara 2001) are:

- Residue curves do not cross or intersect each other.
- The boiling temperature always increases throughout a residue curve (the only exception is given when the steady states where the boiling temperature remains constant because the composition does not change).

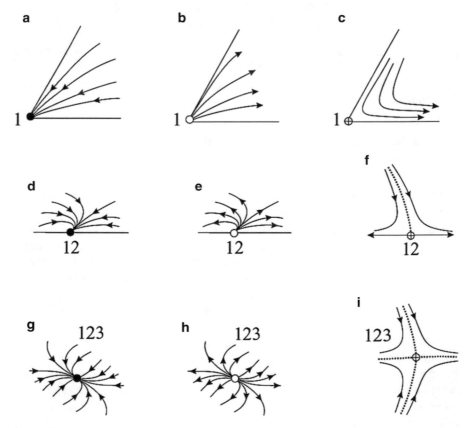

Fig. 6.50 Types of steady points for three component mixtures: (**a**) stable node of a component, (**b**) unstable node of a component, (**c**) saddle node of a component, (**d**) stable node of two components, (**e**) unstable node of two components, (**f**) saddle node of two components, (**g**) stable node of three components, (**h**) unstable node of three components and (**g**) saddle node of three components. The *arrows* indicate the residue curves direction

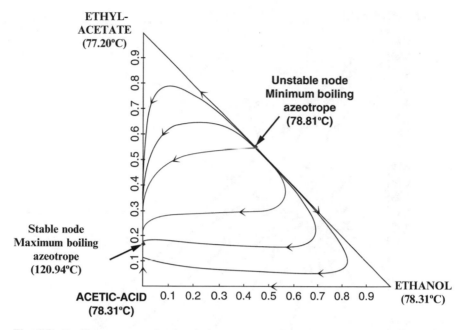

Fig. 6.51 Residue curve map for the ethyl acetate–ethanol–acetic acid system, $P = 1$ atm

- Solutions in steady state of the equations occur in all the pure components and in the azeotropes. In Fig. 6.50 are shown the possible solutions in steady state for a three component system.
- Solutions to stable state are limited to one of the following types: stable node, unstable node, or saddle node.
- Nodes (stable, unstable, and saddles) define a dimensional composition space.
- Residue curves in the nodes are tangent in a common direction, which is determined by the relative volatilities of the components in each node.

The ethanol acetic-acid ethyl-acetate system presented in Fig. 6.51 is more complex in respect to the one from Fig. 6.50, due to the formation of a binary azeotrope between ethyl acetate and ethanol. This system is important to study the reaction of the acetate formation as from acetic acid and ethanol. We can see that the azeotrope is an unstable node because it has the minimum boiling point in the mixture; likewise, ethyl acetate, ethanol, and acetic acid behave as saddle nodes. The presence of the azeotrope generates a particular behavior in residue curves trajectory, and makes the composition triangle to divide into two regions defined by a distillation limit connected between the azeotrope and the stable mode. The regions and limits of distillation generate an additional property of the RCM:

- Each distillation region shall contain a stable node, an unstable node and at least, one saddle node.

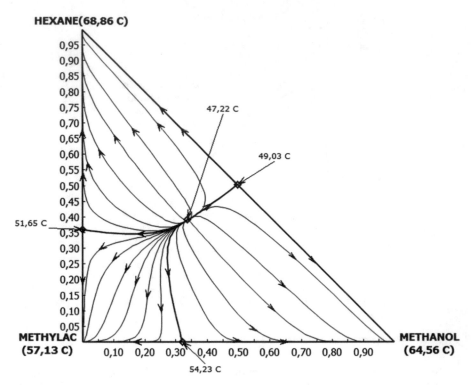

Fig. 6.52 Residue curve map for the hexane–methanol–methyl acetate system, $P = 1$ atm

It is common to see in many ternary systems more than one azeotrope, which can constitute potential distillation regions and, therefore, more complex RCMs. A typical example of this type of maps is shown in Fig. 6.52. Hexane-methyl acetate-methanol mixture has three binary azeotropes of minimum boiling point and a ternary azeotrope of minimum boiling point. Those azeotropes generate three distillation limits and three distillation regions. The ternary Azeotrope of the minimum boiling point is the origin of all the residue curves, and is additionally constituted in the sole unstable node. Depending on the feed composition, the residue curves get to the nodes from each pure component; this behavior is used several times to obtain the lightest components as bottom products of a distillation column (Bekiaris et al. 1994; Ciric et al. 2000; Dennis and Megan 2001b).

For systems where there is partial miscibility, there can also be presence of azeotropes, in the one and two phase regions. Azeotropes contained in the miscible are homogeneous and, likewise, the azeotropes that are within the immiscibility region are heterogeneous and their functioning is restricted to unstable or saddle nodes (Castillo and Towler 1998; Thiele and Geddes 1933; Wahnschafft et al. 1994). An example of this type of systems is shown in Fig. 6.53 for the ethanol–water–benzene mixture. This system has two heterogeneous azeotropes and two homogeneous azeotropes that delimit three distillation regions.

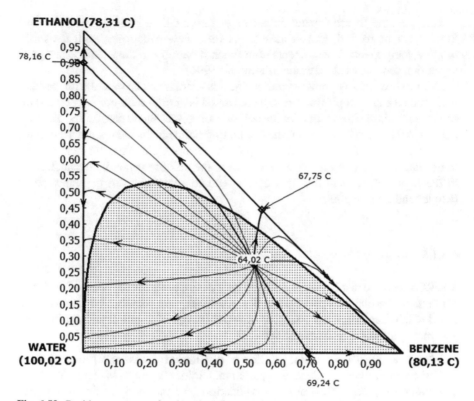

Fig. 6.53 Residue curve map for the ethanol–water–benzene system, $P = 1$ atm

6.6.1.4 Selecting the Separation Agent

To select a separation agent, keep in mind, among other things, the type of Azeotrope to be separated. When a binary azeotropic mixture of minimum boiling point is separated, without any separation agent, it is likely to obtain a pure component in the column bottom, while the composition profile is approximated to the composition of the same enrichment section (Tolsma 1999; Müller and Segura 2000). In the same manner, when a binary azeotropic mixture is wanted to be separated from a maximum boiling point, the composition is obtained in the stripping section (Perry 1992).

To separate the minimum point azeotropes, a high boiling point separation agent shall be fed near the top of the column; this process corresponds to the classic extractive distillation, where the critical separation occurs in the enrichment section (Dhole 1992). In maximum boiling point azeotrope separation, the situation is completely opposite. When a heavy component is used as separation agent, at the beginning it can be fed with the azeotropic mixture because, in this case, only the stripping section is needed.

It is possible to use separation agents of intermediate boiling point, although these are hard to find, and in most cases form new azeotropes with the initial mixture components. These agents offer much flexibility in the adequate design of separation sequences (Bieker and Simmrock 1994).

When a low boiling point separation agent is used, it can be mixed with the main feed when the azeotropic mixture to be separated is of minimum boiling point; if the azeotropic mixture is of maximum boiling point, the agent can be fed near the bottom of the column. However, the low boiling point agents are not much accepted in the separation of maximum boiling point azeotropes, since the nonideal interactions required to modify relative volatilities of the mixture generally occur stronger in the liquid phase, and the light agents tend to accumulate in the vapor phase (Müller and Segura 2000).

6.6.1.5   Residue Curve Map Applications

The RCM and the distillation regions are very useful tools for the comprehension of continuous distillation operations and per batches, specifically if these are combined with other information such as bimodal curves of liquid–liquid equilibrium. The most important applications are:

1. Identification of the system: location of distillation limits, azeotropes, distillation regions, likely products and partial miscibility regions (Malagón 2010).
2. Laboratory data evaluation: data validation and thermodynamic models and confirmation of ternary azeotropes.
3. Process synthesis: conceptual development and construction of flow diagrams for new processes and modification or redesigning of existing processes (Ciric et al. 2000; Lewis and Matheson 1932; Smith et al. 1997).
4. Process modeling: identification of column specifications that could cause difficulties in the convergence of the simulation. Determination of initial estimates of the column parameters, as location of the feeding plate, number of states, reflux ratio, and products compositions (Dennis et al. 2000; Krishnamurthy and Taylor 1985a, b).
5. Control: Profile analysis and column balances that favor the control system design (Dennis and Megan 2001a; Malagón 2010).
6. Process adjustments: operation and malfunction analysis of the separation systems, review of composition profiles and detection of impurity traces that involve corrosion problems.
7. Cost analysis: consideration of different alternatives (Dyk and Nieuwoudt 2000).

As well as in other separation operations, in which triangular composition diagrams are used, as distillation curve maps, mass balances which are graphically represented by straight lines that connect the corresponding compositions. Global flow rates are found through the application of the lever rule. Mass balance lines for the distillation have two restrictions:

1. Bottom compositions, distillate and fee shall always be on the same straight line.
2. Bottom compositions and distillate are, with a very close approximation, on the same residue curve. As, by definition, residue curves do not cross a distillation limit, the bottom compositions and distillate are on the same distillation region, and the intrinsic mass balance and line to the residue line in both sites.

6.6.2   Extractive Distillation

6.6.2.1   Problem Description

In order to illustrate the simulation procedure from an extractive distillation, the ethanol–water system separation is used, using glycerol as separating agent (Leiva 2003).

The distillation process objective is to obtain 300 000 L/day of fuel ethanol (99.5 % molar) from an azeotropic mixture of ethanol and water (88.0 % molar in ethanol). For this purpose it has two distillation columns and the necessary equipment to carry out the feeding conditioning for each column and product conditioning. Water extraction from the azeotropic mixture is made with high purity glycerin (99.7 % molar).

The process flow diagram is the one shown in Fig. 6.54.

The process starts with the input of ethanol and water azeotropic mixture to a heat exchanger in which it is preheated with the glycerol stream recovered from the second column. The recovered glycerol, by its part, is cooled up to 15 °C to store it and prevent loss from evaporation.

By convention, the first distillation column (TD-101 in the diagram) is called dehydration column since this is separated from the anhydrous ethanol, and the second column (TD-102 in the diagram) is called recovering column, since in it the water solvent is separated.

Glycerol and azeotrope streams enter into the dehydrating column; where by top, the anhydrous ethanol is obtained which is further cooled for its storage, and by bottom, a glycerol and water mixture which is pumped toward the recovering column. In this last column both components are separated, obtaining by top relatively pure water and by bottom glycerol that is recirculated.

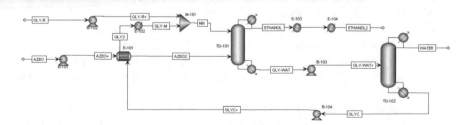

Fig. 6.54 Process flow diagram for the extractive distillation

6.6.2.2 Simulation in Aspen Plus®

First a simulation is started in Aspen Plus® selecting the option _General with Metric Units_ to use metric units. Then the components listed on Table 6.8 are introduced.

The property package for the mixture must be NRTL, according to a previous analysis of the mixture polarity, operation conditions, and availability of interaction parameters for the model. Figure 6.55 shows the _T–x–y_ equilibrium diagram

Table 6.8 Components to introduce in the simulation at Aspen Plus®

Component	Type	Component name
Ethanol	Conventional	Ethanol
Water	Conventional	Water
Glycerol	Conventional	Glycerol

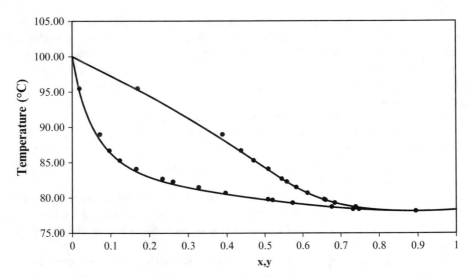

Fig. 6.55 Equilibrium diagram of _Txy_ phases of the Ethanol (1)—Water (2) system at 1 atm

calculated with the NRTL model and compared to the experimental information for the binary ethanol–water mixture. Check that the binary interaction parameters for the components are within the NRTL model and that the experimental information allows to rely on the appropriate selection of the property model.

With this information the system and property model are completely defined. Now input the streams and equipment sets of the flow diagram.

The _Data Browser_ window is closed, once again having the process flow diagram window or _Process Flowsheet Window_ where the process streams and units are set up.

Add a stream called _AZEO_ which corresponds to the ethanol and water azeotropic mixture which is fed then to the dehydrating column. Enter the following information (Table 6.9).

Table 6.9 Operation conditions of the stream *AZEO*

Variable	Value
Temperature	15 °C
Pressure	0.747 bar
Molar flow	250 kmol/h
Mole fraction	
Ethanol	0.88
Water	0.12
Glycerol	0.00

Table 6.10 Operation conditions of stream *GLY±*

Variable	Value
Temperature	153 °C
Pressure	1.334 bar
Mole flow	99 kmol/h
Mole fraction	
Ethanol	0.00
Water	0.01
Glycerol	0.99

Now add the pump B-101 from the menu *Model Library* ≥ *Pressure Changers* ≥ *Pump*, which increases pressure for the stream *AZEO* to *1.334 bar* for the input in the dehydrating column. The stream from this equipment is called *AZEO±*.

The following equipment to be entered is a heat exchanger of two process streams; this is the HeatX module, which is located in Model Library > Heat Exchangers > HeatX. It should be named E-101. The AZEO+ stream is the tube side process stream, the tube outlet is AZEO2.

As a two process stream exchanger was selected, a new stream should be created. This stream has the adequate information to initialize the calculations, because it will be connected to the output stream with recovered solvent. For that a stream named GLY+ should be created with the following information (Table 6.10).

Connect the stream *GLY±* to the fluid intake by the casing (called *Hot* at the time of connecting) and to the outlet create a stream called *GLY2*. Now specify in the exchanger E-101 that the temperature of the cold fluid outlet is to be 54 °C. For this, in the route *Blocks* ≥ *E-101* ≥ *Setup* in the tab *Specification* select the option *Cold stream outlet temperature* and enter the value of 54 °C (Fig. 6.56).

Now select a *Heater* module from the route *Model Library* ≥ *Heat Exchangers* ≥ *Heater* and call it *E-102*. This module shall reduce the stream *GLY2* temperature to 78.2 °C without any pressure drop. The outlet stream of this module is called *GLY-M*.

The glycerol recirculating stream is now defined; however, we still have to define the glycerin recovery stream. For this, create a stream named *GLY-R* with the following operation conditions (Table 6.11).

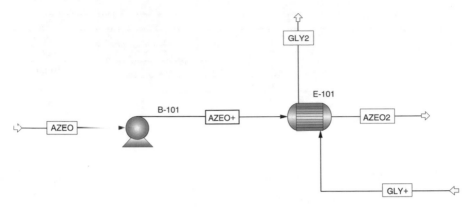

Fig. 6.56 Process flow diagram after adding the exchanger *E-101*

Table 6.11 Operating conditions of stream *GLY-M*

Variable	Value
Temperature	15 °C
Pressure	0.747 bar
Molar flow	0.001 kmol/h
Molar fractions	
Ethanol	0.00
Water	0.003
Glycerol	0.997

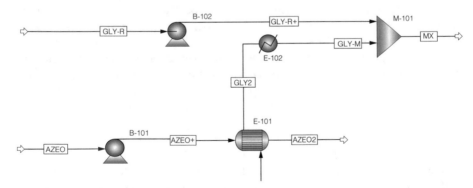

Fig. 6.57 Flow diagram of the process after adding the mixer *M-101*

This stream is connected to a pump (*B-102*) where discharge pressure shall be 1.334 bar in order to be at the same pressure as the stream *GLY-M*. Name the outlet stream, *GLY-R±*. Then select a *Mixer* module from the route *Model Library ≥ Mixers/Splitters ≥ Mixer* and connect as inlets the streams *GLY-M* and *GLY-R±*. The outlet for the said module is called *GLY*. This module is called *M-101* and the outlet stream, *Mix* (Fig. 6.57).

Table 6.12 Specifications for the distillation column *TD-101*

Configuration	
Type of calculation	Equilibrium
Number of stages	24
Condenser	Total
Reboiler	Kettle
Valid phases	Vapor–Liquid
Convergence	Azeotropic
Specifications	
Distillate flow (molar)	217.5 kmol/h
Reflux ratio (molar)	0.6
Streams	
Name	*Stage*
MIX	3
AZEO2	16
Pressure	
Sight	Top/Bottom
Pressure stage 1	0.747 bar
Pressure drop from column	0.016 bar

Now specify the dehydrating column. For this, select the module *RadFrac* from the route *Model Library* ≥ *Columns* ≥ *RadFrac* and enter the specifications from Table 6.12.

The top stream is called *ETHANOL* and the bottom streams, *GLY-WAT*. Once this information is entered, it is possible to carry out the first simulator calculation. Keep in mind that you can see the results as soon as the calculations have been made in the route *Results Summary* ≥ *Streams*.

In the column module you can see flow, concentration, and temperature profiles throughout the column, which can be useful for the column operation analysis.

Figure 6.58 shows the concentration profile throughout the column. In this figure we can see that there are column points where the concentrations vary considerably corresponding to the stages where the feed *MIX* and *AZEO2* are entered. Table 6.13 reported the results from product streams of the dehydrating column.

With this information it can be concluded that the product obtained does meet the specifications; however, a lot of ethanol is lost by the column bottoms. Below is the calculation for ethanol losses.

$$F_{E, \text{ fondos}} = 0.025 \times 131.501 \frac{\text{kmol}}{\text{h}} = 3.287 \frac{\text{kmol}}{\text{h}} \tag{6.9}$$

$$F_{E, \text{cima}} = 0.996 \times 217.5 \frac{\text{kmol}}{\text{h}} = 216.63 \frac{\text{kmol}}{\text{h}} \tag{6.10}$$

which corresponds to:

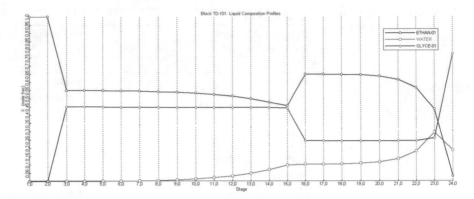

Fig. 6.58 Concentration profile of column *TD-101*

Table 6.13 Conditions of the product streams from *TD-101*

Stream	Ethanol	GLY-WAT
Temperature	70.8 °C	130.7 °C
Pressure	0.747 bar	0.763 bar
Mole flow	217.5 kmol/h	131.5 kmol/h
Mole fraction		
Ethanol	0.996	0.025
Water	0.004	0.230
Glycerol	13 PPB	0.745

Table 6.14 Design specification for column *TD-101*

Specifications	
Type	Mole recovery
Objective	0.995
Components	
Selected components	Ethanol
Product streams	
Selected streams	Ethanol
Varying...	
Distillate flow rate	200 kmol/h
	230 kmol/h

$$\%_{E,\, lost} = \frac{3.287\frac{kmol}{h}}{216.63\frac{kmol}{h} + 3.287\frac{kmol}{h}} \times 100 = 1.49\,\% \tag{6.11}$$

$$\%_{E,\, rec} = 100\,\% - 1.49\,\% = 98.51\,\% \tag{6.12}$$

This loss, although it seems small, is not good for the operation. To guarantee a 99.5 % recovery a design specification on the column shall be included with the parameters established in Table 6.14.

With this specification, the ethanol quality is to be guaranteed after closing the glycerol recycle in the flow diagram. We can see that the distillate flow rate changes slightly from 217.5 to 219.68 kmol/h, meeting the required recovery (Fig. 6.59).

To the stream coming from the bottom of column *TD-101* a pump (*B-103*) must be installed to take pressure to 0.0227 bar, which is the operation pressure for the recovering column of the solvent *TD-102*. The pump outlet is called *GLY-WAT*±.

For the other stream coming from the dehydrating column it must be cooled down to 15 °C for its storage; for this purpose two exchanger types *Heater* (*E-103* and *E-104*) are installed. The first takes the stream to a 30 °C temperature, and the second to the final temperature. It is considered that both exchangers do not have a pressure drop.

The following equipment to install is the distillation column to recover the solvent (*TD-102*). This column has as feed the stream *GLY-WAT*±; to produce by top pure water (stream *WATER*) and by bottom the glycerol for the recirculation (stream *GLYC*). The column specifications are shown in Table 6.15.

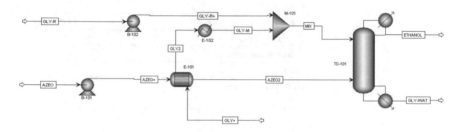

Fig. 6.59 Flow diagram after adding column *TD-101*

Table 6.15 Specifications for the distillation column *TD-102*	Configuration	
	Type of calculation	Equilibrium
	Number of stages	8
	Condenser	Total
	Reboiler	Kettle
	Valid phases	Vapor–Liquid
	Convergence	Standard
	Specifications	
	Distillate flow (molar)	32 kmol/h
	Reflux ratio (molar)	0.05
	Streams	
	Name	*Stage*
	GLY-WAT+	5
	Pressure	
	View	Top/Bottom
	Pressure stage 1	0.02 bar
	Column pressure drop	0.0047 bar

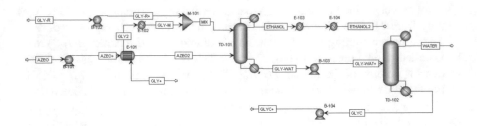

Fig. 6.60 Flow diagram after adding pump *B-104*

Table 6.16 Design specifications for column *TD-102*

Specifications	
Type	Mole purity
Objective	0.995
Components	
Selected components	Glycerol
Product streams	
Selected streams	GLYC
Varying...	
Distillate flow rate	26 kmol/h
	33 kmol/h

Finally, we have to install a pump (*B-104*) to take the glycerol obtained in this column to the adequate conditions for the mixture with the recirculation. For this specify the discharge pressure as 1.334 bar and outlet stream, *GLYC*±.

Figure 6.60 shows the process diagram to the time the pump *B-104* is installed.

Likewise as with the first column, a design specification shall be made to guarantee that when connecting to the glycerol recycle, the product quality will be maintained. For this case the glycerol purity that is to recirculate is the concern, since it affects the ethanol separation which is carried out in the first distillation column.

Table 6.16 shows the information to enter for the designing specification.

In this point we can carry out the calculation to obtain Profiles from the recovering column.

To the present time we have the process simulation in which there is not any solvent recirculation, but, as logical, the solvent recirculation is necessary as from the economic point of view and of the process. For this, follow the procedure described below:

6.6.2.3   Glycerol Recirculation

The following step to construct the simulation is close the glycerol recycle and carry out a design specification on the process, to guarantee a constant glycerol flow in the dehydrating column. For this purpose, delete the stream GLY+ and in its place connect the stream GLYC+. To this moment do not carry out any calculation, since we still have to configure a specification to adjust the glycerol resetting to keep the flow constant.

In the navigation tree, go to the route *Flowsheeting Options* ≥ *Design Spec* where a window appears; there you have to click on the button *New* and select as name of the specification *DS-1*, which is the name by default. The window unfold is similar to the one represented in Fig. 6.61.

In this window the variables set are added; for this case, a variable: *FLOWM*, which corresponds to the resetting glycerol flows and to the glycerol mixture introduced in the dehydrating column. Click on the button New, enter the name *FLOWM* and click on *Accept*.

In the *Edit selected variable* section you have to specify the molar flow of the stream *MIX*. In the section *Category* select the option *Streams* because the variable is from a stream of the process. Then in the tab *Type* select the option *Stream-Var*, which means once again the variables relevant to the process streams. After selecting this option appears the tab *Stream* where the stream *MIX* is selected. Two new tabs appear: *Substream* and *Variable*; from the last select the option

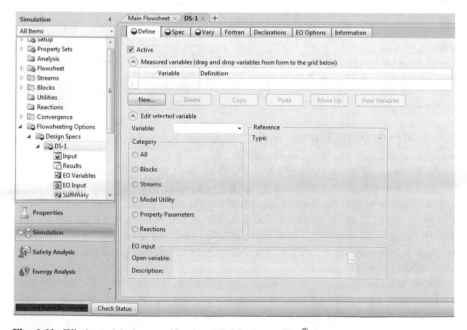

Fig. 6.61 Window of design specification *DS-1* in Aspen Plus®

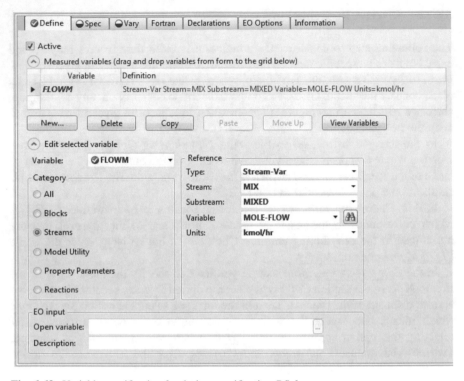

Fig. 6.62 Variable specification for design specification *DS-1*

Table 6.17 Design specification for the glycerol recycle

Specification	
Specification	FLOW-M
Objective	100
Tolerance	0.01

MOLE-FLOW. Verify that the variable units are in *kmol/h*. In this manner the variable is defined, as shown in Fig. 6.62.

As soon as defining the variables, click on the tab *Spec*. In this window the information reported in the following table is specified (Table 6.17).

This indicates that the flow of the stream *MIX*, represented here as the variable *FLOWM*, must have a value of 100 kmol/h and that values within 99.99 and 100.01 kmol/h are admitted. Now on the tab *Vary* select the variable to be modified to meet the specification and the limits on which it flutters. Enter the information reported below (Table 6.18).

With this information the design specification is defined on the glycerol recirculation, now you can make the simulator carry out the corresponding calculations and will have the complete flow diagram.

Table 6.18 Information to be entered in the tab *Vary*

Manipulated variable	
Type	Stream-var
Stream	GLY-R
Substream	MIXED
Variable	MOLE-FLOW
Variable limits	
Lower	0.01
Top	5.5

Table 6.19 Result from the product streams from the extractive distillation

Stream	Ethanol	Water
Mole flow rate (kmol/h)	219.87	30.182
Molar compositions		
Ethanol	0.996	0.036
Water	0.004	0.962
Glycerol	13 PPB	0.02

The following product streams are obtained (Table 6.19).

We can see that a small portion of the fed glycerol is lost in the water stream and this makes it necessary to reset the solvent.

6.7 Nonequilibrium Models

Distillation problems for over 100 years have been approached through equilibrium stage models, based on the assumption that liquid and vapor phases that exit from the stage are in thermal and thermodynamic equilibrium. In practice, the separation processes rarely reach equilibrium, since chemical potential gradients and temperature processes involving mass and heat transfer are generated. The equilibrium model is a simple and very manageable concept as from the mathematical point of view, which has worked as basis for many commercial simulators and which have been widely used in simulation and design of many actual columns.

The equations that base the equilibrium model are known as *MESH* to make reference to the equation types that are used in the model (Eckert and Vanek 2001; Treybal 1996):

- *M* refers to mass balance equations.
- *E* refers to equilibrium equations (assumption between the liquid and vapor streams that abandon a stage).
- *S* refers to the summation equations of the molar fractions of the components present in each of the phases.
- *H* refers to the heat or enthalpy balances.

A traditional method is the process shaping by stages which is the one used by the stage efficiency concept. There are several stage efficiency definitions used in distillation column simulation, including the global efficiency, vaporization efficiency, among others, but the one mostly known and used is the Murphree component efficiency (Eckert and Vanek 2001; Glasser et al. 2000; Thomas 1991), which is defined as:

$$\varepsilon = \frac{y_{ij} - y_{i,j-1}}{y_{ij}^* - y_{i,j-1}} \tag{6.13}$$

Where y_{ij} is the composition of the component i in the vapor stream outgoing from the stage j, $y_{i,j-1}$ is the composition of the incoming vapor from the previous stage $j-1$, y y_{ij}^* is the composition of the vapor that is in equilibrium with the liquid exiting the stage j. This stage efficiency reflects the ratio between actual mass transfer and the one that would occur in the case of reaching equilibrium in the stage. For packed columns an analogue concept to stage efficiency is used, known as HETP (Height equivalent to Theoretical plate). In practice, efficiency and HETP values are estimated from the experience gained with similar processes. However, for new processes, this approach is not totally valid and forces the development of efficiency estimate methods and mechanisms and HETP.

The different types of efficiency try to represent the deviation from the equilibrium, found in each of the stages or in the complete column. Likewise, the HETP is an easily managed number in designing a column. However, there are different issues when this type of efficiencies is used in a simulation based on an equilibrium model (Seider et al. 2004):

- There is not any general consensus on which of the efficiency definitions is the best (although many experts in distillation admit certain preference for the Murphree efficiency).
- Efficiencies vary from component to component and from one stage to another, in a multicomponent mixture. This fact is almost never taken into account in a simulation using an equilibrium model.
- Just like efficiencies change from one plate to another, the HETP is a function of the height of the packed bed. This efficiency and the HETP behavior is not taken into account in the conventional column simulation software.

In the last 10 years distillation and absorption processes have been simulated as an operation based on the mass and heat transfer rates, using models known as *nonequilibrium models* (Boston and Sullivan 1974; Doherty and Malone 2001; Kister 1991; Kiva et al. 2003; Ludwig 1979). The blocks required for the construction of a nonequilibrium model must include mass balances, power balances, equations that relate the thermodynamic equilibrium and mass and energy transfer models. In a nonequilibrium model, balance equations are written for each phase by separate. The conservation equations for each phase are related through the mass balance surrounding the interface, where it is assumed that the material lost in the vapor phase is gained in the liquid phase. Likewise the energy balance is worked both in each phase and in the interface.

In nonequilibrium models the equilibrium equations are used to relate compositions from each phase in the interface; values K are evaluated in the compositions and in this point. For this reason, the temperature and composition shall be determined as part of the column simulation.

Although the nonequilibrium models developed to the present date are several, all of them present some common and fundamental features that are described in this section to illustrate their usefulness and versatility.

A nonequilibrium model was developed by Taylor and his collaborators in his book where he includes a detailed description (Seider et al. 2004). This model can be used to simulate both plate columns and packed columns; the only difference is that different expressions are used to estimate the binary coefficients of mass transfer and interface areas. Specifically, packed columns are simulated with segments representing a discrete integration throughout the packed bed. The larger number of segments, the better the integration and the obtained results are more accurate. Figure 6.63 shows a schematic diagram of a nonequilibrium stage or segment. The vertical waved line that is located in the middle of the diagram represents the interface between the two phases that can be liquid and vapor (distillation), gas and liquid (absorption), or two-liquid phases (extraction). In this representation, the vapor from a lower stage contacts the liquid that comes down from a higher stage; there a mass and energy exchange occurs through a common interface represented by the waved line. In this model feed streams in liquid and

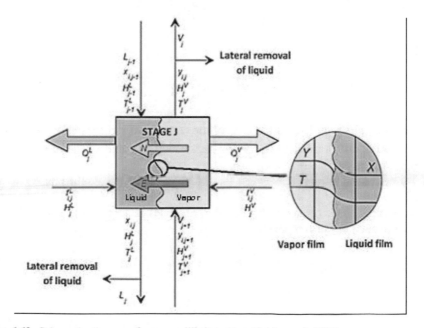

Fig. 6.63 Schematic diagram of a nonequilibrium stage (Seider et al. 2004)

vapor phase are additionally considered, as well as partial withdrawals from these two phases that could or not be considered within the calculations.

Aided by Fig. 6.63, nomenclatures used further in equations that describe the nonequilibrium model are introduced. Vapor and liquid phase flows that exit from the jth stage is denoted by V_j and L_j, respectively. Molar fractions in these streams are y_{ij} and x_{ij}. Molar fluxes from the component i in the stage j are indicated by N_{ij}. Additionally, it is assumed that the liquid and vapor phase temperatures are not equal, so that it is likely there is heat and mass transfer through the interface.

Mass balance equations per component for each phase are the following:

$$M_{ij}^V = \left(1 + r_j^V\right) V_i y_{ij} - V_{j+1} y_{i,\,j+1} - f_{ij}^V + N_{ij}^V = 0 \quad i = 1, 2, \ldots, c \quad (6.14)$$

Where f_{ij}^V is the feed flow in vapor phase of the component i to the stage j; r_j is the side stream flow ratio. The mass balance per component for the liquid phase is:

$$M_{ij}^L = \left(1 + r_j^L\right) L_i x_{ij} - L_{j+1} x_{i,\,j+1} - f_{ij}^L + N_{ij}^L = 0 \quad i = 1, 2, \ldots, c \quad (6.15)$$

Where f_{ij}^L is the feed flow in liquid phase of the component i to the stage j. The term left side of (6.14) and (6.15) represents the net loss or earning of the component i in the stage j due to the interface transportation. Without losing generalities we can define N_{ij}^V and N_{ij}^L as

$$N_{ij}^V = \int N_{ij}^V da_j \qquad (6.16)$$

and

$$N_{ij}^L = \int N_{ij}^L da_j \qquad (6.17)$$

Where N_{ij}^V and N_{ij}^L are the moral fluxes of the component i in a specific point in the dispersion of both phases and da_j is the differential elements of the interface area through which the fluxes pass.

By convention, transfer between the vapor phase to the liquid phase has the positive sign.

Total mass balance for both phases is obtained adding (6.14) and (6.15) in respect to each of the components i.

$$M_{ij}^V = \left(1 + r_j^V\right) \times V_j - V_{j+1} - F_j^V + N_{ij}^V = 0 \qquad (6.18)$$

$$M_{ij}^L = \left(1 + r_j^L\right) \times L_j - L_{j+1} - F_j^L + N_{ij}^L = 0 \tag{6.19}$$

Where F_j indicates the total feed flow for the stage j, and can be calculated with the following equation:

$$F_j = \sum_{i=1}^{c} f_{ij} \tag{6.20}$$

Energy balance for the vapor phase is

$$E_j^V = \left(1 + r_j^V\right) V_j H_j^V - V_{j+1} H_{j+1}^V - F_j^V H_j^{VF} + Q_j^V + \xi_j^V = 0 \tag{6.21}$$

Where V_j is the total vapor flow that leaves the stage j and H_j is the total molar enthalpy. The energy balance for the liquid phase is

$$E_j^L = \left(1 + r_j^L\right) L_j H_j^L - L_{j+1} H_{j+1}^L - F_j^L H_j^{LF} + Q_j^L + \xi_j^L = 0 \tag{6.22}$$

The left side terms from (6.21) and (6.22) represent the energy loss or gain due to the interface transfer. We can define ξ_j^V and ξ_j^L through the following equations:

$$\xi_j^V = \int E_j^V \, da_j \tag{6.23}$$

and

$$\xi_j^L = \int E_j^L \, da_j \tag{6.24}$$

Where E_j is the energy flux in a specific point of dispersion and ξ_j is the energy transfer rate in each phase.

The nonequilibrium model uses two groups of rate equations for each phase through the mass and energy models. Component molar fluxes in each phase are given by

$$N_i^V = J_i^V + N_i^V y_i^V \tag{6.25}$$

$$N_i^L = J_i^L + N_i^L x_i^L \tag{6.26}$$

Where y_i^V is the molar fraction of the component i in the vapor bulk and x_i^L is the molar fraction of the component i in the liquid bulk. Diffusion fluxes are given by (6.27) and (6.28).

$$\left(J^V\right) = c_t^V \times \left[k^V\right]\left(y^V - y^I\right) \tag{6.27}$$

$$\left(J^L\right) = c_t^L \times \left[k^L\right]\left(x^I - x^L\right) \tag{6.28}$$

Combining (6.25) and (6.27) and multiplying by the available interfacial area for the mass transfer, it is possible to obtain the mass transfer rate expression in the vapor phase:

$$\left(R_j^V\right) = c_{tj}^V\left[k_j^V\right]a_j\left(y_j - y_j^I\right) + N_{tj}^V\left(y_j\right) \tag{6.29}$$

In this manner, for the liquid phase an analogue ratio is obtained:

$$\left(R_j^L\right) = c_{tj}^L\left[k_j^L\right]a_j\left(x_j^I - x_j\right) + N_{tj}^L\left(x_j\right) \tag{6.30}$$

Coefficient matrixes of multicomponent mass transfer $[k^V]$ and $[k^L]$ are calculated by (6.31), where $[R^P]$ is the mass transfer resistance and Γ_{ij} is the thermodynamic factor matrix. Additionally, empirical correlations are used to estimate binary coefficients from mass transfer, and the corresponding activity models or state equations for the thermodynamic factor matrix. It is important to keep in mind that the binary coefficients for mass transfer obtained as from the empirical correlations are function of the plate geometry or of type, and size of package, as well as from operation conditions. This means you have to be aware of the equipment design parameters in order to solve the nonequilibrium model equations or have to be determined during the resolution carrying out design calculations simultaneously with the solution of all the model equations.[2]

$$\left[k^P\right] = \left[R^P\right]^{-1}\left[\Gamma^P\right] \tag{6.31}$$

The energy flux can be written for each of the phases through the following equations:

$$E_j^V = h_j^V\left(T_j^V - T_j^I\right) + \sum_{i=1}^{c} N_{ij}\overline{H}_{ij}^V \tag{6.32}$$

and

$$E_j^L = h_j^L\left(T_j^I - T_j^L\right) + \sum_{i=1}^{c} N_{ij}\overline{H}_{ij}^L \tag{6.33}$$

---

[2] For a better comprehension of the calculation of each of these terms, you can look up the text *Multicomponent Mass Transfer* by Ross Taylor, Chapters 8 and 12.

Where h_j is the heat transfer coefficient, H_{ij} is the partial molar enthalpy of the component i in the stage j and T^V, T^L, and T^I are the vapor, liquid, and interface temperatures, respectively.

Likewise, energy transfer rates in liquid and vapor phases are obtained multiplying energy fluxes by the interfacial area a_j.

$$\xi_j^V = h_j^V a_j \left(T_j^V - T_j^I \right) + \sum_{i=1}^{c} N_{ij} \overline{H}_{ij}^V \tag{6.34}$$

$$E_j^L = h_j^L a_j \left(T_j^I - T_j^L \right) + \sum_{i=1}^{c} N_{ij} \overline{H}_{ij}^L \tag{6.35}$$

Heat transfer coefficients can be estimated through correlations and analogies. One of the analogies most frequently used to calculate heat transfer coefficients in the vapor phase is the one by Chilton-Colburn between mass and heat transfer:

$$Le = \frac{\lambda}{DC_p \rho} = \frac{Sc}{Pr} \tag{6.36}$$

$$h^V = k\rho C_p Le^{2/3} \tag{6.37}$$

To calculate heat transfer coefficients in the liquid phase you can use a penetration model:

$$h^L = k\rho C_p \sqrt{Le} \tag{6.38}$$

Where k is the average mass transfer coefficient and D is the average diffusion coefficient.

In the nonequilibrium model of Krishnamurthy and Taylor (Kiva et al. 2003; Kreul et al. 1999), pressure is specified in all the stages in the same way it is done in simulation with equilibrium models. However, pressure drop in the column is a function from the package type (or plate) as the column design and the operation conditions; therefore, this information shall be available in the solution of nonequilibrium model equations. In this manner, you can establish a hydraulic equation that is added to the set of existing equation sets for each of the stages and make the pressure or each stage or package section be an unknown variable. It is assumed that in each stage the mechanical equilibrium is given, and for this reason:

$$p_j^V = p_j^L = p_j \tag{6.39}$$

In the second generation model, the top pressure is specified as the typical pressure drop through a condenser. Pressure in stages under the top stage is calculated as from the pressure information in the superior stage and the specified pressure drop by stage or package section. If the column has condenser (which is numbered as stage 1), the hydraulic equations are written as:

$$P_1 = p_c - p_1 = 0 \tag{6.40}$$

$$P_2 = p_{\text{spec}} - p_2 = 0 \tag{6.41}$$

$$P_j = p_j - p_{j-1} - \left(\Delta p_{j-1}\right) = 0 \qquad j = 3, 4, \ldots, n \tag{6.42}$$

Where p_c is the specified pressure for the, p_{spec} is the specified pressure for the top of the column and Δp_{j-1} is the pressure drop per stage or package section. In general we can consider that the pressure fall is a function from inner flows, densities from fluids, and the equipment design parameters.

Phase equilibrium is assumed only to occur in the interface, and molar fractions in both phases are related through the following expression:

$$Q_{ij}^I = K_{ij}\left(x_{ij}^I - y_{ij}^I\right) = 0 \quad i = 1, 2, \ldots, c \tag{6.43}$$

Where K_{ij} is the equilibrium relation for the component i in the stage j. K_{ij} values are calculated from temperature, pressure, and molar fractions in the interface. The restriction corresponding to the molar fraction summation shall be complied in each phase and interface, which shall be equivalent to one. For the interface case, this restriction is written as follows:

$$S_j^{VI} = \sum_{i=1}^{c} y_{ij}^I - 1 = 0 \tag{6.44}$$

$$S_j^{LI} = \sum_{i=1}^{c} x_{ij}^I - 1 = 0 \tag{6.45}$$

Equilibrium and nonequilibrium models require many similar specifications. Feed flows and their corresponding thermal condition shall be specified in both models, as well as the column configuration (number of stages, location of feeding stages and side streams, etc.). Additional specifications that are similar for the two simulation models include reflux ratios, product flows, among others. Pressure specification is required in each stage when we do not plan to carry out a pressure drop calculation; otherwise, it is only necessary to specify top pressure.

If the nonequilibrium model is solved with the Newton method, initial estimates are required for all the variables. This is done with an initialization routine that is used in equilibrium model simulation in which with the column bottom flow and the reflux ratio specification the column is solved using Wilson ideal model. Vapor, liquid temperatures, and interface are all initialized with the same value, taking as reference values the one obtained in the equilibrium model. Mass and energy transfer rate are initialized in zero and molar fractions of the interface are initialized with similar values to the ones from the molar fractions in the mixture sine, which are also obtained from said initialization. Finally, pressure drops are initially assumed as equivalent to zero.

In the nonequilibrium model there are $6c + 8$ unknown variables per stage, which are defined in the following manner:

- Liquid and vapor flows (V_j, L_j; 2)
- Liquid and vapor phase compositions (y_{ij}, x_{ij}; 2c)
- Liquid and vapor temperatures (T_j^V, T_j^L; 2)
- Liquid and vapor composition in the interface (y_{ij}^I, x_{ij}^I; 2c)
- Temperature in interface (T_j^I, 1)
- Mass transfer rate (R_j^V, R_j^L; 2c)
- Energy transfer rates (ξ_j^V, ξ_j^L; 2)
- Stage pressure (P_j, 1)

Consequently, $6c + 8$ equations are required to solve this nonlinear system by Newton method. These equations are referred to as the *MERSHQ* equations and are the following:

- M: mass balance for vapor phase ($c + 1$)
- M: mass balances for liquid phase ($c + 1$)
- M: Mass balances for interface (c)
- E: energy balance equations (3)
- R: mass transfer rate equations ($2c - 2$)
- R: energy transfer rate equations (2)
- S: summation restrictions of the molar fractions in the interface (2)
- H: hydraulic equations (1)
- Q: Equilibrium in interface equations (c)

A nonequilibrium simulation requires several additional specifications, compared to a conventional equilibrium model. Several of the additional modifications required are:

- Type and arrangement of column internals
- Mass transfer coefficient model
- Flux model for both phases
- Desorption and dripping models
- Model for pressure drop
- Models for physical properties

For the estimate of transport properties, the nonequilibrium model requires an evaluation of the additional amount of properties (densities, viscosity, diffusivity, calorific capacity, thermal conductivities, and surface stress) that the equilibrium model does not require.

Added to the information requirements mentioned previously, a nonequilibrium simulation cannot be carried out without knowing the type of column and internal arrangement in order to determine the mass transfer coefficients, interfacial area, and pressure drop. The type of plate and its mechanical information, for example, are necessary to calculate the mass transfer coefficient in every plate. For packed columns, the type of package, dimension and material shall be known for this same purpose. The arrangement or *layout* is specified for every section of the column, where a section is represented by one or more plates (or packed bed). Generally, standard *layout* for plate columns and packed columns are stored at online libraries which can be easily accessed.

6.7.1 Nonequilibrium Model Example

To introduce the topic proposed before, an exercise is developed on the industrial application column which separates a methanol–water mixture. First, a simulation is effected in order to evaluate the possibility of making adjustments and verify the operation behavior.

The column has a 68 % aqueous solution feed in methanol weight. It is packed with 1 in Raschig rings. Total packed height is 9 m and the diameter is 37.5 cm. In the base case 250 kg/h are fed in order to obtain a solution of 88 % in methanol weight. It is assumed that the feed is entered at 1 m height measured from the bottom. The molar reflux ratio is 1. The distillation column is located in Bogotá and operates at atmospheric pressure. The column has several operating problems which are commented below, and that are expected to be solved, at least in its greater part, using computer tools.

- Column pressure drop is very high and the cause is unknown.
- Reflux ratio is very small due to the current condenser installation.
- The top composition obtained is only near 88 wt% of methanol when a 98 % is required as minimum.

Below is the information available for the package installed in the column (Table 6.20).

Initially it is considered that the column has nine ideal stages; however, because for the nonequilibrium model segments must be defined to calculate heat and mass transfer rates, the same number of segments is considered. Furthermore, there is a 9 m height, so every stage will be associated to a meter height.

Thus, we have the complete information to enter the system in Aspen Plus® for the column simulation using the nonequilibrium model since the other specifications are made in a similar way to the calculation using the equilibrium model.

For this, start a simulation in Aspen Plus® with metric units. Then add two *RadFrac* modules as explained previously, one for the equilibrium model and another for the nonequilibrium model. Use a *Dupl* module to duplicate the feed stream. The simulation flow diagram is represented in Fig. 6.64.

As soon as completing the corresponding connections, click on the button *Next*. Input the necessary components and use *NRTL* as property model. The *Feed* stream conditions are summarized below (Table 6.21).

Table 6.20 Main characteristics of the packing used for the nonequilibrium model

| Type | Material | Size | | Thickness | | CF | CF BD | Weight | Ap | ε (%) | #/m³ |
		(in)	(mm)	(in)	(mm)			(kg/m³)	(m²/m³)		
Raschig rings	Ceramic	1	25	0.125	3.2	155	150	673	190	74	47,700

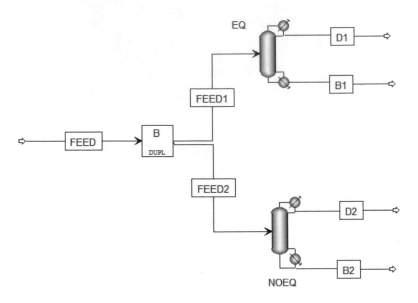

Fig. 6.64 Diagram entered in Aspen Plus® for the nonequilibrium model calculation module

Table 6.21 Operating conditions for the stream *Feed*

Variable	Value
Pressure (bar)	1
Temperature (°C)	70
Mass flow (kg/h)	250
Mass fraction	
Methanol	0.68
Water	0.32

Now configure the distillation column. For this case a shortcut calculation is not required, since what we want to see is the behavior of the installed column. For this we modify the option *Calculation type*, that is in *Blocks* ≥ *COL* ≥ *Setup*, where we find *Equilibrium* by default, and *Rate-Based*. Select the option *Rate-Based* in the Calculation type box as shown in Fig. 6.65.

The segment number corresponds to a division that is made in this module and that would correspond to the column number of stages; for this case this value is the one calculated previously in addition to two stages that correspond to the condenser and the reboiler, that is to say 11. The condenser is Total. The reflux ratio is 0.3. The distillate flow corresponds to 175 kg/h. With this information the module basic information is defined. Now we have to add, as well as done in the equilibrium module, the input stage information for streams and pressure profile (Fig. 6.66).

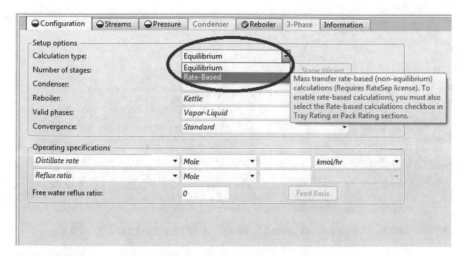

Fig. 6.65 Rate-based calculation type in Aspen Plus®

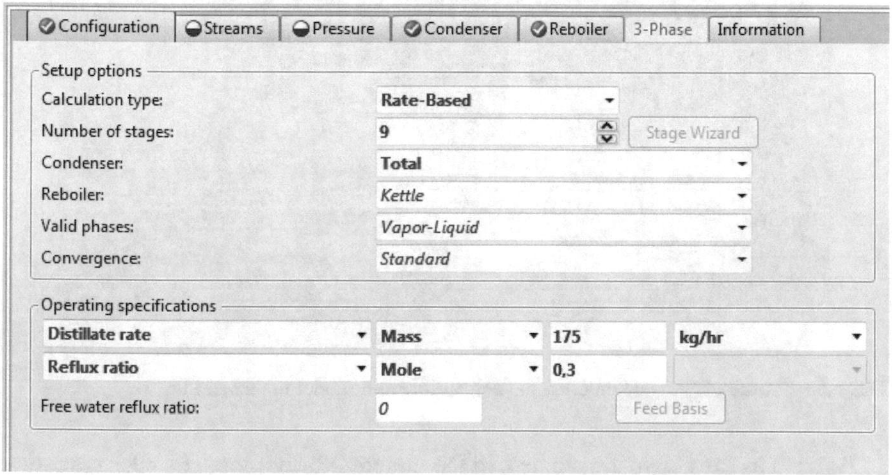

Fig. 6.66 Basic information entered in the nonequilibrium model in Aspen Plus®

Then the segment in which the feed stream is input is estimated, since according to the provided information, it enters at 1 m height measured from the bottom; therefore, and assuming each stage corresponds to said height, the stream *Feed2* must be entered in the segment 8. Streams *D1* and *B1* exit by the column both ends, that is to say by segments 1 and 9, respectively (Fig. 6.67).

As to the pressure profile, it is specified that all the column operations are at a same pressure, since when the package evaluation is made the pressure drip is calculated. Therefore it is specified that pressure in stage 1/condenser is 560 mmHg (Fig. 6.68).

Fig. 6.67   Information of streams entered in the nonequilibrium model in Aspen Plus®

Fig. 6.68   Profile of pressure entering the nonequilibrium model in Aspen Plus®

As soon as the above is specified, in the nonequilibrium model a package rating or plate shall be made in order to obtain the information required to carry out the calculation based on the mass and heat transfer rate.

According to the information provided, the column is packed with 1 in Raschig Rings; therefore, click on *Packing Rating* on the lower left part of the screen. Simultaneously, enter the information corresponding to plates in the option *Tray Rating* (Fig. 6.69).

In this window displayed, select the option *New*... and then *OK* to enter a new pack section. You can only enter one section because the column has only one type of pack and only one diameter (Fig. 6.70).

Then the window that allows entering packing information is open. Here is defined that the packed section begins in stage 2 (because stage 1 represents the condenser) to stage 8 (because stage 9 represents the reboiler). In the option

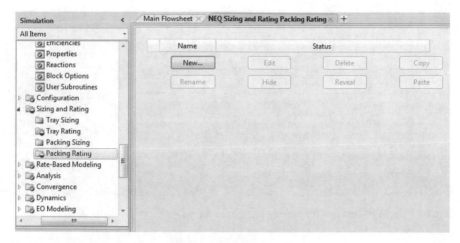

Fig. 6.69 Main Window for *Packing Rating* in Aspen Plus®

Fig. 6.70 Window to enter segments in *Packing Rating* in Aspen Plus®

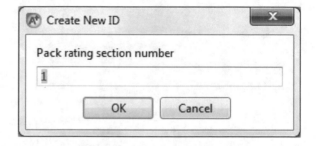

Starting Stage enter the value of 2 and for the option *Ending Stage* a value of 8. In the box *Type* select the option *RASCHIG*. In the box *Vendor* select *Generic*; in *Material* select *Ceramic* and finally in *Dimension*, the option *1-IN OR 25-MM*. On the right side of the screen enter into the box *Section diameter* a 37.5 cm value, which corresponds to the column diameter.

Now, to complete the necessary information, specify on the lower part the equivalent height of the plate (HETP) or the packed section length. For this, select the option *Section packed height* and assign it a 9 m value. The final configuration entered is shown in Fig. 6.71.

With the button *Update Parameters* you can edit the pack properties based on the information available in the literature. In order to develop this exercise, default parameters will be used. Lastly, select the option *Rate-based* from the tree on the screen left side in packing section 1. A window is displayed where you must activate the option *Rate-based calculations* in order for the simulator to carry out the calculations corresponding to this module (Fig. 6.72).

In this section you can select the parameters to solve the nonequilibrium mode, and in the tab *Correlations* you can select the empirical correlation for the

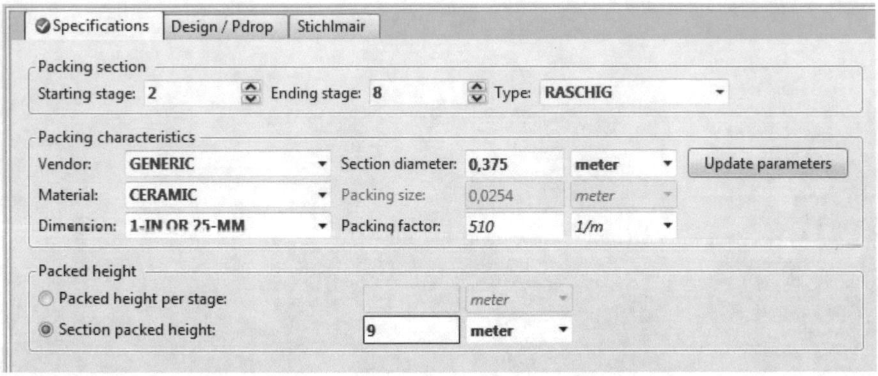

Fig. 6.71 Input information for packing segment in Aspen Plus®

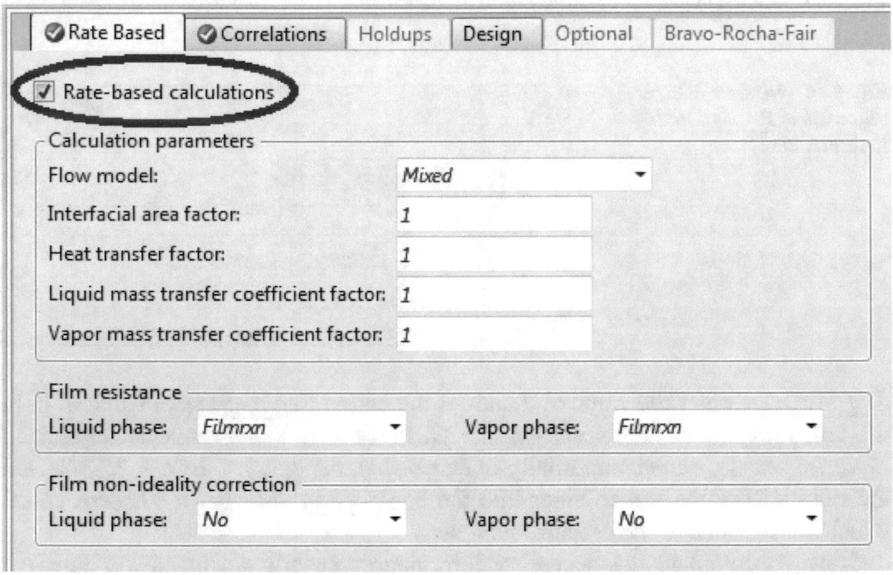

Fig. 6.72 Main window of _RateSep_ in Aspen Plus®

calculation required for mass transfer coefficient, of heat and for the problem resolution in the interface (Fig. 6.73).

In this way, all the elements to carry out the simulation are specified. Click on the button _Next_ and see the results. Make sure that the mass compositions are reported; for this, in case of not being reported, in the route _Setup_ ≥ _Report Options_ select the tab _Stream_ and activate the option _Mass_ in the section _Fraction Basis_. Once again click on the button _Next_ (Fig. 6.74).

Here you can see that a top stream is obtained with 87.6 % methanol concentration, and by bottom a stream with 77.8 % water concentration. These results are

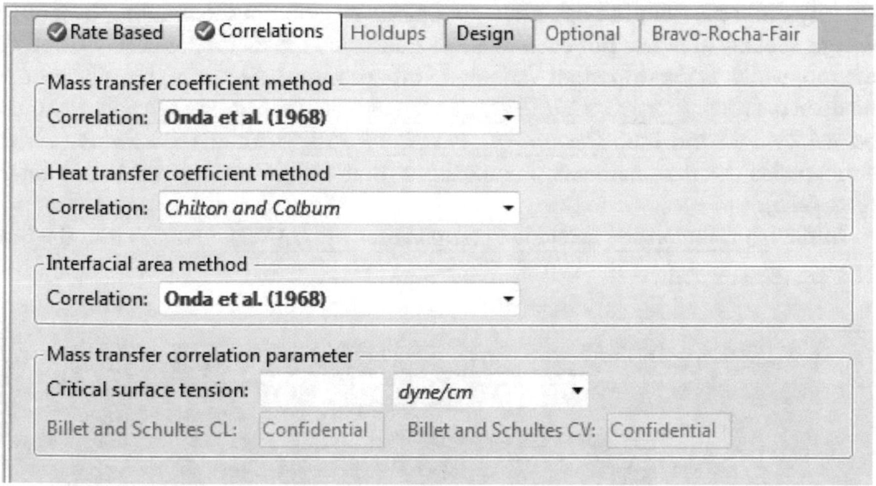

Fig. 6.73 Correlations available in *Rate-based* calculation in Aspen Plus®

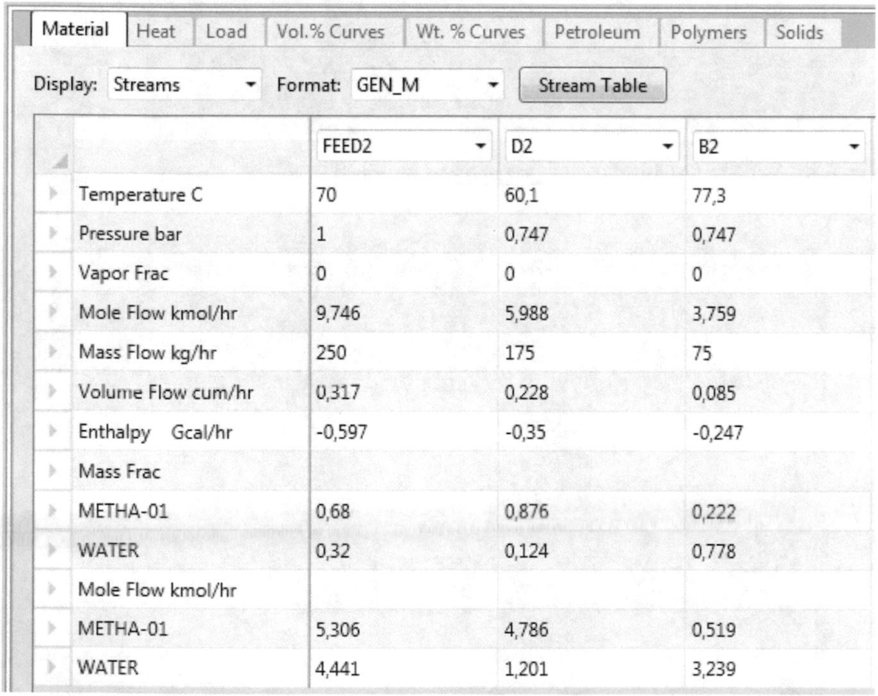

Material | Heat | Load | Vol.% Curves | Wt. % Curves | Petroleum | Polymers | Solids

Display: Streams Format: GEN_M Stream Table

	FEED2	D2	B2
Temperature C	70	60,1	77,3
Pressure bar	1	0,747	0,747
Vapor Frac	0	0	0
Mole Flow kmol/hr	9,746	5,988	3,759
Mass Flow kg/hr	250	175	75
Volume Flow cum/hr	0,317	0,228	0,085
Enthalpy Gcal/hr	-0,597	-0,35	-0,247
Mass Frac			
METHA-01	0,68	0,876	0,222
WATER	0,32	0,124	0,778
Mole Flow kmol/hr			
METHA-01	5,306	4,786	0,519
WATER	4,441	1,201	3,239

Fig. 6.74 Results per streams of the nonequilibrium model in Aspen Plus®

slightly different from the operation conditions reported in the column, because a design specification has not been made to guarantee the methanol composition on the top, which is the important variable in this process. The design specification is made in the route *Blocks ≥ COL ≥ Design Specs* where the option *Mass Purity* must be selected in the box *Design Specification ≥ Type* with a 0.88 value, which corresponds to the methanol composition that exists by top of the column (Fig. 6.75).

In the tab *Components* select the component *METHANOL* and add it to the box *Selected components* (Fig. 6.76).

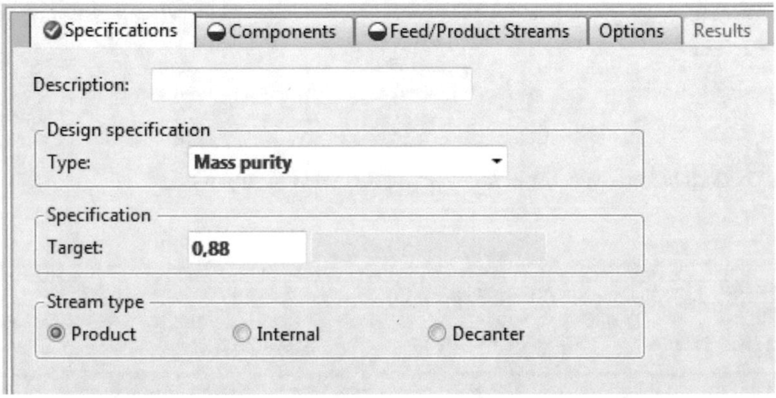

Fig. 6.75 Window specifications of *Design Specs* in Aspen Plus®

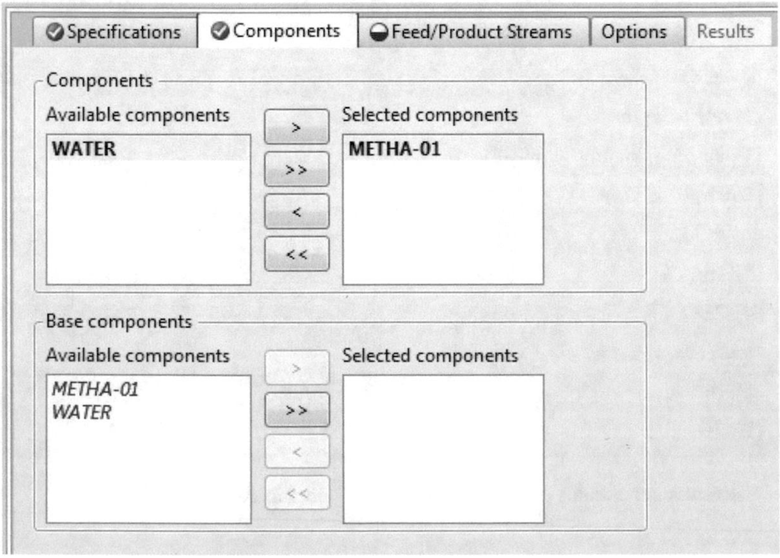

Fig. 6.76 Window components of *Design Specs* in Aspen Plus®

Now in the tab *Feed/Product Streams* select the stream *D2* which corresponds to the stream where the concentration entered previously shall be obtained (Fig. 6.77).

Now, in the route *Blocks* > *COL* > *Vary*, indicate if you want to modify the distillate flow, *Distillate Rate*, in an interval within 160 and 190 kg/h, which is a sufficient interval and which contains the current value of the distillate flow rate, 175 kg/h (Fig. 6.78).

Click on the button *Next* and as soon as the simulator carries out the corresponding calculation, verify the *Dist* stream composition, in order to obtain a mixture with 88 % methanol by the column top.

Figure 6.79 shows the results obtained. Here we can see that the distillate flow was reduced from 175 kg/h to 172.67 kg/h; this is because the concentration was lower initially (approximately 87.8 %), making it necessary to reduce the flow so that the 88 % concentration could be reached.

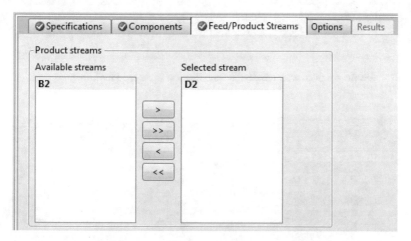

Fig. 6.77 Window *Feed/Product Streams* de *Design Specs* in Aspen Plus®

Fig. 6.78 Window *Specifications* of *Vary* in Aspen Plus®

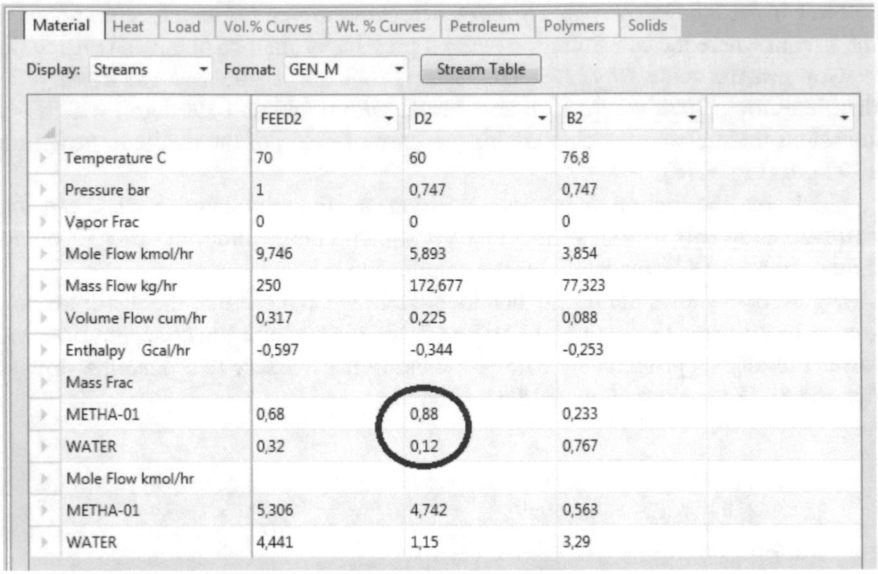

	FEED2	D2	B2	
Temperature C	70	60	76,8	
Pressure bar	1	0,747	0,747	
Vapor Frac	0	0	0	
Mole Flow kmol/hr	9,746	5,893	3,854	
Mass Flow kg/hr	250	172,677	77,323	
Volume Flow cum/hr	0,317	0,225	0,088	
Enthalpy Gcal/hr	-0,597	-0,344	-0,253	
Mass Frac				
METHA-01	0,68	0,88	0,233	
WATER	0,32	0,12	0,767	
Mole Flow kmol/hr				
METHA-01	5,306	4,742	0,563	
WATER	4,441	1,15	3,29	

Material | Heat | Load | Vol.% Curves | Wt. % Curves | Petroleum | Polymers | Solids

Display: Streams   Format: GEN_M   Stream Table

Fig. 6.79 Results obtained after the design specification in Aspen Plus®

Specifications | Components | Results

Type:	MASS DISTILLATE RATE	
Lower bound:	160	kg/hr
Upper bound:	190	kg/hr
Final value:	172,677	kg/hr

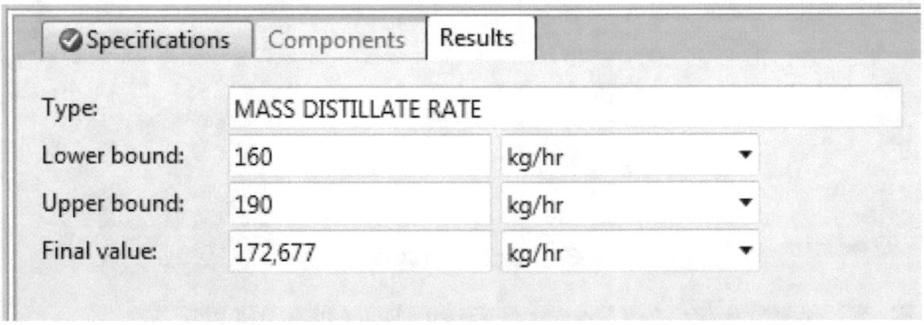

Fig. 6.80 Results from the manipulated variable in the design specification in Aspen Plus®

Additionally we can see that the bottom stream contains methanol, this is why the balance made previously was not complied and a design specification was required to obtain operation conditions nearer to the ones proposed in the problem.

Likewise, observe the results obtained by the design specification in the route *Blocks* ≥ *COL* ≥ *Design Specs* ≥ *1* ≥ *Results* for the composition, and in the route *Blocks* ≥ *COL* ≥ *Vary* ≥ *1* ≥ *Results* for the manipulated variable, in this case, the distillate flow (*Distillate Rate*) (Fig. 6.80).

To generate the graphics of heat and mass transfer rate profiles, activate the corresponding options in the folder *Rate-Based Modeling* in the tab *Rate-based Report* of the nonequilibrium module, as you can see in Fig. 6.81.

Likewise, you can include other variables in the report that may get to be determinant for the operation analysis and functioning, such as: interfacial area,

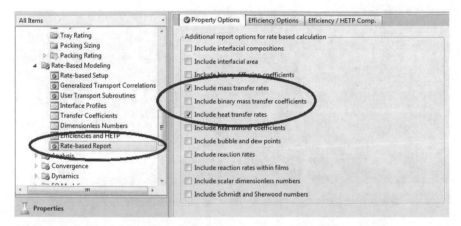

Fig. 6.81 Tab *Rate-based Report* from the nonequilibrium model in Aspen Plus®

Stage	METHA-01	WATER
2	-0,484422	0,396213
3	-0,23271	0,193355
4	-0,100098	0,0844618
5	-0,0405508	0,0344812
6	-0,0160333	0,0135499
7	-0,00641778	0,00494935
8	-1,49904	1,26419

Temperatures | Compositions | Mass Transfer | Heat Transfer | Interfacial Area

Phase: **Liquid**

Component mass transfer rates (positive for transfer from vapor to liquid)
Units: kmol/hr

Fig. 6.82 Results from mass transfer rates for the nonequilibrium model in Aspen Plus®

composition in Interface, binary diffusion coefficients, binary mass transfer coefficients, dew and bubble points, reaction rate (whenever applicable), heat transfer coefficient, among others.

As soon as selecting the variables to be analyzed, once again execute the calculation for the simulator to load the values in the section *Interface Profiles* ≥ *Heat Transfer* or *Interface Profiles* ≥ *Mass Transfer*, as applicable.

Then, in the section *Interface Profiles* ≥ *Mass Transfer* we have the results from the mass transfer rates in respect to the liquid phase or the vapor phase throughout the column, as shown in Fig. 6.82 in *Interface Profiles* ≥ *Heat Transfer* we can

	Stage	Total		Conductive Liquid	Convective Liquid	Conductive Vapor	Convective Vapor
		kcal/hr	▾	kcal/hr ▾	kcal/hr ▾	kcal/hr ▾	kcal/hr ▾
▶	2	461,987		163,958	298,029	179,178	282,809
	3	46,2576		88,4109	-42,1533	86,4023	-40,1447
	4	-54,607		45,968	-100,575	37,3868	-91,9938
	5	-36,9888		20,7205	-57,7092	15,5705	-52,5592
	6	-8,45298		8,57028	-17,0233	7,55118	-16,0042
	7	28,6712		3,40563	25,2655	7,84897	20,8222
	8	-843,102		631,659	-1474,76	497,805	-1340,91

Temperatures | Compositions | Mass Transfer | Heat Transfer | Interfacial Area | Reactions | Holdups

Heat transfer rates (positive for transfer from vapor to liquid)

Fig. 6.83   Results from heat transfer rates for the nonequilibrium model in Aspen Plus®

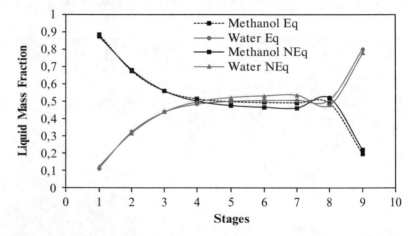

Fig. 6.84   Comparison between composition profiles for equilibrium and nonequilibrium models

observe the results corresponding to the head transfer broken down by each of the four resistances (connective and conductive for each phase) and the total transferred energy, as observed in Fig. 6.83.

Now, with the results obtained for the two modules, we can see a difference between the equilibrium model and the nonequilibrium model. For this we use Figs. 6.84 and 6.86, which show the composition and temperature profiles for both calculation methods. Additionally, Figs. 6.85 and 6.87 report the behavior of heat and mass transfer rates and tray efficiency throughout the column, and finally in Fig. 6.88 the relevant to theoretical plate equivalent height for the packing used.

In these graphics you can see that the behavior is slightly different from equilibrium conditions. This is because it is considered that the liquid and vapor in each stage does not have the same temperature, and therefore there is simultaneous heat and mass exchange within the phases. If we observe Figs. 6.84 and 6.85, we can relate the changes in phase composition with mass transfer rate taking into

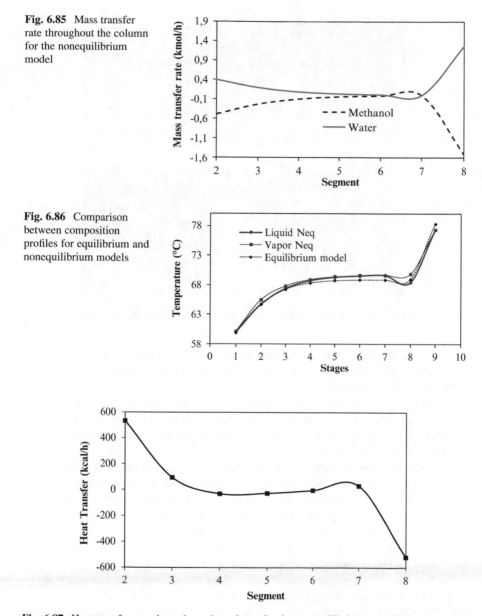

Fig. 6.85 Mass transfer rate throughout the column for the nonequilibrium model

Fig. 6.86 Comparison between composition profiles for equilibrium and nonequilibrium models

Fig. 6.87 Heat transfer rate throughout the column for the nonequilibrium model

account that this value is positive for transfer from vapor to liquid. We can see that in the higher and lower segments, speed is higher, because the concentration change in each phase is different. In the lower part, temperature increase makes the phenomena even more marked. Changes observed in the component profile are explained in the entrance of the feed stream in segment 8.

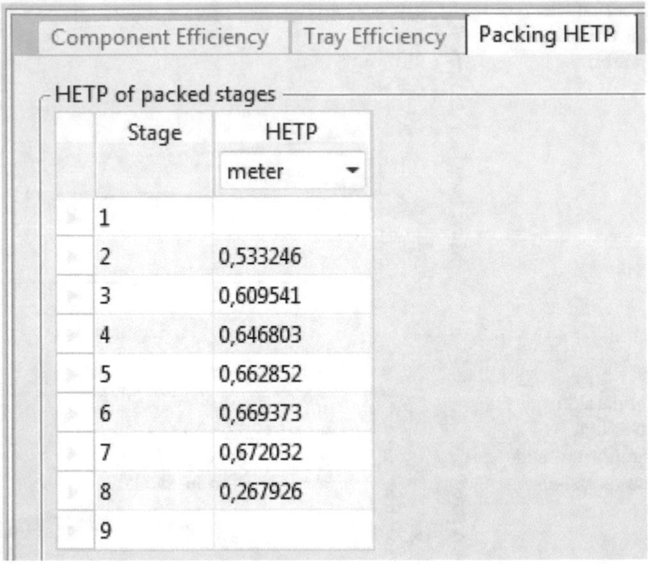

Component Efficiency	Tray Efficiency	Packing HETP

HETP of packed stages

Stage	HETP
	meter ▼
1	
2	0,533246
3	0,609541
4	0,646803
5	0,662852
6	0,669373
7	0,672032
8	0,267926
9	

Fig. 6.88 Results for the height equivalent to theoretical plate for the installed packing

Likewise the Figs. 6.86 and 6.87 can be related since the heat transfer is carried out by the difference of temperatures present in them. Once again we observe the feed stage effect in both profiles.

Additionally the plate efficiency calculation is considered (for the plate column case) or height equivalent to theoretical plate (HETP) (for the case of packed columns) that can be seen in the tab *Efficiencies and HETP* of the corresponding column module (Fig. 6.88).

We can see that segment 8 is the most efficient since it has a lesser HETP in respect to the others. For the other segments we have an increase in the equivalent height as it decreases in the column. Feed in segment 8 once again makes easier the separation between the phases and increases the efficiency of said stretch.

To this moment the process simulation is kept under the current operation conditions. However, modifications can be made that enable us to obtain several responses to the problems present in this equipment. As first operation issue we have that the pressure drop is much more than the normal of operation. For this reason we will study a likely deterioration of the pack that can be simulated using a smaller pack (¼ in) within the Pack Rating carried out in the nonequilibrium module. With the base case simulation a pressure drop of 9.167 mmHg is obtained throughout the column, while for smaller packs the behavior observed in Table 6.22 is obtained.

With this information we confirm that lesser size of pack, the pressure drop increases significantly. That is to say, that if a potential breaking and wear of the original pack of 1″ is considered, we could think that smaller particles found in the column hinder the fluids circulation and, therefore, increase the pressure drop. We

Table 6.22 Pressure drop for different packing sizes

Packing	ΔP (mmHg)
Raschig rings 1 in	9.167
Raschig rings ¾ in	13.193
Raschig rings ½ in	32.471
Raschig rings ¼ in	186.048

Table 6.23 Results obtained for different configurations for the distillation column

Scenario	ΔP (mmHg)	Top (weight %)
NE = 9 RR = 0.3	9.167	87.62
NE = 9 RR = 2	54.537	95.40
NE = 14 RR = 0.3	9.231	91.53
NE = 14 RR = 2	51.762	96.98

can consider changing the worn pack as a solution to the high pressure drop obtained currently, and that makes difficult the separation.

To solve the other two problems, we have to analyze the situation adequately. If the column is not big enough to take the methanol composition on top to 99 % in weight, it is likely that it requires height to be increased. Furthermore, the other variable that affects the product purity is the reflux ratio. Aided by process simulators, we can easily modify the process variable values and observe the equipment behavior. Keep in mind to delete the design specification to prevent setting the top composition in the column.

Table 6.23 summarizes several arrangements and the results obtained.

In this manner the best configuration is the one of 14 stages, with a reflux ratio of 2.0. And additionally the composition is adjusted modifying the distillate flow. This corresponds to a height increase to up 14 m, and it is necessary to make the appropriate changes in the condenser operation in order to set a greater value of the reflux ratio.

6.8   Columns Thermal and Hydraulic Analysis

The thermal and hydraulic analysis of columns is a tool available for *RadFrac*, *MultiFrac*, and *PetroFrac* modules. This option execution can be activated on the sheet *Analysis/Analysis Options* corresponding to each module, as observed in Fig. 6.89.

The column thermal and hydraulic analysis tool of Aspen Plus® is used to identify possible modifications in column designing, in order to reduce service costs, improve energetic efficiency, reduce capital investment (by improvements in driving power) and facilitate the unclogging of equipment.

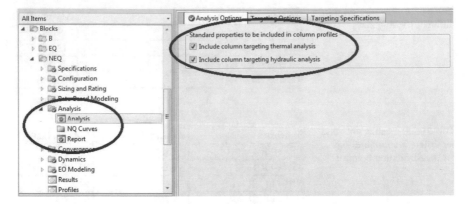

Fig. 6.89 Window of properties that enable the inclusion of thermal and hydraulic analysis

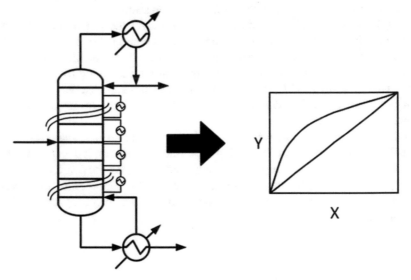

Fig. 6.90 On the right, hypothetic representation of a distillation column operating in MCT; on the left, scheme of operating and equilibrium line in the MCT (overlapped)

6.8.1 Thermal Analysis

Thermal analysis is based on the concept of the minimum thermodynamic condition of a column. This state belongs to the reversible operation of the column with zero thermodynamic loss, which involves that the column operates at the minimum reflux ratio, with an infinite number of stages and with infinite coolers and heaters placed in each state for the operating lines and equilibrium lines to coincide in every point. A scheme of the reversible operation of a column is shown in Fig. 6.90.

Figure 6.89 shows how the window where the option thermal and hydraulic calculation is activated for a *RadFrac* model in Aspen Plus®. In the corresponding module is the option *Analysis* activating the boxes: *Include column targeting thermal analysis* and *Include column targeting hydraulic analysis*.

The MTC (*minimum thermodynamic condition*) is described through a temperature-enthalpy profile. These graphs show how the condenser and reboiler duties are distributed over the entire column temperature interval. It can also be used to identify the chance to install side condensers and reboilers. The temperature-enthalpy diagram is known as *Column Grand Composite Curve* (CGCC).

A column in the MTC requires an infinite number of stages with infinite heat exchange facilities in each stage. In this way, the operating line coincides with the equilibrium curve and its points are overlapped throughout the composition interval, as you can see in Fig. 6.90.

The CGCC evaluation involves the simultaneous resolution of the equations described in the equilibrium curve and the operating line. Aspen Plus calculates the CGCC based on the (*Practical near-minimum thermodynamic condition*) proposed by Dhole and Linnhoff (1992). Enthalpies in each stage are calculated assuming that the operation and equilibrium lines coincide in each stage. The PNMTC takes into account losses and practice inefficiencies inevitable in the column and that are due to the design itself and to the modifications carried out in the column. Main inefficiencies considered are: feed losses, pressure drops, drastic separation, chosen column configuration (multiple products, side strippers, etc.) and loss generated by the separation scheme.

Equilibrium equations and operating line are solved simultaneously in each stage for the key components of the mixture, which are previously selected. The thermal analysis results depend significantly on the correct selection of light and heavy key components of the mixture. Aspen Plus® has four different methods to select key components:

- User defined.
- Based on component split fractions.
- Based on component K-values.
- Based on column composition profiles.

These four options are available on the sheet *Analysis/Targeting Options* from each calculation module (See Fig. 6.91).

The thermal analysis enables the identification of a series of modifications on a distillation column design; these modifications are considered through a methodology that follows the following order:

1. Appropriate feed stage location.
2. Changes in operating pressure and reflux ratio.
3. Feed conditioning (cooling/heating).
4. Installation of side condensers/reboilers.

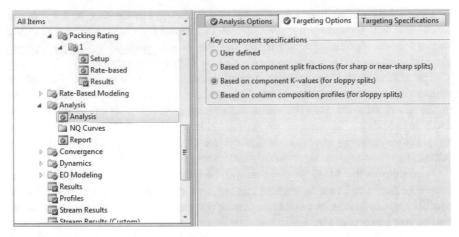

Fig. 6.91   Key component specification in module *RadFrac*

6.8.2   *Hydraulic Analysis*

The hydraulic analysis of a distillation column is a tool of Aspen Plus® used to learn the behavior of liquid and vapor flows, compared to the minimum limits (corresponding to the PNMTC) and maximum (corresponding to flooding). For plate and pack columns, the column flooding calculation is carried out through the maximum limit in the vapor flow. The hydraulic analysis is also used to identify and eliminate clogging situations of the column.

In order to calculate maximum liquid and vapor flows leading to the column flooding, you have to specify the information corresponding to the plate or pack used in the column in all its sections. Furthermore, it is necessary to specify the flooding factor (as total fraction or flooding). These specifications are made on the sheets *Tray Rating* ≥ *Design* ≥ *Pdrop* or *Pack Rating* ≥ *Design* ≥ *Pdrop*. Values used by Aspen by default are: for flooding with vapor 85 % and for flooding with liquid 50 %.

Below is an example that illustrates the usefulness of the column thermal or hydraulic analysis.

6.8.3   *Application Exercise*

Here the nonequilibrium model from the methanol–water mixture exercise is used again. For further information, verify the previous section on the nonequilibrium model.

The first step in developing the thermal or hydraulic analysis is to activate the respective verification boxes, as indicated in Fig. 6.89, and execute once again the simulation. The geometry and geometrical specifications for the column are exactly

Fig. 6.92 Graphic
selection window of the
menu *Plot* for the thermal or
hydraulic analysis curves in
Aspen Plus®

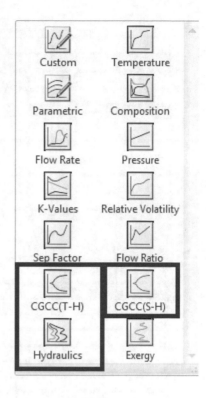

the same from the exercise developed previously as base case. Do not forget to
deactivate the design specification on the column.

As first measure we can see the curves CGCC (thermal analysis) and the flow
ones (hydraulic analysis) of the column current configuration, in other words, the
base case from which the analysis will start. Once in the menu *Plot* (Fig. 6.92), it is
necessary to select that graphic you wish to visualize; for the analysis it is fit to use a
CGCC (*T-H*) curve. Likewise, select the option *Hydraulics* to generate a graphic
with the information corresponding to the column flows. Likewise, check the
analysis data in table format in the routes *Profiles* > *Thermal Analysis* or
Profiles > *Hydraulic Analysis*, as applicable.

In the curves in Fig. 6.93 is displayed the results obtained for the thermal or
hydraulic analysis for the base case. In part (a) of the Figure, the curve CGCC
presents a zone with an important area within the ideal profile and the actual profile.
This zone is located in the highest temperature sections of the column, i.e., near the
reboiler and under the feed section. There you can make modifications to enable a
reduction in this difference making both profiles to be more similar, and therefore,
the thermal behavior of the column to be more efficient.

In the nonequilibrium model, exercise was identified as the problem, the low
value in reflux ratio which does not allow the required composition to be achieved
in the column top. Then, initially the reflux ratio must be increased to solve in some

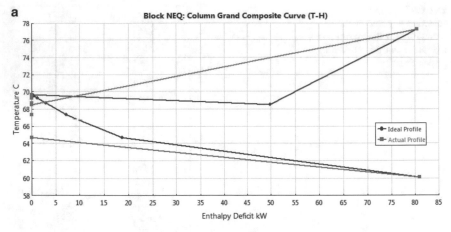

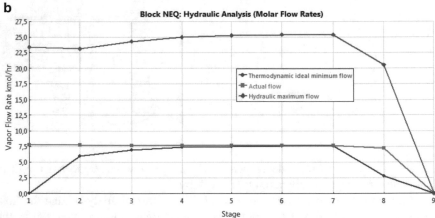

Fig. 6.93 Thermal (**a**) and hydraulic (**b**) analysis for the base case ($RR = 0.3$ and $N = 9$) in Aspen Plus®

form the operation problems and once again carry out the thermal and hydraulic analysis in respect to the new reflux condition. Use a value of 0.8 to repeat the analysis.

As we can see in Figs. 6.93, 6.94, and 6.95, varying the reflux ratio we find that a marked effect occurs both in the thermal and hydraulic behavior. We suggest to use a reflux ratio of 1.0 that allows us to obtain a greater methanol composition on top. It is also recommendable to increase the number of stages to improve the contact between the two phases and the methanol purity. For this study we will use 14 segments and the pack height is defined then in 14 m.

To select appropriately the modifications that can be implemented to upgrade the separation functioning, it is very useful to analyze the information provided by the thermal or hydraulic analysis and modify several of the operation conditions. Below are shown the proposed modifications and the corresponding analysis for each one.

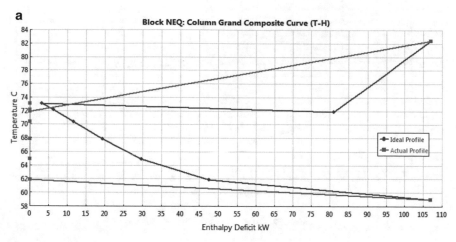

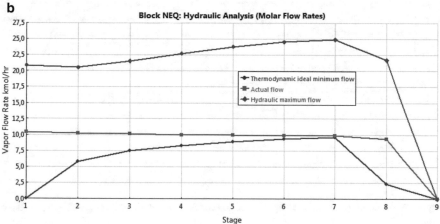

Fig. 6.94 Thermal (**a**) and hydraulic (**b**) analysis for the scenario 2 ($RR = 0.8$ and $N = 9$) in Aspen Plus®

6.8.3.1 Feeding Stage

We observe that when the feeding stage is varied for the stage 11 to 9 and further to 7, the CGCC curve is affected because the feed is every time higher and its temperature is greater to any of the stages reported in Fig. 6.96; in this manner affecting the column temperature profile. Simultaneously, the hydraulic behavior is affected since the entrance of the feed stream represents an increase in the column liquid and vapor flow. In Figs. 6.97, 6.98, and 6.99 are shown the hydraulic Profiles (or of flows) throughout the column.

You can see that as the feed enters in segments each time higher in the column, there is a variation in the curves and a substantial change in the behavior. The minimum ideal vapor flow decreases, and the maximum vapor flow increases just in

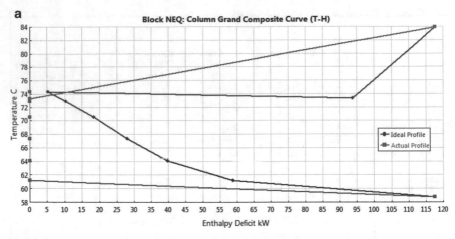

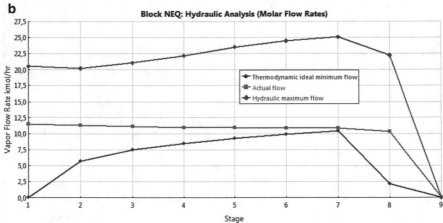

Fig. 6.95 Thermal (**a**) and hydraulic (**b**) analysis for the scenario 3 ($RR = 1$ and $N = 9$) in Aspen Plus®

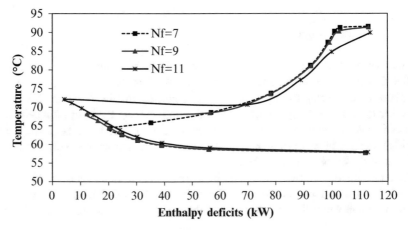

Fig. 6.96 CGCC from the distillation column for a change in the feeding stage in Aspen Plus®

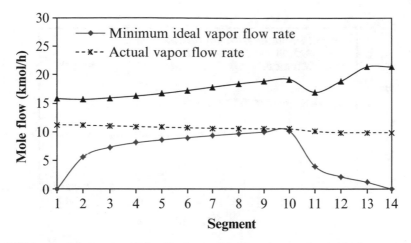

Fig. 6.97 Hydraulic analysis of the distillation column for the base case with $N = 14$, $N_f = 11$, and $RR = 1$ in Aspen Plus®

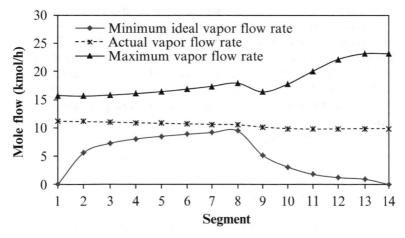

Fig. 6.98 Hydraulic analysis of the distillation column for the base case with $N = 14$, $N_f = 9$, and $RR = 1$ in Aspen Plus®

the feed segment. The objective of this type of analysis lies on that the line that corresponds to the current vapor flow is the nearest possible to the ideal minimum vapor flow curve, since in this manner the column functions very near the ideality condition, and therefore the efficiencies are higher. In general, we can say that the change in the feed stage affects the hydraulic behavior, but in any case allows the operating condition to be nearest to the ideal and relatively distant from the maximum condition, which indicates that the column could be operated with slightly higher loads without any problem.

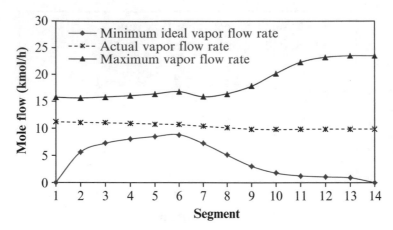

Fig. 6.99 Hydraulic analysis of the distillation column for the base case with $N = 14$, $Nf = 7$, and $RR = 1$ in Aspen Plus®

From these two analyses we can see that when segment 11 is fed, the column hydraulic behavior is nearer to the ideal one than when it is fed at higher height. As to the thermal behavior, we can see that when it is fed in said segment, the curve has a behavior closer to the ideal one that would be the situation in which there is not any enthalpy deficit throughout the column, and which is represented with a green line for the case in which the graphics are generated by the simulator. Additionally, the modification of the feed state allows us to observe a displacement in the pinch point location of the graphic, which moves away more from the zero enthalpy deficit axis when it is fed in a stage closest to the column top. This also justifies the stage or segment 11 as appropriate location for the feed.

6.8.3.2   Feeding Temperature

In the base case of this exercise we set the temperature in 70 °C; however, we wish to quantify the effect of said variable in the column thermal and hydraulic behavior, and in this manner select the appropriate feeding temperature.

In Figs. 6.100, 6.101, and 6.102 we can see the obtained results. Regarding to the base case, the feed temperature variation within 60 and 70 °C does not have any significant incidence. However, as expected, when we have a 60 °C feed temperature the enthalpy deficit increases in the column as consequence of the need for a greater amount of energy, which is not used in the mixture separation but in the thermal adapting of the feed stream. Likewise, entering feed at 80 °C we have that the enthalpy deficit drops notably and also the flows inside the column. This reduction is because probably the feed in this condition enters as a liquid–vapor mixture which causes a large part of what enters the column to immediately vaporize to rise by the column and, therefore, not accumulate in the lower zone of the column.

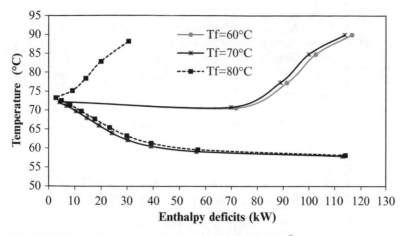

Fig. 6.100 CGCC for a change in the feeding stage in Aspen Plus®

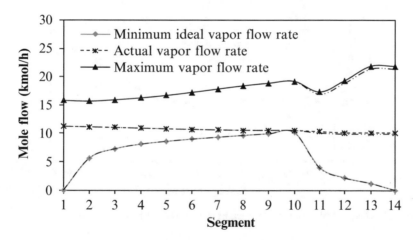

Fig. 6.101 Hydraulic analysis from the distillation column with $N = 14$, $N_F = 11$, and $T_F = 60\,°C$ in Aspen Plus®. On the dotted line, the base case with $T_F = 70\,°C$

6.8.3.3 Packing Type

For the case where the possibility to carry out a change in the pack type is considered, you have to analyze the thermal behavior, although we expect it does not change in a very significant manner, and the hydraulic, in which the maximum vapor flow will change in a significant manner. Keep in mind that the maximum flow calculation depends on the relevant characteristics of the pack and on the approximation to the maximum flooding capacity. For this reason it is likely it changes from one pack to another.

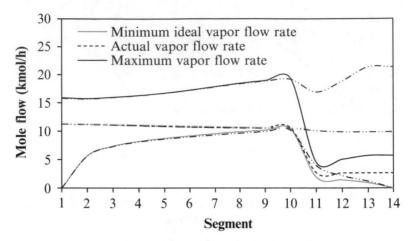

Fig. 6.102 Hydraulic analysis of the distillation column with $N = 14$, $N_F = 11$, and $T_F = 80\,^\circ$C in Aspen Plus®. On the dotted line the base case with $T_F = 70\,^\circ$C

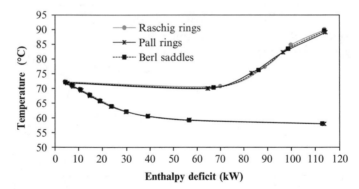

Fig. 6.103 CGCC for a change in the packing type, all of a 1 in size in Aspen Plus®

Figures 6.103 and 6.104 show the effect the type has on the thermal and hydraulic behavior of the column. For it, the possibility of using other two packs with the same dimension as the base case pack is considered (Pall Rings and Berl saddles). The effect on the CGCC curve profiles is practically null, thus for the three cases the enthalpy deficit throughout the column is the same. For the hydraulic case, obviously the maximum flows increase for the cases where Pall rings and Berl saddles are used, packs with different parameters as to vacuum fraction and density that allow working with higher capacities.

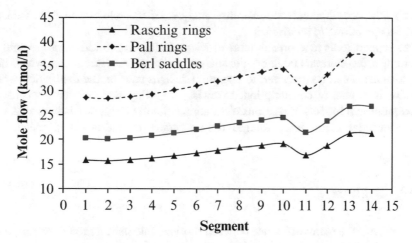

Fig. 6.104 Maximum vapor flow of the distillation column for a change in the packing type, all in a 1 in size in Aspen Plus®

6.8.3.4 Additional Remarks

As conclusion from the thermal and hydraulic analysis of the methanol distillation column, we can see that the initial configuration deviates from an efficient operation which allows obtaining a product with the expected qualities. The reflux ratio shall be adjusted to 1 to improve the separation and accurately know the energy consumption from the column. On the other hand, an appropriate location of the feed state allows energy consumptions to be reduced as consequence of an improvement in the separation. The feed temperature adjustment shall have a notable affect if you keep in mind that the energy is the separation agent of the distillation, and fundamentally this is the objective of the energy addition to the column. For this reason, the feed temperature evaluation affects considerably the hydraulic response and the enthalpy deficit of the column. Finally, as seen in the nonequilibrium model and in this thermal and hydraulic analysis section, the appropriate selection of the pack type and its dimensions has significant consequence on the pressure drop and the hydraulic of the system, which finally is translated to efficiency in the separation and better results in the operation energy consumption.

6.9 Summary

Aspen Plus® and Aspen HYSYS® have a strong calculation modules corresponding to distillation columns; this is due to distillation being one of the most important separation operations in process engineering. Additionally, in order to provide a robust and sound model, the use of computing tools to reduce calculation time as well as the development of sensibility analysis of the relevant variables on the process operation are required.

As support tools to approach more efficient designs for liquid–vapor separation operations, the thermal and hydraulic analysis utilities were introduced, with which improvements can be proposed to operation designs such as the distillation which involve high energy consumption. Likewise the utility that can have the use of nonequilibrium models in rigorous distillation column calculation is shown as well as the advantages and disadvantages of these facing the traditional approach.

6.10   Problems

P6.1 What is the main difference between shortcut calculation models and rigorous models?

P6.2 What is the nonequilibrium model for distillation and absorption columns based on? What advantages does it have?

P6.3 What is the importance of carrying out a thermal and hydraulic analysis of the distillation columns? What information can be obtained from said analysis?

P6.4 One of the absorption operation applications at industrial scale is the natural gas dehydration using glycol; this process is important in the natural gas processing since water presence in transportation lines can cause obstructions due to the formation of solid hydrates.

The stream to be dehydrated has the conditions shown in Table P6.1. Initially, according to the heuristics introduced by Kidnay, the usual amount of glycol, specifically TEG (triethylene glycol) to use in the absorption is 3 gal/lb H_2O, and the gas outlet specification is 0.04 lb H_2O/MMscf, because is the one used in natural liquefied gas, process that cannot bear water amounts over the one mentioned.

Table P6.1   Specifications of the natural gas stream to be dehydrated

Component	Mole flow rate (lbmol/h)
Water	10.920
Methane	12,028.034
Ethane	1034.652
Propane	617.161
n-Butane	227.722
n-Pentane	102.310
Nitrogen	2417.488
Helium	74.257
T [°F]	93
P [psi]	2000

Carry out the necessary calculation to enter the corresponding data in the simulation. Compare the results obtained in Aspen Plus® and Aspen

HYSYS®. Why do results vary? What results are more correct? What would you recommend to guarantee the data is reliable?

P6.5 Figure P6.1 shows the data of the feed stream in a column used to recover propane along with the operation pressures and wanted recovery percentages. Design the required column applying a shortcut method and later a rigorous method. Then propose a pack and calculate the column diameter.

P6.6 You are required to separate a methanol–ethanol–water mixture which is fed to a tray distillation column at a ratio of 100,000 kg/h. Feeding pressure is 5 bar and temperature 86 °C. The mixture contains 79.8 % weight of methanol, 0.2 % in ethanol weight, and 20 % in water weight. The configuration and main specifications of the column currently installed are:

- Condenser total, Distillate flow: 3 kg/h
- Reboiler type Kettle, reboiler duty: 55 MMkcal/h
- Feed is introduced in tray 48
- A lateral liquid stream is extracted in tray 4 at a ratio of 78,500 kg/h
- A second lateral liquid stream is extracted in tray 66 at a ratio of 1500 kg/h
- Pressure in condenser: 1.3 bar
- Pressure drop in condenser: 0.2 bar; pressure drop in the rest of the column: 0.4 bar
- The column as 78 valve trays; the overflow height is 0.015 m
- The column diameter is 5.4 m

The objective of the exercise is to calculate the column in an equilibrium model and a nonequilibrium model and compare the results obtained by the routes. The information provided is from the equipment actual construction. This means that for the equilibrium model it is necessary to make an efficiency assumption and readjust the parameters for the simulation. Finally, an analysis is required concerning the information that can be withdrawn from the nonequilibrium model results.

P6.7 A deisobutanizer column is required with a saturated liquid feed of 500 lbmol/h of isobutane and 500 lbmol/h of butane. The distillate is to be 99 mol% isobutene and the bottoms 99 mol% n-butane.

1. Select an adequate operating pressure.
2. Use a shortcut method to determine the minimum number of stages and minimum reflux ratio.
3. Use a rigorous equilibrium method to size the column, using a reflux-to-minimum-reflux ratio of 1.1.
4. Determine the condenser and reboiler duties.

P6.8 Distillation is used to separate pentane from hexane. The feed flow rate is 100 kmol/s and has a mole ratio pentane/hexane $= 0.5$, $P = 1$ atm. The bottom and top products have pentane compositions $x_B = 0.05$ and $x_D = 0.98$. The feed, at the bubble point, enters the column at stage whose temperature is equal to the feed temperature. The distance between the trays amounts to 0.50 m.

(a) Calculate via a shortcut method the minimum reflux ratio and define an operation reflux ratio
(b) Calculate the feed temperature
(c) Calculate the vapor stream from the reboiler
(d) Calculate the required energy in the reboiler
(e) Construct the y–x diagram
(f) Construct the operating lines and locate the feed line
(g) Determine the number of equilibrium stages
(h) Determine the height of the column

References

Aspen Technology, Inc. (2001) Aspen Plus 11.1. Unit operation models. Columns. Aspen Technology, Cambridge (Chapter 4)

Bekiaris N, Meski G, Radu C, Morari M (1994) Design and control of homogeneous azeotropic distillation columns. Comput Chem Eng 18:15–24

Bieker T, Simmrock K (1994) Knowledge integrating system for the selection of solvents for extractive and azeotropic distillation. Comput Chem Eng 18:825–829

Black C (1980) Distillation modeling of ethanol recovery and dehydration processes for ethanol and gasohol. Chem Eng Prog 76:78–85

Boston JF, Sullivan SL (1974) A new class of solution methods for multicomponent, multistage separation processes. Can J Chem Eng 52:52–63

Castillo F, Towler G (1998) Influence of multicomponent mass transfer on homogeneous azeotropic distillation. Chem Eng Sci 53(5):963–976

Ciric A, Mumtaz H, Corbett G, Reagan M, Seider W, Fabiano L, Kolesar D, Widagdo S (2000) Azeotropic distillation with an internal decanter. Comput Chem Eng 24:2435–2446

Dennis Y, Megan J (2001a) Multicomponent homogeneous azeotropic distillation 1. Assessing product feasibility. Chem Eng Sci 56:4369–4391

Dennis Y, Megan J (2001b) Multicomponent homogeneous azeotropic distillation 2. Column design. Chem Eng Sci 56:4393–4416

Dennis Y, Megan J (2001c) Multicomponent homogeneous azeotropic distillation 3. Column sequence synthesis. Chem Eng Sci 56:4417–4432

Dennis Y, Castillo F, TOWLER G (2000) Distillation design and retrofit using stage-composition lines. Chem Eng Sci 55:625–640

Diamond D, Hahn T, Becker H, Patterson G (2004) Improving the understanding of a novel complex azeotropic distillation process using a simplified graphical model and simulation. Chem Eng Process 43:483–493

Dhole VD, Linnhoff B (1992) Distillation column targets. Comput Chem Eng 17:549–560

Doherty M, Malone M (2001) Conceptual design of distillation systems. McGraw-Hill, New York, p 568

Dyk V, Nieuwoudt I (2000) Design of solvents for extractive distillation. Ind Eng Chem Res 39 (5):1423–1429

Eckert E, Vanek T (2001) Some aspects of rate-based modelling and simulation of three-phase distillation columns. Comput Chem Eng 25:603–612

Glasser D, Hildebrandt D, Hausberger B (2000) Costing distillation systems from residue curve based designs. Comput Chem Eng 24:1275–1280

Henley E, Seader J (1998) Separation process principles. Wiley, New York, pp 527–543

Henley E, Seader J (1981) Equilibrium stage separation operations in chemical engineering. Wiley, New York

Kister H (1991) Distillation design. McGraw-Hill, New York

Kiva V, Hilmen E, Skogestad S (2003) Azeotropic phase equilibrium diagrams: a survey. Chem Eng Sci 58:1903–1953

Kreul L, Górak A, Barton P (1999) Dynamic rate-based model for multicomponent batch distillation. AIChE J 45(9):1953–1961

Krishnamurthy R, Taylor R (1985a) A nonequilibrium stage model of multicomponent separation processes. Part I: model description and method solution. AIChE J 31:449–456

Krishnamurthy R, Taylor R (1985b) A nonequilibrium stage model of multicomponent separation processes. Part III: the influence of unequal component efficiencies in process design problems. AIChE J 31:1973–1985

Leiva F (2003) Estudio de una columna de destilación reactiva sobre un simulador de procesos químicos, Graduate work to obtain the title of Master in Chemical Engineering. Universidad Nacional de Colombia, Bogotá

Lewis WK, Matheson GL (1932) Studies in distillation design of rectifying columns for natural and refining gasoline. Ind Chem Eng 24:494

Ludwig EE (1979) Applied process design for chemical and petrochemical plants, vol 2, 2nd edn. Gulf Publications, Houston

Malagón F (2010) Optimización de la destilación extractiva de etanol con glicerina en estado estable. Graduate work to obtain the title in Chemical Engineering. Universidad Nacional de Colombia, Bogotá

Manan Z, Bañares-Alcántara R (2001) A new catalog of the most promising separation sequences for homogeneous azeotropic mixtures I. Systems without boundary crossing. Ind Eng Chem Res 40(24):5795–5809

Müller N, Segura H (2000) An overall rate-based stage model for cross flow distillation columns. Chem Eng Sci 55:2515–2528

Perry RH (1992) Perry's chemical engineers' handbook, 7th edn. McGraw-Hill, Mexico, pp 13-56–13-81

Pham H, Doherty M (1990) Design and synthesis of heterogeneous azeotropic distillations—II. Residue curve maps. Chem Eng Sci 45(7):1837–1843

Rodríguez I, Gerbaud V, Joulia X (2001) Entrainer selection rules for the separation of azeotropic and close-boiling—temperature mixtures by homogeneous batch distillation process. Ind Eng Chem Res 40:2729–2741

Rooks R, Julka V, Doherty M, Malone M (1998) Structure of distillation regions for multicomponent azeotropic mixtures. AIChE J 44(6):1382–1391

Scenna N (1999) Modelado, simulación y optimización de procesos químicos. Universidad Tecnológica Nacional, México, pp 373–434

Seider WD, Seader JD, Lewin DR (2004) Product & process design principles, 2nd edn. Wiley, New York

Smith J, Van Ness H, Abbott M (1997) Introducción a la termodinámica en Ingeniería Química, 5th edn. McGraw-Hill, Mexico, p 857

Tao L, Malone M, Doherty M (2003) Synthesis of azeotropic distillation systems with recycles. Ind Eng Chem Res 42(8):1783–1794

Taylor R, Krishna R (1993) Multicomponent mass transfer. Wiley, New York

Thiele EW, Geddes RL (1933) Computation of distillation. Apparatus for hydrocarbon mixtures. Ind Chem Eng 25:289

Thomas M (1991) A study in the synthesis of separation processes for multicomponent azeotropic mixtures. Thesis submitted to the Graduate School in partial fulfillment of the requirements for the degree Master of Science in Chemical Engineering. Carnegie Mellon University, Pittsburgh

Tolsma J (1999) Analysis of heteroazeotropic systems. Submitted to the Department of Chemical Engineering in partial fulfillment of the requirements for the degree of Doctor of Philosophy in Chemical Engineering. Massachusetts Institute of Technology

Treybal R (1996) Operaciones de transferencia de masa, 2nd edn. McGraw-Hill, México (Chapter 9)

Urdaneta R, Bausa J, Brügemann S, Marquardt W (2002) Analysis and conceptual design of ternary heterogeneous azeotropic distillation processes. Ind Eng Chem Res 41(16):3849–3866

Van Winkle M (1967) Distillation. McGraw-Hill, New York, p 684

Villiers W, French R, Koplos G (2002) Navigate phase equilibria via residue curve maps. Chem Eng Prog 98:66–71

Wahnschafft O, Köhler W, Westerberg A (1994) Homogeneous azeotropic distillation: analysis of separation feasibility and consequences for entrainer selection and column design. Comput Chem Eng 18:531–535

Wang J, Henke G (1966) Tridiagonal matrix for distillation. Hydrocarb Process 45(8):155–163

Wankat P (2008) Ingeniería de procesos de separación, 2nd edn. Prentice Hall, México

Wasylkiewicz S, Kobylka L, Castillo F (2000) Optimal design of complex azeotropic distillation columns. Chem Eng J 79:219–227

Wasylkiewicz S, Kobylka L, Castillo F (2003) Synthesis and design of heterogeneous separation systems with recycle streams. Chem Eng J 92:201–208

Widagdo S, Seider W (1996) Azeotropic distillation. AIChE J 42(1):96–130

Chapter 7
Process Optimization in Chemical Engineering

7.1 Introduction

In chemical processing units, optimization is the method that seeks to solve the problem of minimizing or maximizing an objective function that relates the variable to optimize with the design and operating variables. The criteria for analysis of the economic objective function involve fulfilling a process criteria restrictions, conditions, design equations, and respecting the limits of the variables (Seider and Warren 2003). Most problems in chemical engineering processes have many solutions, in some cases becoming endless. The optimization is related to the selection of an option that is best in a variety of efficient options but being the only one that comes closest to an economic optimum performance and operation.

Plants operating profits are achieved by optimizing performance of the valuable products or reduction of pollutants, reducing energy consumption, improving processing flows, decreasing operation time, and minimizing plant shutdowns. To this end it is useful to identify the objective, constraints, and degrees of freedom in the process, reaching benefits such as improving the design quality, ease of troubleshooting, and a quick way to make correct decisions. However, the argument for the implementation of the optimization process is not well supported if the formulation of the optimization process has an uncertainty in the mathematical model describing the process. Although the mathematical model is a description of reality, an optimization on the mathematical model does not guarantee optimization modeling phenomenon due to the difference between the mathematical model and the actual phenomenon. In the case of chemical engineering problems, most processes and operations are well represented by mathematical models with some complexity, this leads by ensuring mathematical model optimization is going to optimize the process.

Optimization can take place at any level within an organization, from a complex combination of plants, distribution facilities to each floor, units combinations, individual equipment, subsystems of a piece of equipment or smaller units. Process

© Springer International Publishing Switzerland 2016
I.D.G. Chaves et al., *Process Analysis and Simulation in Chemical Engineering*,
DOI 10.1007/978-3-319-14812-0_7

design and equipment specifications is usually performed by taking decisions that affect the whole life of the plant or process, which is why the right decisions are important and can be based on results optimizing both the design and daily operation of the process (Edgar et al. 2001).

7.2   Formulation of Optimization Problem

The formulation of the objective function is based on the mathematical model of the phenomenon being optimized. In the chemical industry, cost minimization or profit maximization is held as usual objective function, in which case the design equations and describing the process should relate to the operation cost or invest-ment. In other cases it is important that an objective function is raised in order to maximize the operation performance or choose the most appropriate diameter for specific equipment, in such a case the equations describing the operation must be related to the process efficiency and unit sizing. Depending on what you want to optimize, more rigorous and complex models need to be used. Each optimization problem has three main categories:

1. A target function (at least) to be optimized.
2. Equality constraints.
3. Inequality constraints.

To achieve a feasible solution for the optimization problem, the problem vari-ables must satisfy the restrictions to the degree of accuracy required, so a workable region is obtained in the problem where the optimum point is found. In some cases the optimal solution is a single point, but in other cases the problem becomes indeterminate (Edgar et al. 2001).

7.2.1   Degrees of Freedom

In design and control calculations, information is important and user must remove redundant equations before running calculations. In the case where there is no debug information the solution of the problem cannot be obtained and the infor-mation should be reviewed again. The degrees of freedom in a model are the number of variables that can be specified independently and are defined as:

$$N_f = N_v - N_e \tag{7.1}$$

where

N_f: degrees of freedom number
N_v: total number of variables
N_e: number of independent equations including specifications

An analysis of the degrees of freedom for modeling problems separated into three categories:

1. $N_f = 0$. The problem is exactly determined. The number of independent equations is equal to the number of process variables and established equations have a unique solution, in which case the problem is not an optimization problem. If the equations are linear, single solution is obtained, but if they are not linear can be obtained no real solutions or multiple solutions.
2. $N_f > 0$. The problem is indeterminate or is underspecified. In this case there is more process variables independent equations in the problem. At least one variable can be optimized.
3. $N_f < 0$. The problem is over is overdetermined or specified. There are fewer process variables independent equations in the problem and consequently established equations have no solution.

In the operating and design variables, study of chemical engineering operations shows that, by varying one of the operating conditions, can change the value of the objective function both directly and inversely reaching set point wherein the objective function is optimized. An example would be that if the reflux ratio of a distillation column is decreased, the diameter of the column is increased by decreasing the energy required in the system; if the reflux ratio is increased, reducing the number of plates but the energy is increased in the reboiler and condenser of the column. So in this case is an example in which it can minimize the operation cost.

7.2.2   Objective Function

The formulation of the objective function is one of the crucial steps in the application of optimization to a practical problem. In the chemical industry the objective function is generally expressed in monetary terms, since the goal of companies is to minimize production costs and maximize profits subject to variations in the constraints. In other cases, the problem to be solved is to maximize the performance of a component in a reactor or minimizing service flow heat exchanger. As formulated, the mathematical model of each problem the complexity of the equations that represent, at present, has the advantage of having software capable of solving highly nonlinear functions must be analyzed.

The ability to understand and apply concepts of cost analysis, profit analysis, budgets, and balances are key to assess opportunities. Economic decisions are made at various levels of detail and the more detailed study; it takes longer and requires more resources. To formulate the objective function and which are related to economic parameters must take into account two types of costs (Aspen Technology, Inc. 2005).

1. Costs associated with the mass flows, such as purchase costs of raw materials or revenues for products.
2. Costs associated with operating variables in the model, such as electric power, steam, fuel gas, etc.

7.2.3 Classification of Optimization Problems

Optimization problems in chemical engineering have various types and therefore requires to properly classify, considering that there are specific methods for each type of them.

The simplest problems encountered in process engineering can be linear as the blending of two or more batches of products. These problems can be represented graphically with three optimization variables. Typically, optimization problems have constraints that limit the feasibility area of the solution. In most cases, these restrictions are given by problem physics; for example the sum of compositions of a mixture must always add up to 1, or purity of any outlet must be less than or equal to 1. The same applies to the recovery rates and process efficiencies.

There are different strategies for optimization of engineering problems; basically consist of reduce or limit the area of possible answers being careful not to leave out the potential global maximum or minimum.

7.2.3.1   Linear Programming Problems

The simplest classification of problems depends on the nature of both objective function and constraints. If the equations are linear, calculation is performed more easily in order to search for the optimum. They are called linear programming problems.

Linear programming is the term used to describe the optimization techniques in which the mathematical model that represent a process can be characterized as linear equations. The linear nature of the set of equations makes these methods widely implemented and an effective tool for solving optimization problems.

The most common examples solved using linear programming methods are:

1. Schedule to improve the use of labor, increasing productivity, and worker safety.
2. Optimization of profits through the proper use of raw materials and production of the highest paid product.
3. Delivery time minimization and distribution network optimization.

To representing processes using mathematical models involve variables, equations, and constraints for a single problem. It is desirable that the solution is not only to find the points that satisfy the system but to find the optimal points that minimize or maximize the objective function (Tarquin and Blank 2001).

7.2.3.2 Nonlinear Problems and Sequential Quadratic Programming

In most problems in chemical engineering, models describing the unit operations (chemical and phase equilibrium, transport phenomena, etc.) are highly nonlinear. These problems are called nonlinear programming problems.

The Sequential Quadratic Programming (SQP) method is a method design to find the minimum of a nonlinear n-dimensional function where n variables are related to nonlinear constraints. Can be expressed mathematically as follows:

$$
\begin{aligned}
&\text{Min} && f(x) \\
&\text{s.t.} && g(x) = b
\end{aligned}
\tag{7.2}
$$

where $f(x)$ is a nonlinear function. X contains the n design parameters related to m nonlinear equality and/or inequality constraints.

The SQP method locally approaches the nonlinear function f by a quadratic function q and the nonlinear *constraints*. To find the minimum of a function of this type with linear constraints, the active set method is used which solves quadratic subproblems with equality constraints having analytical solution. Finding the minimum of this approximate quadratic function with linear constraints approach it is required to go back again to another function and constraints on anew point. This continues until some stopping criterion is met.

7.2.3.3 Mixed Integer Nonlinear Programming

Another important classification refers to the nature of the variables. There are two types of variables: Continuous and discrete variables. Discrete variables are those that take only integer values; in the case of process engineering would be appropriate to: number of process equipment, number of tube passes in a heat exchanger, number of stages on a gas–liquid separation, etc. These problems are known as mixed integer programming problems.

In design and operation of plants, some problems involve nonlinear relationships, binary or integer variables, and continuous variables. Continuous variables generally represent process variables such as flow, and integer variable represents a decision (Gomez and Esteban 2006). Some problems in plant operation and activities programming involve variables that are not continuous and often are integer values. Integer variables can take binary values to represent decisions define equipment configuration, or to define whether or not a specific stage feed of a distillation column. Other integer variables can take values from the set of natural numbers (positive integers) and can be useful if you want to define the number of stages of evaporation in a train of evaporators or the number of steps in a heat exchanger. The solution of such problems involves mathematical variables and the use of more resources for which methods are designed to make the simplest problem complexity. And the problems of continuous variable integer variable issues can also be classified as linear (MILP) and nonlinear (MINLP) depending on the type of equations that characterize the objective function or constraints (Edgar et al. 2001; Grossmann 2002).

The formulation of the problem when there are integer variables is similar to what is done when the variables are real except for the inclusion of a new term that characterizes the integer variables. The formulation of the problem is:

$$
\begin{aligned}
\min \quad & Z = f(x, y) \\
\text{s.t.} \quad & g_j(x, y) \leq 0 \quad j \in J \\
& x \in X, \quad y \in Y \quad y \in \mathbb{Z}
\end{aligned} \tag{7.3}
$$

This formulation is general and each problem can develop your own equations describing the problem.

One method for solving linear and nonlinear problems with integer variables is the "Branch and Bounds" (B&B) method using linear relaxations. This method can be used in both linear and nonlinear problems with integer variables. You can consider using the B&B method when all integer variables are binary but with other by this method the solution is more complex. Relaxation method of discrete variables is based on the fact that the discrete variables can be viewed as continuous if a regression of the discrete variable data is performed. By solving the problem with integer variables can be obtained relaxed variables. For relaxed variables you need to create a new LP (branching problem) in which the variables are evaluated at each of the nearest integer values and select which will provide better value to the objective function. For each subproblem limits, the relaxed variables (bounding) are set and the behavior of this variable is checked to see if you need further consideration (Gomez and Esteban 2006).

The second method of solving the MILP and MINLP problems is "outer approximation" (OA). Each iteration of the OA method involves two subproblems: A nonlinear programming problem with continuous variable and a linear problem with mixed variables. In general, the method is based on solving the problem by the lower limit of NLP MINLP problem. The second subproblem, the MILP, takes into account the whole and continuous variables, all nonlinear functions in the linearized rated range. MILP problem optimization generates the upper limit of the optimal solution MINLP. At each iteration, the step size analysis of the problem is reduced to that of a finite number of iterations and errors depending on the accepted solution to optimal MINLP (Edgar et al. 2001) is located.

7.3   Optimization in Sequential Simulators

One of the great advantages of process simulators lies in the ability to integrate with design and optimization process. Due to the implementation of powerful numerical methods, simulators allow solving larger problems and a lot of variables; however, some of these variables are not behind the calculations of each operation so that problems rarely have more than 100 degrees of freedom. Among the many advantages of integrated process simulation optimization is the possibility of carrying out analysis on the different variables in a more intuitive and efficient manner.

The first attempts to integrate the modular simulation optimization processes were developed based on the methods of "black boxes." In these optimization models uses the process simulation results, so the process had to be solved several times, and errors or small differences in the convergence did not allow optimization would be carried out. However, from the eighties, the modular integrated process simulation optimization has become a widely used tool in industry. This has been possible thanks to three major advances in the implementation:

1. SQP strategy requires evaluating few functions and works quite well for optimization problems with few functions to evaluate.
2. The internal cycles of convergence, such as design specifications and recycle streams can be incorporated as constraints in the optimization problem.
3. Since strategy SQP is a Newton-type method, this can be incorporated into the simulation as an additional block, such as a convergence block recycle, allowing the simulation structure to stay the same.

Thus optimization algorithms have been easily incorporated into the modular simulation. By including the optimization algorithm as a convergence block is able to reduce the calculation speed by an order of magnitude with respect to the method of black boxes.

Currently simulators Aspen Plus® and Aspen HYSYS® have different methods and optimization options depending on the type of problem. Table 7.1 summarizes the methods present in the simulators. Aspen Plus® has only the complex method and the SQP method, while Aspen HYSYS® has all the methods except the complex method.

7.3.1 General Aspects

Here are some tips on using optimization simulators which are presented below.

Most simulations have the option of using design specifications. Iterative calculation related to the specifications commonly found in recycle loops and convergence shall be achieved in the specifications for each recycle iteration. In other instances the specifications are implemented as external loops, where the convergence should be achieved in recycles for each iteration of the outer loop. It is also possible to solve both loops simultaneously, usually using only one pass through the units in the recycle loops and specification.

When optimizing a process is often preferable to include design specifications at first, then when the optimization algorithm must be removed include those specifications. Thus the optimization algorithm replaces those specifications.

It is recommended that prior to any optimization algorithm to conduct preliminary searches varying the most important variables. The more used to perform this

Table 7.1 Important methods present in process simulators (Aspen Technology, Inc. 2005)

Method	Description	Restrictions
Fletcher-Reeves	Corresponds to the Polak Ribiere modification of the Fletcher-Reeves conjugate gradient scheme	This scheme is efficient for minimization without restrictions. This method does not handle constraints
Quasi Newton	It refers to the method of Broyden-Fletcher-Goldfarb-Shanno	In limited applicability is similar to the Flechter-Reeves method. This scheme is efficient for minimization without restrictions. This method does not handle constraints
		In both limited applicability and is similar to the method Flechter-Reeves
BOX	This method corresponds to a sequential search problem solver with nonlinear objective functions	Handles inequalities constraints, but not equalities. Usually requires a large number of iterations, however it is a fairly robust method
SQP	It is the Sequential Quadratic Programming (SQP) method	Considered one of the most efficient methods to minimize linear and nonlinear problems. Allows equality and inequality constraints
Mixed	This strategy seeks to take advantage of the strength of the BOX method and SQP efficiency. Use an initial assessment with BOX method using a low tolerance, then find the solution with tolerance using SQP	Only allows inequality constraints
Complex	Traditional black boxes method. Only available in Aspen Plus	It does not require derivative calculations. It can handle as inequality constraints and bounds on the optimization variables. Blocks should be used for converging external convergence recycles or design specifications. Requires many iterations to converge

procedure is by way sensitivity analysis so that the results will be closer to the optimum and the optimization algorithm is more likely to converge with fewer iterations.

7.4   Introductory Example

To illustrate the optimization case of distillation columns, an extractive distillation of ethanol using glycerol in steady state presented in Chap. 6 is planned to optimize the process taking into account the operational cost and the corresponding fixed cost related to process equipment. So that it will obtain the values of the optimization variables that maximize the annualized earnings of the process.

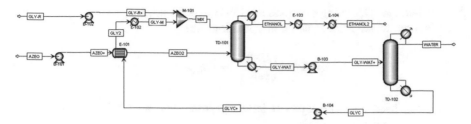

Fig. 7.1 Process flow diagram for extractive distillation

7.4.1   *Aspen Plus*® *Simulation*

In the example of extractive distillation (Gil 2003; Uyazan 2006), the recycle of glycerol specification from column to the heat exchanger E-101 and subsequent admission to the dehydration column should be reviewed. Now it is required to optimize this process to maximize earnings (Fig. 7.1).

The first step in any optimization process is to properly define the objective function and constraints to narrow the search for the optimal operating conditions. In this case it will raise the profit goal or earnings to be having the process for 10 years taking into account the investment costs of the equipment, the cost of raw materials, and the utilities cost consumed in that period of time (Langston et al. 2005). The value of anhydrous ethanol sold product is also taken into account, so that you can calculate the annualized earnings or profit of the process of extractive distillation and solvent recovery.

In this case the objective function is described below:

$$\text{Earnings} = \text{Income} - \text{Expenses} \qquad (7.4)$$

The next step is to determine if the objective function need to be minimized or maximized, since in most cases earnings are used, it should be maximized. In the nomenclature used for the optimization is expressed as follows:

$$\max f(x) = \text{Earnings} = \text{Income} - \text{Expenses} \qquad (7.5)$$

Now you must disaggregate each of the terms to include operating conditions which will be varying during optimization modifications. The only income in this process is due to the sale of product, in this case, anhydrous ethanol.

$$\text{Income} = \text{Anhydrous alcohol sales} \left(\frac{\text{US\$}}{\text{year}}\right) \qquad (7.6)$$

Regarding expenses, there are two types: fixed and operating costs. Fixed costs are the expenses incurred that are not dependent on the amount of product manufactured. Operating expenses are those expenses which are incurred for the production, which are a function of operating time.

$$\text{Expenses}\left(\frac{\text{US\$}}{\text{year}}\right) = C_{\text{fixed}} + C_{\text{oper}} \tag{7.7}$$

In the overhead costs of equipment, buildings, roads, etc. are taken into account, for this case the cost of the equipment.

The operational costs include raw materials, defined as substances that enter the process, utilities costs, mainly steam, electricity, instrument air for elements, etc.; however, equipment costs have an important weight in this term of the equation. When replacing (7.7) such terms are converted into:

$$\text{Expenses}\left(\frac{\text{US\$}}{\text{year}}\right) = C_{\text{raw materials}} + C_{\text{utilities}} + C_{\text{equipment}} \tag{7.8}$$

Now each of the terms is decomposed depending on their nature. The cost of raw materials for this process corresponds to:

$$\text{Raw materials cost}\left(\frac{\text{US\$}}{\text{year}}\right) = C_{\text{Ethanol}} + C_{\text{Glycerin}} \tag{7.9}$$

The utilities costs correspond to the costs of electricity for pumping steam to raise the temperature in reboilers, cooling water for condensers, and finally refrigerant for heat exchanger E-104. The use of the same is minimized by installing double exchanger to cool the anhydrous ethanol; thus it can reduce the temperature by using water up to 30 °C (while maintaining a temperature difference of 10 °C) and then use to carry coolant to the storage temperature.

$$\text{Utilities cost}\left(\frac{\text{US\$}}{\text{year}}\right) = C_{\text{Electricity}} + C_{\text{Steam}} + C_{\text{Cooling Water}} + C_{\text{Refrig.}} \tag{7.10}$$

Finally, the costs of equipment for this problem are only three types: pumps, heat exchangers, and distillation columns.

$$\text{Equipment cost}\left(\frac{\text{US\$}}{\text{year}}\right) = C_{\text{Pumps}} + C_{\text{Dist. Towers}} + C_{\text{Heat Exchangers}} \tag{7.11}$$

The described terms of the objective function must be calculated by the simulation program and updated with actual process data for each simulation run. For this Aspen Plus has a very useful tool for this purpose. This is the *Calculator tool* which is accessed through the route *Flowsheeting Options/Calculator Data Browser* button. This tool allows the calculation of equations in this case correspond to the cost models of equipment, products, and utilities used in the objective function.

7.4.1.1   Raw Material Cost

The flow of azeotropic ethanol is calculated based on installed capacity and to ensure that the flow of dehydrated ethanol is 300,000 L/day. The price of hydrated ethanol varies between 0.72 and 0.78 US$/L and an average price of 0.76 US$/L is taken. For glycerin must be considered the initial cost to be taken to start the operation and the replacement cost of the glycerin losses. Additionally should be noted that glycerin should be replaced totally after a determined period of time due to the degradation thereof by heating. However, total glycerin amount could be replaced considering losses first that the total programmed replacement is due to degradation.

7.4.1.2   Utilities Cost

The utilities cost is associated with the electricity, steam, and cooling water consumption on each of the process equipment. Prices for each service are taken from (Seider and Warren 2003).

7.4.1.3   Pumps

The cost of a pumping system should include both the cost of the pump and engine that can be estimated with the following correlations:

$$S = QH^{0.5} \qquad (7.12)$$

where

S: size factor
Q: volumetric flow in gpm
H: pump head in ft

$$C_B = \exp\left\{9.2951 - 0.6019 \times [\ln(S)]^2\right\} \qquad (7.13)$$

where

C_B: base cost
S: size factor

$$C_P = F_T \times F_M \times C_B \qquad (7.14)$$

C_P: purchase cost
F_T: pump type factor (Table 16.20 from Seider and Warren (2003))
F_M: material factor (Table 16.21 from Seider and Warren (2003))

For pump motor cost:

$$P_C = \frac{P_T}{\eta_P \times \eta_P} = \frac{P_B}{\eta_M} = \frac{Q \times H \times \rho}{33,000 \times \eta_P \times \eta_M} \tag{7.15}$$

where

P_C: power consumed
P_T: theoretical pump power
η_P: pump efficiency
η_M: electric motor efficiency
P_B: pump brake power

$$\eta_P = -0.316 + 0.24015 \times \ln Q - 0.01199 \times (\ln Q)^2 \tag{7.16}$$

within volumetric flows between 50 and 5000 gpm:

$$\eta_M = 0.80 + 0.0319 \times (\ln P_B) - 0.00182 \times (\ln P_B)^2 \tag{7.17}$$

with P_B in a range from 1 to 1500 hp.

Base cost is calculated as follows:

$$C_B = \exp\Big\{5.4866 + 0.13141 \times (\ln P_C) + 0.053255 \times (\ln P_C)^2 + 0.028628$$

$$\times (\ln P_C)^3 - 0.0035549 \times (\ln P_C)^4\Big\} \tag{7.18}$$

$$C_P = F_T \times C_B \tag{7.19}$$

F_T: motor type factor (Table 16.22 from Seider and Warren (2003))

The sum of purchase cost of the pump and motor, which depends on the flow and hydraulic head, is the total cost of the pumping system, thus the cost of the pump is related to the independent variable in this equipment, which is the flow.

7.4.1.4   Heat Exchangers

According to Seider and Warren (2003) the base cost of a heat exchanger is given by the type of heat exchanger that will be used. In this case, Kettle type heat exchangers will be used as reboilers in distillation columns, and floating heat exchangers will be used for condensers and other exchangers in the process. According to this:

For Kettle exchanger type:

$$C_B = \exp\left\{11.967 - 0.8709 \times (\ln A) + 0.090005 \times (\ln A)^2\right\} \tag{7.20}$$

For floating head exchangers:

$$C_B = \exp\left\{11.667 - 0.8709 \times (\ln A) + 0.090005 \times (\ln A)^2\right\} \tag{7.21}$$

A: heat transfer area in ft^2

The purchase cost can be calculated as follows:

$$C_P = F_P \times F_M \times F_L \times C_B \tag{7.22}$$

F_L: length tube correction factor
F_M: material type factor (Table 16.25 from Ravagnani (2010))

$$F_M = a + \left(\frac{A}{100}\right)^b \tag{7.23}$$

F_P: pressure factor based on shell side pressure:

$$F_P = 0.9803 + 0.018 \times \left(\frac{P}{100}\right) + 0.0017 \times \left(\frac{P}{100}\right)^2 \tag{7.24}$$

This makes it possible to obtain the cost of the exchangers depending on the heat transfer area mainly. You need to evaluate the different configurations of heat exchangers for the required processes: reboilers, condensers, and heat exchangers.

7.4.1.5 Distillation Columns

Since the distillation columns operate under pressure, they must be designed as pressure vessels, thus starts by determining the design pressure, which depends on the operating pressure as follows (Seider and Warren 2003):

$$P_d = \exp\left\{0.60608 + 0.91615 \times (\ln P_o) + 0.0015655 \times (\ln P_o)^2\right\} \tag{7.25}$$

Regarding the design pressure, the wall thickness shall be calculated:

$$t_P = \frac{P_d \times D_i}{2 \times S \times E - 1.2 \times P_d} \tag{7.26}$$

where

D_i: inner diameter in inches
S: maximum material stress allowable
E: welding efficiency

Then, the tower weight is calculated.

$$W - \pi \times (D_i + t_S) \times (0.8 \times D_i) \times t_S \times \rho \tag{7.27}$$

ρ is column construction material density.
Then, the empty vessel cost is calculated including manholes, supports, etc.

$$C_V = \exp\left\{7.0374 + 0.18255 \times (\ln W) + 0.02297 \times (\ln W)^2\right\} \tag{7.28}$$

Then, it is considered the stairs and platforms costs:

$$C_{PL} = 237.1 \, (D_i)^{0.63316} \times (L)^{0.80161} \tag{7.29}$$

Finally, the purchase cost is calculated:

$$C_P = F_M \times C_V + C_{PL} \tag{7.30}$$

Equipment depreciation is important to consider in the equation of the objective function term, and not directly in the objective function that evaluates the annualized profit, but in the time of payback. In Colombia, given the variability in the economy is expected to recover the investment of 3–4 years maximum, so that depreciation is set to 4 years.

Heating, cooling, and power utilities are not available as modules of such exchangers and pumps, but will be used and must be added by the user. Aspen Plus® has the *Utilities* tool by which a wide variety of utilities preset by the user intended shape.

To access the tool click the *Utilities Data Browser* button to detach the navigation tree, select the window called Utilities. Click *New* and enter the following information for electricity (*ELECT*) (Fig. 7.2).

A window for data entry is displayed. In the *Type* tab select the *Electricity Utility* option. Now in the Cost Utility section, enter a value of 0.04 U/kW-h in the *Purchase Price* tab. In this way the service is defined (Fig. 7.3).

Now refrigerant service is defined, called *REFRI*, with the information reported in Table 7.2.

To enter the values of the operating variables of the *Calculation Options* select the *Specify inlet/outlet conditions* section option. Then go to the *State* variables tab and enter the appropriate information (Fig. 7.4).

Now, enter the medium pressure steam utility, called *STEAMM*, with the information reported in Table 7.3.

Fig. 7.2 Utilities specification window in Aspen Plus®

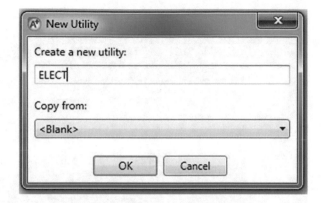

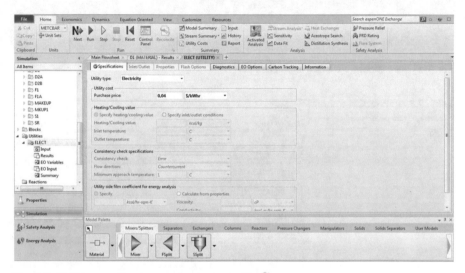

Fig. 7.3 Utilities specification window in Aspen Plus®

Table 7.2 Information to be entered for *REFRI* utility

Variable	Value
Utility type	Refrigeration
Price	14 $/kg
Inlet conditions	
Temperature	5 °C
Pressure	1 bar
Outlet conditions	
Temperature	20 °C
Pressure	1 bar
Composition	
Glycerol	1

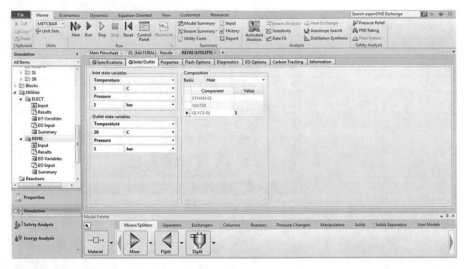

Fig. 7.4 State variables window in Aspen Plus®

Table 7.3 Information to be entered in the *STEAMM* utility

Variable	Value
Utility type	Steam
Price	0.044 $/lb
Inlet conditions	
Temperature	358.42 °F
Vapor fraction	1
Outlet conditions	
Temperature	358.42 °F
Vapor fraction	0

With the data entered for this facility, the medium pressure steam is used which condenses isothermally at a temperature of 358.42 °F (181.34 °C) thus make use only, its latent heat of vaporization.

The next service to enter the cooling water is called *CW*, with the information shown in Table 7.4.

A utility that does not correspond to any particular application, called *U-1*, which will be useful later, is added. The information on this utility is shown in Table 7.5.

To enter this information keeps the option *Specify heating/cooling* selected in the *Calculation* section.

Once finished entering utilities, should be included in each unit as required. To this end, each equipment block in *Setup Utility* is a tab called where the relevant service is entered. In Table 7.6 the relevant utilities for each unit is shown.

Note that the mixer and the heat exchanger *E-101* utilities required by the calculation (Figs. 7.5 and 7.6).

Table 7.4 Information to be entered in the *CW* utility

Variable	Value
Utility type	Water
Price	1 $/t
Inlet conditions	
Temperature	25 °C
Pressure	1 bar
Outlet conditions	
Temperature	120 °F
Pressure	1 bar

Table 7.5 Information to be entered in the U-1 utility

Variable	Value
Utility type	General
Price	1 $/kg
Energy requirements and temperature specifications	
Mass heat capacity	1 kcal/kg
Inlet temperature	10 °C
Outlet temperature	15 °C

Table 7.6 Utilities associated to each process unit

Equipment	Service
B-101	ELECT
B-102	ELECT
B-103	ELECT
B-104	ELECT
E-102	CW
E-103	REFRI
E-104	REFRI
TD-101 Condenser	CW
TD-102 Condenser	CW
TD-101 Reboiler	STEAMM
TD-102 Reboiler	STEAMM

To enter the proper utility in the distillation columns should be directed to the Condenser and Reboiler of each module tabs and enter the information in the *Utility* tab.

The cost model is associated to each item with a set of empirical equations describing the cost based on the same design variables. The variables must be known and must be updated within each calculation simulator. To this end, Aspen Plus provides the *Calculator tool* which has the ability to program in Fortran or Excel equations describing such cost models. For this case Excel is used to calculate the costs and the objective function.

To access the Calculator tool, follow the *Flowsheeting Options > Calculator* route in the *Data Browser* menu. A new calculator block is activated with the *New* button and the default name, *C-1* is left. It speaks directly to the *Calculate tab in Excel* Calculation method is selected (Fig. 7.7).

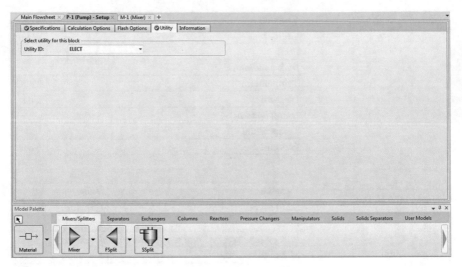

Fig. 7.5 *Utility* tab of utilities specification in Aspen Plus®

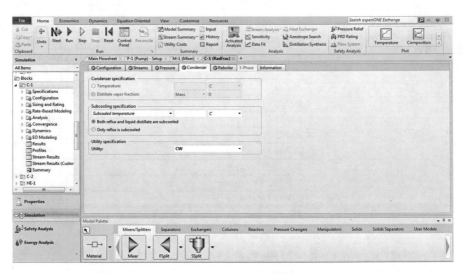

Fig. 7.6 *Utility* tab of utilities specification in Aspen Plus®

Clicking the *Open Excel Spreadsheet* button, the window version of Microsoft Excel® installed on the computer is displayed. This is completely defined as the *Calculator* tool because it takes the values of the spreadsheet Microsoft Excel. Then the pop-up window is displayed (Fig. 7.8).

Microsoft Excel automatically in a new tab bar tool called *Aspen* (in newer versions) appears. If the menu not appears, check in the complements menu that the Aspen Plus tool (Fig. 7.9) is activated.

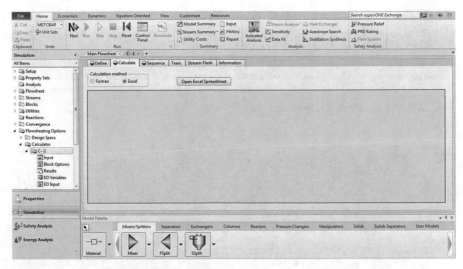

Fig. 7.7 *Calculator* window in Aspen Plus®

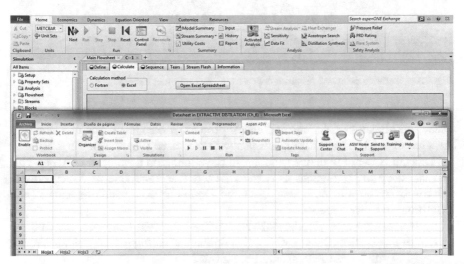

Fig. 7.8 Integration of *Calculator tool* with Microsoft Excel®

Fig. 7.9 *Aspen* tab in Microsoft Excel®

Fig. 7.10   Complement tab in Microsoft Excel®

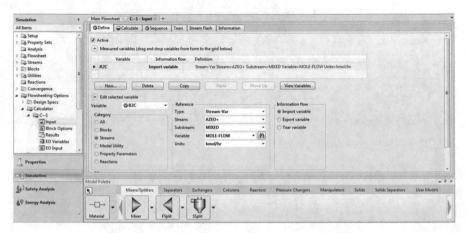

Fig. 7.11   Variable definition tab from Aspen Plus® in Microsoft Excel®

In this tab you can integrate Aspen Plus calculation engine with Excel spread-sheet and allows calculations and check properties databases. For this exercise, Aspen tool can be found in the *Complement* tab is used (Fig. 7.10).

This tab is useful to define variables in the simulation developed in Aspen Plus and use them in calculations as required. To define a variable of Aspen Plus in Microsoft Excel you must first select the *Define* button, which appears in the definition variables window.

In this case, for example, the molar flow of defined *Azeo* + stream is selected. As an entering data sheet, Microsoft Excel is defined as an *Import variable*, if a result obtained in the spreadsheet should be exported to Aspen Plus, *Export Variable* option must be selected as is the case of the objective function value (Fig. 7.11).

Closing the *Variable Definition* tab and if the variable does not already appear in the cell in which it is defined, it is necessary to refresh the Microsoft Excel sheet using the *Refresh* button on the *Complements* tabs and then run the simulation from Aspen Plus. This procedure must be done to define each variable that you want to import from or export to Aspen Plus® using Microsoft Excel®.

Thus all imports of information are performed from Aspen Plus to Microsoft Excel and equipment costs, raw materials, utilities, product, and objective function.

The value of the objective function is exported to Aspen Plus using the cost of the *U-1* utility, Microsoft Excel sheet cools, and Aspen Plus simulation runs. The *U-1* utility is defined in the *Utilities* section.

The cost of the *U-1* utility takes the value of the objective function as Aspen Plus has no direct way to relate the calculated heat of the objective function in Microsoft Excel with a variable whose name is in Aspen Plus. As the cost of *U-1* utility is used, any variable can be used in a stream, block, or usefulness that is completely independent and does not affect the development of the simulation or process. The variable takes the value of the objective function is then used to perform sensitivity analysis and optimization.

7.4.2 Sensitivity Analysis

7.4.2.1 Process Variables

The results of sensitivity analysis are considered for determination of the operating conditions of the process and subsequently the target responses (called earnings function) for the optimization problem performed for this example.

The sensitivity analysis shows that a low recovery of solvent (95 %) and low solvent feed ratio, none of the independent variables analyzed has great influence. Differences in energy requirements at each analysis reboiler to 95 % recovery are shown, because the amount of water in the reboiler to increase the purity requirement on top ethanol is increased (Fig. 7.12).

The sensitivity analysis in which varies the glycerol flow, temperature, and the reflux ratio with a solvent feed ratio of 2.2, introduces substantial changes to what had been observed with a solvent realción 0.2 feed. In the solvent temperature analysis, Fig. 7.13 shows that there is stability in the ethanol composition at 150 °C, where the temperature of the glycerol can evaporate water extracted and increase in distillate product fraction. In Fig. 7.12 the lower solvent to feed ratio must be to carry out water extraction shown in ethanol, the solvent feed ratio is about 0.4. In Fig. 7.14, a large area in which the reflux ratio composition maintains high ethanol

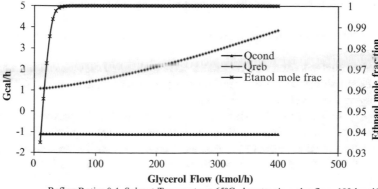

Reflux Ratio: 0.4. Solvent Temperature: 65°C. Azeotropic molar flow: 100 kmol/h

Fig. 7.12 Sensitivity analysis on glycerol flow (95 % rec)

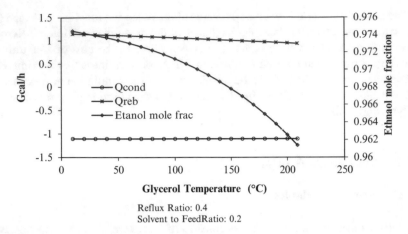

Reflux Ratio: 0.4
Solvent to FeedRatio: 0.2

Fig. 7.13 Sensitivity analysis on glycerol feed temperature (95 % rec)

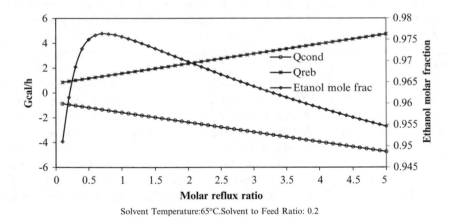

Solvent Temperature:65°C.Solvent to Feed Ratio: 0.2

Fig. 7.14 Sensitivity analysis on reflux ratio from first tower (95 % rec)

in the distillate product of high solvent ratio under which feed the process simulation is observed.

A similar analysis for a recovery of 99 % is realized. These sensitivity analyses help to establish the conditions and limits of process variables when performing sensitivity analyses with objective functions.

Similarly to the variables shown, it is recommended to perform sensitivity analyses on all optimization variables (or influences on the objective function), in order to determine the ranges of these so as to properly identify the area of response objective function and gain access to a global (or local) maximum easier to calculate.

7.4.3 Results

As results of the above process, optimization on the objective function behavior is analyzed, in this case the process annual income (Fig. 7.15). Then the optimization variables, the initial values, and the optimum values shown (Table 7.7).

7.4.3.1 Objective Function

When comparing the results obtained from the sensitivity analysis developed by Gil and Uyazán (2003) with the values found by the optimization process you may notice that it is not far from each other. This indicates conducting sensitivity analysis on the process parameters on the same optimal solution. It is expected that the number of iterations will be reduced from the point obtained by this method.

However, the greatest changes occurred in the reflux ratio of the extractive column and the operating pressure of the regeneration column. For the first, you may notice that by optimizing the system as a whole, the number of stages is increased by one, so it is expected that equipment with as many stages can make the same separation with less reflux. Also the column initial cost increased slightly

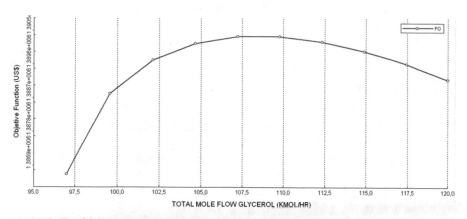

Fig. 7.15 Sensitivity analysis results varying solvent flow on tower TD-10

Table 7.7 Column parameters before and after optimization

Parameter	Initial	Optimum
Solvent to feed molar ratio	0.6	0.6626
Theoretical stages of extractive column	24	25
Theoretical stages of regeneration column	8	7
Reflux ratio of extractive column	0.6	0.3094
Regeneration column bottom pressure (bar)	0.02	0.0322

considering operating costs due to solvent flow. Operating cost reduce by half considering the varying in reflux flow.

For the case of regeneration column, the number of stages reduced. This shows that the operation of the extraction column has improved such that it allows performing the recovery with fewer steps, due to increased solvent flow. Also operating costs are reduced because the pressure that must operate this column is a little higher as in the base case.

7.5   Summary

Aspen Plus and Aspen HYSYS® have powerful optimization tools into their interfaces that allow optimization of both simple and complex systems. The example developed allowed us to observe the recommendations, specifically the development of case studies on process units, are bringing the system to the optimal solution.

In the example, it was possible to see how to optimize the system as a whole, the total system performance increases. So the optimization should be considered an important role of process design aspect, since that improves the operation of each process unit considering its interaction with other operations. This is one reason why the computational tools can help you perform complex calculations related to process engineering.

7.6   Problems

P7.1 What importance has the restrictions in an optimization problem? Which operating criterion is fixed?

P7.2 What is the difference between process and optimization variables?

P7.3 A mixture containing paraffins, from n-C5 to n-C9 is fed to the distillation tower shown in Fig. P7.1 with 25 stages (including the condenser and reboiler) in stage 15 counting from the reboiler. The goal is to adjust the process conditions to obtain a distillate (D) C5 concentrate, a side product (S1) in stage 20 C6 concentrate, a second side product (S2) in stage 10 C7 and C8 concentrate, and bottom product (B) concentrated on NC9. No costs are involved. The operating conditions to be adjusted are the reflux and distillate streams and product side, this achieves the formulation of a nonlinear problem in which the stages of extraction feed and product stages are fixed.

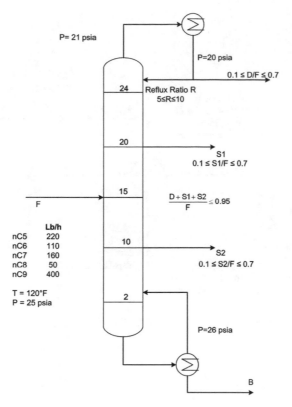

Fig. P7.1 Distillation column with side products

The objective function is:

$$F_{obj} : D_{C5} + S1_{C6} + S2_{C7} + S2_{C8} + B_{C9}$$

The restrictions are shown in Fig. P7.1.

P7.4 In Fig. P7.2 a flow diagram is shown where liquid toluene needs to be heated from 105 to 400 °F, while liquid styrene at 290 °F is cooled to 105 °F. Additionally, there are also included the E-2 and E-3 exchangers, which use steam and cooling water, respectively, to meet the temperature specifications in the event that temperature is not satisfied by E-1. The process should be optimized with respect to minimum approach temperature in exchanger E-1, which should be between 1 and 50 °F. The temperature of the outlet streams from exchanger E-1 must be less than or equal to 200 °F for styrene stream and less than or equal to 300 °F for toluene stream. The annualized cost needs to be minimized regarding the return on investment, r, equal to 0.5. All necessary data are reported in Fig. P7.2 (Adapted from Seider and Warren (2003)).

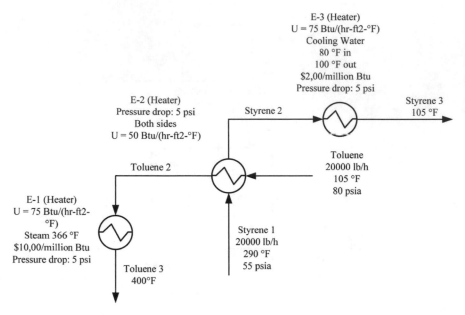

Fig. P7.2 Heat exchanger network described in problem P7.4

References

Aspen Technology, Inc. (2005) Aspen HYSYS simulator help. Aspen Technology, Cambridge

Edgar TF, Himmelblau DM, Lasdon LS (2001) Optimization of chemical processes, 2nd edn. McGraw Hill, New York

Gil I, Chasoy W (2010) Determinación experimental de datos de Equilibrio Líquido-Vapor. Sistema Agua-Glicerina, VI Congreso Argentino de Ingeniería Química

Gil ID, Uyazán AM (2003) Simulación de la deshidratación de etanol azeotrópico por destilación extractiva. Thesis in Chemical Engineering, Department of Chemical and Environmental Engineering, Universidad Nacional de Colombia

Gomez P, Esteban H (2006) El método SQP de optimización con restricciones. XXI Simposium Nacional de la Unión Científica Internacional de Radio

Grossmann IE (2002) Review of nonlinear mixed-integer and disjunctive programming techniques. Optim Eng 3(3):227–252

Langston P, Hilal N, Shingfield S, Webb S (2005) Simulation and optimisation of extractive distillation with water as solvent. Chem Eng Process 44(3):345–351

Lee FM, Pahl RH (1985) Solvent screening study and conceptual extractive distillation process to produce anhydrous ethanol from fermentation broth. Ind Eng Chem Process Des Dev, Phillips Petroleum Company 24(1):168–172

Miller-Klein Associates (2006) Impact of biodiesel production on the glycerol market

Pagliaro M, Rossi M (2010) The future of glycerol, 2nd edn, RSC green chemistry series. Royal Society of Chemistry, Cambridge

Ravagnani MS (2010) Anhydrous ethanol production by extractive distillation: a solvent case study. Process Saf Environ Prot 88(1):67–73

Seider JD, Warren D (2003) Product & process design principles: synthesis, analysis and evaluation, 2nd edn. Wiley, Somerset

Taha HA (1998) Investigación de operaciones, 6th edn. Pearson Education, Harlow

Tarquin AJ, Blank LT (2001) Ingeniería económica. McGraw Hill, New York

Uyazán AM (2006) Producción de alcohol carburante por destilación extractiva: simulación del proceso con glicerol. Revista Ingeniería E Investigación 26(1):45–50

Chapter 8
Dynamic Process Analysis

8.1 Introduction

Dynamic analysis has been increasingly gaining importance over the last few years in the field of process design. Largely because novel process designs are more efficient, if the controllability is considered during the detailed engineering stage. By means of dynamic simulation, it is possible to monitor the behavior of the main process variables when subjected to disturbances typical of an industrial plant operation.

Furthermore, the possibility of suggesting different control strategies and assess their effect on the operability makes possible to study several scenarios in a relatively short time. Generally speaking, new plant designs and control strategies must guarantee product quality, process safety, equipment protection, and compliance to environmental regulations. All these fields can be considered when developing dynamic models of existing processed or new designs.

In this chapter, a series of case studies is presented; in these, some of the above mentioned process parameters are analyzed, willing to provide the reader with the fundamentals for the analysis of future simulations and also aiming to provide a tool to expand the understanding of fundamental process control principles.

8.2 General Aspects

The behavior of a process with time is subject to the nature and the way the system feeds are applied. The study of the dynamic characteristics of a process allows to determine what the best control strategy is considering that many of the systems present in industry are highly nonlinear in nature. The challenge in process control lies in that most of the times there is a dependence of the process response with time; this causes the process variables to experiment a delay on the response time.

© Springer International Publishing Switzerland 2016
I.D.G. Chaves et al., *Process Analysis and Simulation in Chemical Engineering*,
DOI 10.1007/978-3-319-14812-0_8

This dependence with time is known as the process dynamics and it is necessary to be familiar with it before approaching the problem of process control.

The dynamic behavior characteristics of a system, mechanic, chemical, thermic, or electric are defined by any of the following effects: inertia, capacitance, resistance, and dead time.

Consider a system where a tank receives a liquid stream Fin that accumulates until a certain level and an outlet stream at a rate of F_{out}. When developing a mathematical model that allows describing the behavior of the tank level the following is stated:

The mass balance in the tank can be written as:

$$\frac{dV}{dt} = F_{in} - F_{out} \tag{8.1}$$

Then, it can be written as a function of the height h of liquid in the tank keeping in mind that the cross sectional area is constant.

$$A\frac{dh}{dt} = F_{in} - F_{out} \tag{8.2}$$

Likewise, it can be stated that the outlet flow rate F_{out} is defined by

$$F_{out} = \frac{h}{C} \tag{8.3}$$

where:

V: liquid volume in the tank
h: liquid height
A: cross sectional area of the tank
F_{in}: inlet flowrate
F_{out}: outlet flowrate
C: flow coefficient (show the dependence of the outlet flow with the liquid height)

Equation 8.2 can be reordered as:

$$A\frac{dh}{dt} = F_{in} - \frac{h}{C} \tag{8.4}$$

$$A\frac{dh}{dt} + \frac{h}{C} = F_{in} \tag{8.5}$$

$$A\frac{dh}{dt} + \frac{h}{C} = F_{in} \tag{8.6}$$

Equation 8.6 corresponds to the general first-order differential equation, it can be written in terms of a steady state gain K and a time constant τ. Having:

$$\tau \frac{dy}{dt} + y = K_u(t) \tag{8.7}$$

where $y(t)$ is the output function and $u(t)$ is the process input that when modified affects the output. As shown in Fig. 8.1

In this way, (8.6) can be finally written as

$$\tau \frac{dh}{dt} + h = KF_{in} \tag{8.8}$$

where $\tau = AC$ and $K = C$.

It can be seen that the time constant τ is determined by the cross sectional area of the tank, in other words, by the geometry of the tank. Also, the gain of the process is given by the characteristics of the outlet valve of the system.

In general the process gain K is defined as the relationship between the change in the process output and a change in the process input. In this example F_{in} corresponds to the input and the height h to the output. In steady state, the first term of (8.8) equals zero and it can be solved for K.

The time constant represents the response speed of the system. For first-order systems, when a step type disturbance occurs, there is a time interval τ in which the process variable changes 63 % of the total change, see Fig. 8.2.

Process dead time (θ) of a process is defined as the time between a change in an input variable and the detection of a change in the output variable. In some systems, the change in the output variable when a disturbance occurs is almost instantaneous thus the dead time is close to zero. However, with most of the physical and chemical systems dead times can be significant depending on the process nature and the location of the measurement instruments. Frequently dead time is caused by mass and heat transfer gradients in different points of the process.

8.2.1   Process Control

In the processing industry, there is an increasing interest for properly standardized processes as well as for ensuring product quality. Furthermore, environmental regulations and process safety require proper control mechanisms that allow for the fulfillment of these requirements.

Given the ever changing economic conditions and the competitiveness of the market, process control becomes important to provide efficient processes and high quality products. The operability of more integrated processes with fewer degrees of freedom makes room for error smaller. Hence, more efficient control strategies are required (Luyben2002; Perry1992).

Take a fermentation reactor as an example, initially it is required that all the variables are monitored to verify correct operation: Temperature, pH, dissolved oxygen, liquid level, feed flow rate and agitation speed; all these variables, in a way,

Fig. 8.1 Scheme of a
reservoir tank for liquids

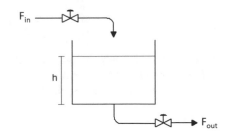

Fig. 8.2 First-order system
response

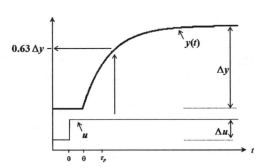

define the reaction performance. In this case it is required to keep the reaction temperature (controlled variable) at a specific level (set point) to control the reaction rate. This temperature can be adjusted by modifying the cooling water flow rate that flows along the reactor jacket (manipulated variable). In a conventional control loop, temperature is measured by a sensor that sends a signal to a controller. In the controller, a control algorithm determines the action to take by comparison between the measured value and the set point. The difference between these values is known as error. The execution of the action is applied by the final control element, usually a control valve. This element receives the information from the controller and changes the valve opening percentage. For the above example a control valve would regulate the cooling water flow rate to the reactor.

In the case of the temperature controller, detecting the main four elements of a control loop is not hard: the sensor or primary element, the transmitter or secondary element, the controller and the final control element. These final control elements can be speed regulators for pumps, electrical motors, electrical heating elements, etc.

Some relevant terms for process control, fundamental for the definition of a control loop are described next.

Process: A set of equipment and operations limited by a boundary; together with the corresponding material and energy streams flowing between vessels as well as the ones crossing the boundary. In the reactor example, valves, the reaction vessel, the stirring system and all the piping are part of the process.

Disturbance variable: Part of the input variables group is a variable that it is not manipulated and its changes cause an effect on the process performance. In the example, the temperature at which the reacting mixture enters the reactor can be considered a disturbance variable.

Controlled variable: Part of the output variables group, it is used to verify the desired operation of the process. For example, the product temperature or the liquid level in the reactor.

Set point: It is the desired value for the controlled variable.

Manipulated variable: Part of the input variables group, used to adjust a controlled variable and drive it to the set point. In the fermenter, the manipulated variable is the cooling water flow.

Final control element: A device that adjusts the manipulated variable.

Measured variable: Any variable recorded over time, usually a controlled variable.

Sensor: Measurement instrument. Usually grouped by pressure flow, level, temperature, or composition. It measures a variable over time.

Controller: A device that takes the information from the sensor and processes it to determine the deviation of a variable from the set point and based on that sets an action for the final control element to carry out, for instance opening or closing a valve. Its function is based on the control algorithm supplied.

Open-loop operation: Also known as manual operation, when the controller is not providing feedback to the final control element. Hence no action can occur to adjust the controlled variable.

Closed-loop operation: Operation where the controller is connected and takes decisions executed by the final control element.

8.2.2   Controllers

Feedback control is the traditional and most widely used type of control, it is characterized by its simplicity and versatility. In this algorithm, a controlled variable is kept at the desired value according to the calculated error value. The main advantage of feedback control is that it compensates any disturbance present in the process, its main disadvantage is that the control action is only done when the disturbance has occurred and deviations may have propagated in the process (Fig. 8.3).

The equation or a feedback controller PID is given by:

$$OP(t) = K_C E(t) + \frac{K_C}{T_i} \int E(t) dt + K_C T_d \frac{dE(t)}{dt}$$ (8.9)

where

OP(t): controller output at t
$E(t)$: error at t

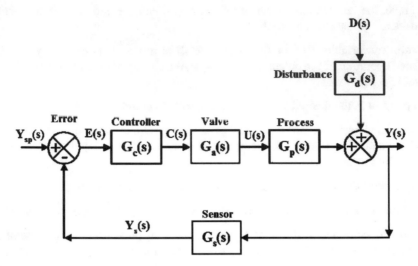

Fig. 8.3 Block diagram of a feedback control loop

K_C: proportional gain of the controller
T_i: integral time
T_d: derivative time

Each element of the control loop impacts the controller performance. Particularly, the final control element determines the effectiveness of the mitigation of the process deviation. The control valve is the most common final control element. It counts with an orifice of variable restriction that allows for the control of the fluid by manipulation of the pressure drop. The flow rate through a valve is also dependant on the valve type and opening percentage (Fig. 8.4).

The relationship between the flow rate and the valve opening is known as the *characteristic curve* of the valve. There are three main types of valves: fast opening, equal percentage or linear. A fast opening valve allows high flow rates at low opening percentages; the flow rate increase is lower with further increases on the opening percentage. An equal percentage valve allows low flow rates at low opening percentages and the highest flow increases occur when the opening is close to 100 %. In a linear valve there is a direct correlation between the opening and the opening percentage. Figure 8.4 shows the characteristic curves of these valves.

A PID feedback controller uses three tuning parameters that must be adjusted to obtain a satisfactory performance and an operation within the admissible ranges of a specific system. When tuning a controller, it is important to know the control objective as well as the existing restrictions e.g., error limits, response time, acceptable transient state time, among others.

Controller tuning proposes a set of parameters that are suited for a narrow and very precise control of the variable, at the expense of rough and sharp changes in the manipulated variable (close to the instability zone) or that allows the variable to be

Fig. 8.4 Valves
characteristic curves

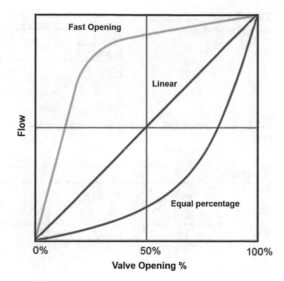

somewhat apart from the set point value (higher variability of the controlled variable) but with enough robustness to respond to different disturbances.

Some rules of thumb in this regard are:

If a steady state error (offset) in the controlled variable is acceptable, the use of a proportional (P) controller is advised.

If the system has signal noise or dead times, the use of a proportional-integral (PI) controller is recommended.

If the signal noise is negligible, a proportional-integral-derivative (PID) controller is recommended.

The selection of tuning parameters is also function of the controlled variable. Experience shows that, for example, flow control responds rapidly due to the proximity between the measurement and the final control element. Moreover, a flow rate signal is usually accompanied by noise; suggesting the use of a PI controller with a low proportional gain.

On the other hand, liquid level control can be approached in two different ways: When dealing with a system like a buffer tank where its purpose is to control disturbances to the process, strict level control is not required and a proportional controller is enough. When precise level control is required a PI control is recommended.

Pressure control in a liquid is analogous to liquid flow control, as well as pressure control of a gas is analogous to liquid level. Hence, the same recommendations made above apply for this case.

Finally, temperature control in industry is widely done with PI control. Nevertheless, when the feedback loop is slow, it is possible to include the derivative action to enhance the response time. Table 8.1 shows some recommended tuning parameters for common control variables in process engineering.

Table 8.1 Typical control
tuning parameters

Variable	K_c	T_i [min]	T_d [min]
Flow	0.4	0.3	0
Level	2	2	0
Pressure	2	10	0
Temperature	10	20	0

Table 8.2 Summarizes the open-loop tuning rules for the Ziegler-Nichols and Cohen-Coon
methods

Controller	K_c		T_i [min]		T_d [min]	
	ZN	CC	ZN	CC	ZN	CC
P	$\frac{\tau}{Kp\theta}$	$\frac{\tau}{Kp\theta}\left[1+\frac{\theta}{3\tau}\right]$	–	–	–	–
PI	$\frac{0.9\tau}{Kp\theta}$	$\frac{\tau}{Kp\theta}\left[0.9+\frac{\theta}{12\tau}\right]$	3.3θ	$\frac{\theta[30+3\theta/\tau]}{9+20\theta/\tau}$	–	–
PID	$\frac{1.2\tau}{Kp\theta}$	$\frac{\tau}{Kp\theta}\left[\frac{4}{3}+\frac{\theta}{4\tau}\right]$	2θ	$\frac{\theta[32+6\theta/\tau]}{13+8\theta/\tau}$	0.5θ	$\frac{4}{11+2\theta/\tau}$

Methods for controller tuning are classified into two groups; open and closed
loop. Open-loop methods consist in setting the controller to manual mode and cause
a step type disturbance. By doing so, and with the assumption that the process is
approximately first order with dead time, the process gain, time constant, and dead
time are calculated. With these parameters, the gain, integral time, and derivative
time are calculated. The calculation of the PID parameters is based on one of the
following tuning rules: Ziegler-Nichols, Cohen-Coon, IMC, IAE, ISE, or ITAE
(Table 8.2).

In the closed-loop tuning methods the first task is to obtain the ultimate gain
value as well as the ultimate period of oscillation. Initially the integral and derivative actions are disabled. Then, the gain is increased until the response oscillates
with a constant amplitude. When this condition is achieved, the ultimate gain and
ultimate oscillation period are obtained.

Among the existing tuning methods, one of the most widely used due to its
simplicity and quickness is the Auto Tuning Variation (ATV). This method consists
in generating an oscillatory cycle between the manipulated and controlled variable.
The procedure is as follows:

1. Determine a reasonable value for the variation of the opening percentage of the
   valve (h% = percentage of the change of valve position). Usually this value is 5–
   10 % of the nominal controller output.
2. Perform a change in the negative direction −h%.
3. Wait for the controlled variable to change and then make a positive change +2 h%.
4. When the process variable crosses the set point value, make a negative change
   −2 h%.
5. Continue making alternate changes every time the variable crosses the set point
   value until a cycle is obtained.
6. Determine the response amplitude a, and the oscillation period Pu.

Fig. 8.5 Closed-loop controller tuning using ATV method

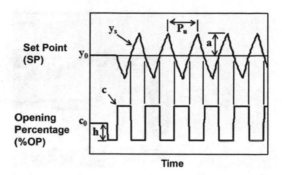

7. Determine the tuning parameters as:

$$\text{Ultimate gain}: \ K_U = \frac{4h}{a\pi} \tag{8.10}$$

Ultimate Period: As obtained above.

$$\text{Controller gain}: \ K_C = \frac{K_U}{3.2} \tag{8.11}$$

$$\text{Integral time}: \ T_i = 2.2 \times P_U \tag{8.12}$$

Figure 8.5 shows the ATV tuning method graphically.

8.3 Introductory Example

Establishing a control system of a process demands the appropriate selection of manipulated-controlled variable pairs. Occasionally, identifying the interaction in each pair is not trivial and requires the development of dynamic models to assess it. Moreover, the inclusion of some control loop elements such as sensor dead time, control valve characteristics, and control algorithm parameters are important for the dynamic analysis done by dynamic process simulation.

In the following example, the methodology to follow for the development of a dynamic state simulation using Aspen Hysys Dynamics® is presented. Through this test a series of different steady state applications in Aspen Hysys® have been studied. Now, the required steps for the execution of a dynamic state simulation and the setting of simple feedback control loops are shown.

The process comprises two phase separators connected by a couple of heat exchangers as shown in Fig. 8.6. The system is fed with a hydrocarbon mixture at a rate of 3200 lbmol/h, pressure of 900 psia, and a temperature of 32 °F. Mixture composition is provided in Table 8.3.

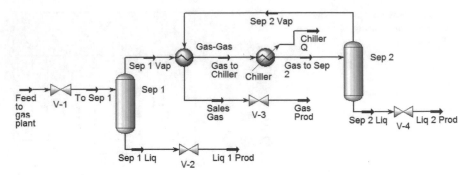

Fig. 8.6 Process flow diagram for the hydrocarbons mixture separation

Table 8.3 Molar composition of the *To gas plant* stream

N₂	H₂S	CO₂	C1	C2	C3
0.0066	0.0003	0.0003	0.7576	0.1709	0.0413
i-C4	n-C4	i-C5	n-C5	C6	H₂O
0.0068	0.0101	0.0028	0.0027	0.0006	0.0000

Both separators operate adiabatically and their pressure drop is negligible. The valve *V-1* in the feed stream has a pressure drop of 4 psi. For the gas–gas heat exchanger, the model used is *Simple Weighted*. The inlet to the tube side is *Sep 1 Vap* and the outlet *Gas to Chiller*, the pressure drop is 5 psi. The inlet to the shell side is *Sep 2 Vap* with unknown conditions and the outlet is *Sales Gas*, the pressure drop is 1 psi. In this heat exchanger a design specification must be provided, a minimum temperature difference approach of 10 °F. The *Chiller* heat exchanger has a pressure drop of 5 psi and requires the outlet temperature to be −4 °F. This information is sufficient to run the steady state simulation. The only step missing is to install the valves *V-2*, *V-3*, and *V-4*.

8.3.1 Dynamic State Simulation

The dynamic state simulation configuration requires some adjustments to guarantee flow-pressure conditions in the process. Aspen HYSYS Dynamics® verifies that there is a pressure difference that guarantees flow through the different vessels. That is why it is necessary to install valves that create the pressure difference and also regulate the flow rate. Install *V-2*, *V-3*, and *V-4* as per Fig. 8.6. Define the outlet pressure of *V-2* and *V-4* as 875 psia and of *V-3* as 870 psia.

Subsequently, the selection and sizing of the control valve is necessary. Aspen HYSYS Dynamics® includes information from some control valve manufacturers as characteristic curves and correlations for the flow coefficient. This information is available in the valve configuration window *Rating > Sizing (dynamics)*. Figure 8.7 shows the specification and results of valve *V-1*. All four valves are set as linear and

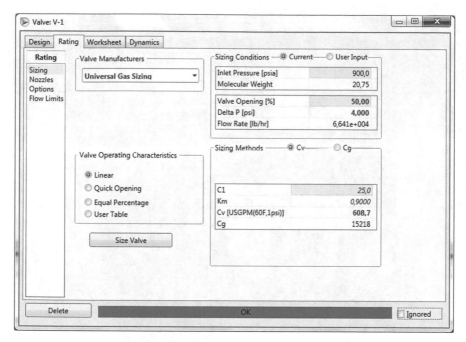

Fig. 8.7 Specification and results for valve *V-1*

the manufacturer selected is *Universal Gas Sizing*. The opening percentage is set to 50 % and then the valve is sized by clicking on *Size Valve* so that the flow coefficient is computed (C_v or C_g).

The definition of the phase separators volumes is also required; since this parameter sets the time constant, which in turn affects the response time of the system.

Vessel volume is supplied in the tab *Dynamics > Specs* in the option *Vessel Volume*. In this case, both separators have a volume of 70 ft^3. Note also that the dimensions (length, diameter) can also be supplied, if these are not available the simulator assumes a length to diameter ratio and back calculate the dimensions. Finally, an initialization option for the separators is set. For the *Sep 1* choose *Initialize from Products*, for the *Sep 2* select *Dry Startup*, as shown in Fig. 0.0.

Now, the flow rate specification of the feed stream needs to be removed. This value is not fixed anymore and is determined by the pressure drop through *V-1*, defined by the opening percentage. To remove this specification go to *Dynamics > Specs > Flow Specification* and uncheck the *Active* checkbox, Fig. 8.9. This procedure is also done automatically by the simulator when the dynamic mode is active.

Finally, the process control loops are created. For this example, a level control for *Sep 1* and a pressure control for *Sep 2* are installed. The level control is tied to *V-2* as the manipulated variable, and the liquid level percentage as controlled variable. The pressure control is tied to *V-3* as manipulated variable, and vessel pressure as controlled variable. Both controllers are direct action, and the default set

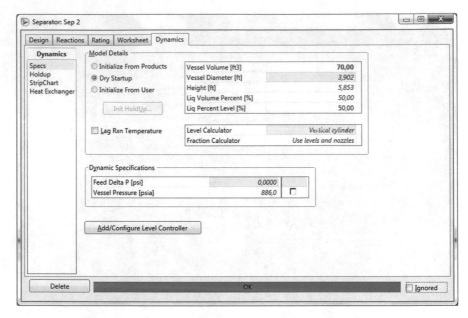

Fig. 8.8 Dynamic parameters definition for the Sep 2 vessel

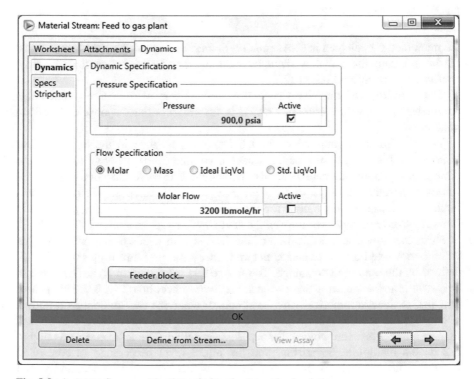

Fig. 8.9 A stream flow specification window for dynamic simulation

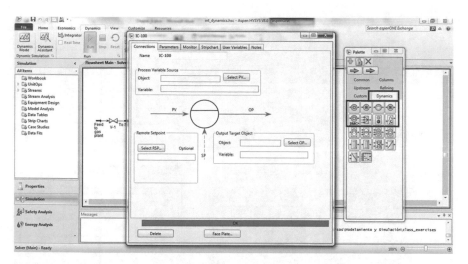

Fig. 8.10 Installation and configuration of a controller in Aspen HYSYS Dynamics®

points correspond to the steady state values for the variables. The high and low limits are user defined, for the liquid level set 0–100 % as limits, and for the pressure control set 800–950 psia.

The level controller has a proportional gain of 2 and an integral time of 5 min. The pressure controller has a gain of 2 and an integral time of 2 min.

A control loop setting starts with the selection of an appropriate control model. At the bottom of the simulator toolbar, there is an icon called *Control Ops*, by clicking on it, the five available control models are displayed: *Split Range, Ratio, PID, MPC (multivariable predictive control), and DMC (Dynamic Matrix Control)*. Click on the PID option and install it on the flowsheet as shown in Fig. 8.10. Then, click on the icon to open the controller configuration window. In the first tab, *Connections*, input the information related to the process variable PV (*Process Variable Source*) and the controller output signal OP (*Output Target Object*). On the *Parameters* tab, the values for proportional gain, integral time, and derivative time, maximum and minimum values, controller action, and operation mode, are supplied. By doing so, the control loops are configured and the *Face Plate* where the *PV, SP*, and *OP*, can be visualized. Also, in the *Stripchart* tab a trend graph can be configured.

After the entire configuration is completed, arrange the screen to simultaneously visualize all the information available, as shown in Fig. 8.11.

The next step is to click the *Dynamics Assistant*, available in *Dynamics* tab, to verify that all required parameters for a dynamic simulation are met. The result is shown in Fig. 8.12. Note that the simulator detects necessary changes and suggests to perform these automatically. Note that the first suggested change, is to disable the flow specification on streams and the third is to assume volumes for the separation vessel. This information has been supplied beforehand. The other two suggestions expedite the pressure calculation in the simulation. Click on *Make Changes* and then click on the dynamic mode button in the toolbar.

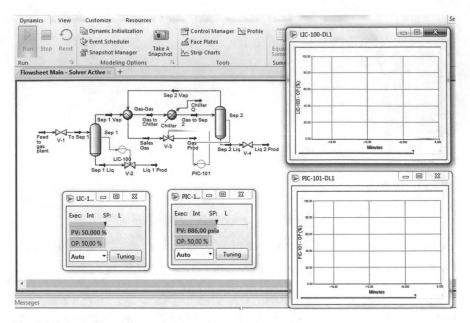

Fig. 8.11 Control loops for the hydrocarbon separation system

Fig. 8.12 Results from the *Dynamic Assistant in* Aspen HYSYS Dynamics®

When the simulation is changed to dynamic mode, the integrator is deactivated waiting to establish the zero time of the simulation. Since every parameter is already set, the integrator can be activated. Immediately, the trend lines start showing the progress of the simulation as a steady behavior over time. As a next step, make changes in the set point of the level controller to observe the system response over time, as well as the effect that the tuning parameters have on it. Modify the level controller set point to 60 % and wait for stabilization. Then, take it back to 50 % and again wait for stabilization Fig. 8.13 (a). In order to observe the effect of the integral time on the control, stop the integrator and change the integral time to 1.5 min. Perform a change in the set point to 60 %. It is evident that the response of the controlled variable is more oscillatory (Fig. 8.13 (b)), and hence, generates a higher instability in the control loop compared to the original settings. This is attributed to the short integral time causing that at short time intervals, the proportional action is duplicated. Finally, change the integral time setting to 50 min and perform the change in the set point; now, the response is overdamped, the oscillation disappears and the stability is reached faster (Fig. 8.13 (c)).

Now, consider the effect of changes in the set point in the pressure controller. Change it from 886 to 875 psia. The control valve opens quickly to knock the pressure down. However, after reaching 100 % opening, the pressure is not reduced. After some time, the valve opening and the pressure start fluctuating (Fig. 8.14); this behavior is due to the saturation of the valve due to a small pressure drop across V-3. To eliminate the oscillation increase the set point value to 880 psia; this causes the valve to close and regulate the pressure, Fig. 8.14.

An introduction to the Aspen HYSYS® dynamic tools has been illustrated with the previous example, showing the effect of the controller tuning parameters and pressure drop across a valve on the stability of the control loop. Ahead in this chapter, additional case studies are developed to elaborate on some other concepts of dynamic process analysis.

8.4 Gasoline Blending

Currently, some countries are implementing policies towards the reduction of fossil fuel consumption. One of these policies, is the ethanol alternative as a component in gasoline blends. By blending ethanol in the gasoline the use of fossil fuel is reduced and a cleaner combustion takes place. Let us develop a simulation example for the blending of these two substances.

8.4.1 Steady State Simulation

This simulation comprises a mixing tank as the main process vessel. Additionally, control valves are included on the inlet and outlet streams, to guarantee the flow-pressure ratios and establish control loops. The process flowsheet is shown in Fig. 8.15.

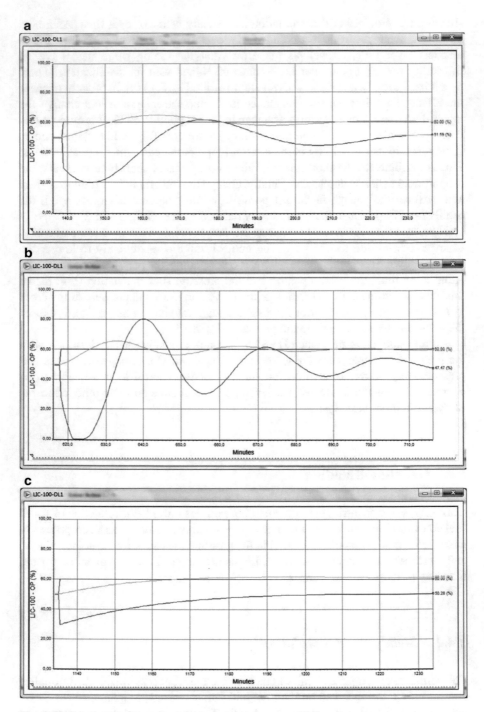

Fig. 8.13 Level control response with a setpoint change to 60 %

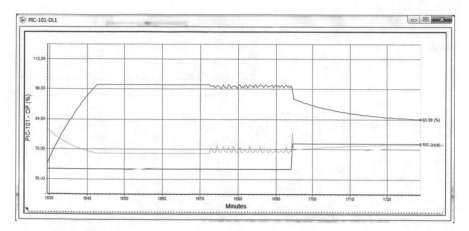

Fig. 8.14 Pressure response with a setpoint change for the pressure controller of the *Sep 2* separator

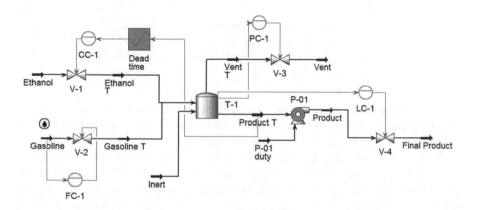

Fig. 8.15 Process flowsheet for the gasoline blending process

The components included in the simulation are: ethanol, an inert gas to blanket the tank and avoid oxidation (nitrogen), and gasoline, as a hypothetic hydrocarbon mixture defined by its distillation curve. The ethanol stream has a small water content that also needs to be accounted for.

The property package to be used is Wilson-Ideal. Finally, the distillation curve information for regular gasoline, available from Ecopetrol S.A. has to be provided to the simulation, Table 8.4.

In order to supply distillation curve information to Aspen HYSYS®, the *Properties* section has in the main menu the *Oil Manager* tab in order to define information about a new assay. The window shown in Fig. 8.16 appears. Click on the *Input Assay* button.

Table 8.4 ASTM D-86 distillation curve for regular gasoline

% distilled volume	Temperature [°C]
10	77
50	121
90	190
100	225

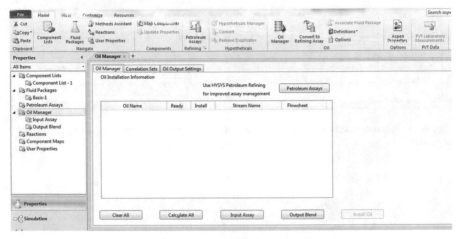

Fig. 8.16 *Oil Manager* window in Aspen HYSYS®

In the *Input Assay* window (Fig. 8.17), petroleum streams can be specified by providing physical properties information; such as, light end composition, distillation curves, viscosity, density, molecular weight, etc.

In this window, click on *Add…* to include a new assay that represents a petroleum stream; another window appears where the type of information to be provided is defined. By this method, information of several oil streams can be provided to simulate oil blending or any other operation involving crude oil products.

In the *Assay Data Type* cell select *ASTM D86*, this opens four additional options:

- *Light Ends*: This is the stream fraction composed by light compounds. Usually C1–C4 gases as well as CO_2, N_2, among others.
- *Molecular Wt. Curve*: To include the average molecular weight dependence on temperature. This information is seldom available.
- *Density Curve*: Curve of the density behavior with temperature.
- *Viscosity Curves*: Curve of the viscosity behavior with temperature. Usually available.

For this example, given that no additional information is known, select *Not Used* for each of them, except for light ends which are specified as *Ignore*.

The distillation curve information is displayed on the right side of the window. Click on *Edit Assay* and supply the distillation profile information. Then, click *OK* (Fig. 8.18).

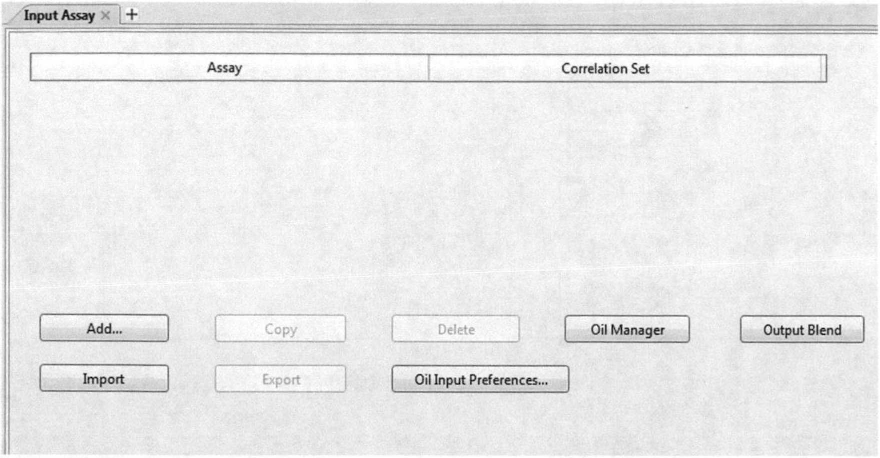

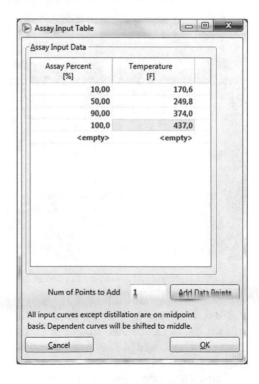

Fig. 8.17 *Input Assay window* in Aspen HYSYS®

Fig. 8.18 Distillation curve data input in Aspen HYSYS®

Then, click on the *Calculate* button in the *Assay* window (Fig. 8.19). In this manner the gasoline specification is complete. Other tabs in this window allow to check the calculated properties.

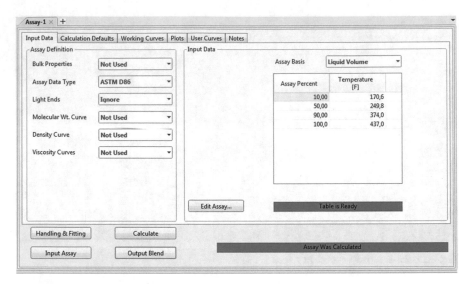

Fig. 8.19 Assay window in Aspen HYSYS®

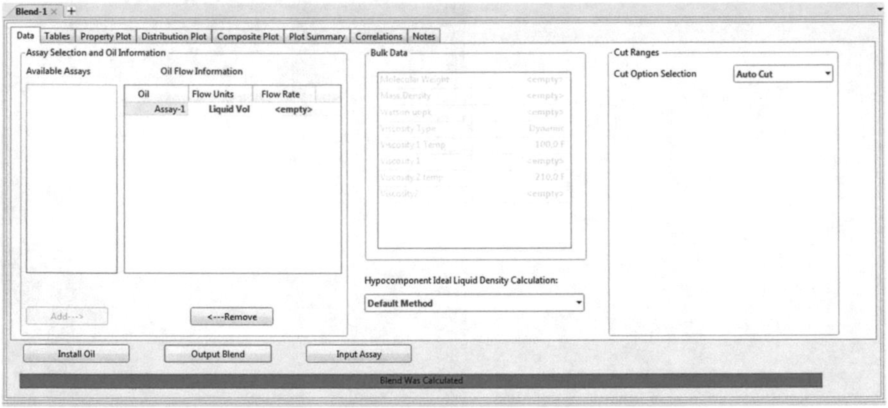

Fig. 8.20 Cut/Blend window in the Oil Environment, Aspen HYSYS®

Now, click on the *Output Blend* button and a window where the blending information of the defined assay can be provided appears. In this window, create a new cut and assign the *Assay-1* previously configured by clicking on *Add* (Fig. 8.20).

Now click on the *Install Oil* button to add the blend to an actual material stream in the flowsheet. In the window that is opened (Fig. 8.21), type *Gasoline* in the *Stream Name* column and then in the *Install* button. In this way, when entering the simulation environment the *Gasoline* stream appears, containing the properties calculated in the *Oil Manager*.

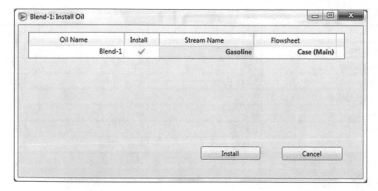

Fig. 8.21 *Install Oil* window of the *Oil Manager* in Aspen HYSYS®

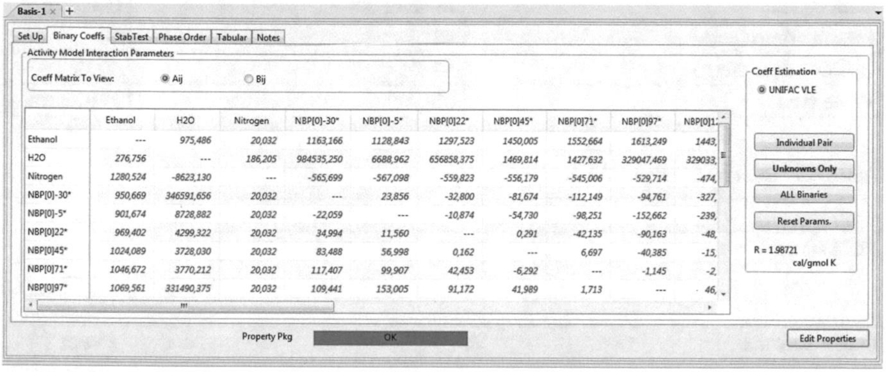

Fig. 8.22 Binary parameter estimation in Aspen HYSYS®

Finally, click on the *Basis-1* folder under *Fluid Packages* to estimate the thermodynamic model missing parameters. This is done after defining the *Gasoline* Stream in order to include it in the calculation of binary coefficients.

Next, the missing parameters are estimated. For this purpose, in the property package selection window click on the *Binary Coeffs* tab where the binary coefficients are displayed. The missing parameters can be estimated using the UNIFAC method. Click on the *Unknowns Only* button. The window shown in Fig. 8.22 is displayed.

Then, the feed streams can be defined according to the information in Table 8.5.

The valves associated to the *Ethanol* and *Gasoline* streams are defined; their specifications are outlined in Table 8.6.

Remember to click on the *Size Button* when specifying the valves.

The parameters for the tank and pump are shown in Tables 8.7 and 8.8.

Once the tank and pump are defined, the rest of the control valves can be specified and sized as well (Table 8.9).

Table 8.5 Feed stream conditions for the gasoline blending example

	Ethanol	Gasoline	Inert
Temperature [°F]	95	95	95
Pressure [psia]	150	150	150
Flowrate	5221 lb/h	46,650 lb/h	25 lbmol/h
Mass composition			
Nitrogen	0	–	1
Ethanol	0.997	–	0
Water	0.003	–	0

Table 8.6 Control valve specifications

Name	*V*-1	*V*-2
Inlet stream	Ethanol	Gasoline
Outlet stream	Ethanol T	Gasoline T
Pressure drop [psi]	40	40
Opening percentage	50 %	50 %
Valve type	Linear	Fast opening

Table 8.7 Blending tank conditions

Name		T-1
Feed streams		Ethanol T
		Inert
		Gasoline T
Outlet streams		Vent T
		Product T
Pressure drop		0 psi
Volume		1000 gal
Vessel disposition		Vertical

Table 8.8 Process pump specification

Name		P-01
Feed stream		Product T
Outlet stream		Product
Pressure increase		80 psi

Table 8.9 Control valve specifications

Name	*V*-3	*V*-4
Feed stream	Vent T	Product
Outlet stream	Vent	Final product
Pressure drop [psi]	90	40
Opening percentage	20 %	50 %
Valve type	Linear	Linear

Table 8.10 Dead time parameters for the composition control loop

Name	Dead time
Input element	Ethanol mass fraction, Product T
Output element	–
PV range	0–0.25
OP range	0–0.25
K	1
Dead time [min]	3

Table 8.11 Controller parameters

Name	FC-1	CC-1	PC-1	LC-1
Input element	Mass flow, gasoline	Output signal (PV) dead time	Tank Pressure, T-1	Tank level, T-1
Output element	Valve V-2	Valve V-1	Valve V-3	Valve V-4
Set point (SP)	46 650 lb/h	0,1 (mass fraction)	110 psia	50 %
PV minimum	20 000 lb/h	0.0	50 psia	0 %
PV maximum	65 000 lb/h	0.25	200 psia	100 %
Action	Reverse	Reverse	Direct	Direct
K_C	0.5	1	2	2
T_i [min]	0.3	–	10	–

With the supplied data, most of required information for the steady state simulation has been provided. However, before moving to the dynamic state simulation, the controller parameters and dead time have to be specified. For this case a dead time of 3 min is assumed corresponding in the delay caused by the composition analysis and feedback.

Care must be taken during the configuration of the transfer function. Ethanol mass composition in the *Product T* stream must be set as input both in the transfer function and the PID controller. In order to specify the dead time, go to *Delay* in the *Parameters* tab of the transfer function window. Enter the parameters shown in Table 8.10 and make sure that $G(s)$ *enabled* is active, this activates the transfer function.

After specifying the transfer function, complete the setting on the remaining controllers with the parameters shown in Table 8.11. It is important to note that the *CC-1* controller input corresponds to the dead time module output as proves value (PV). In this way the 3 min delay in the composition is effective (Fig. 8.23).

Click on the *Dynamics Assistant*, no errors should appear at this point (Figs. 8.24 and 8.25).

Make the prompted changes and start the dynamic mode.

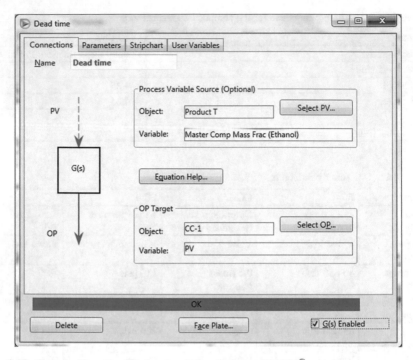

Fig. 8.23 *Transfer Function Block* main window in Aspen HYSYS®

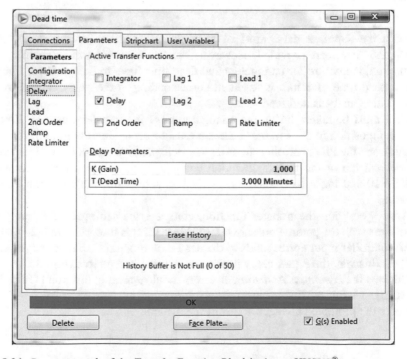

Fig. 8.24 *Parameters* tab of the *Transfer Function Block* in Aspen HYSYS®

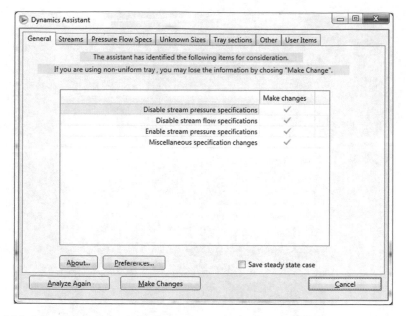

Fig. 8.25 *Dynamics Assistant* window

8.4.2 Dynamic State Simulation

In order to start the dynamic simulation, arrange the display of the trend graphs of the controlled variables (liquid level, pressure, and product composition) as well as the graphs of the controllers (*CF-1 and CC-1*). Then set the action of all controllers to automatic and start the integrator.

The composition controller does not have with tuning parameters yet. Aspen HYSYS® has implemented an algorithm to automatically tune the controller. This is located in the *Autotuner* option in the *Parameters* tab (Fig. 8.26).

Make sure the integrator is active and click on *Start Autotuner*. Right after, the trend graph showing the oscillation and the calculated parameters are displayed. In this case, the tuning is for a PID controller (Fig. 8.27; Table 8.12).

8.4.3 Disturbances

For this example, three different disturbances affecting the controlled variables are done. These are:

As shown in the Figs. 8.28, 8.29, and 8.30, the disturbances are readily assimilated by the system. This is a consequence of proper controller tuning. The velocity at which the disturbances are controlled depends on two parameters: First, a high proportional gain increases the proportional and integral action of the controller. Second, a small integral time value magnifies the integral action of the controller. This joint effect achieves a quick response to the disturbances shown earlier.

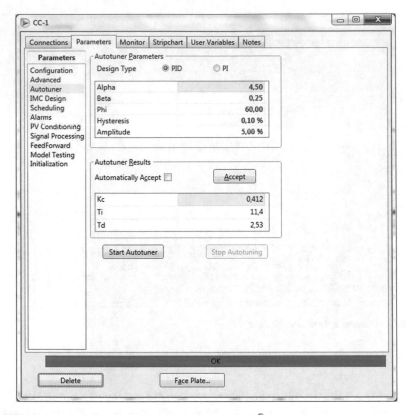

Fig. 8.26 *Autotuner* in the *Parameters* tab in Aspen Hysys®

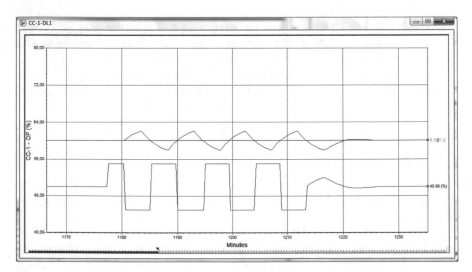

Fig. 8.27 Tuning of the ethanol composition controller CC-1

Table 8.12 *CC-1* Controller tuning parameters

Parameter	Values
K_C	0.412
T_i (min)	11.4
T_d (min)	2.53

Note: The tuning parameters obtained may be different

a

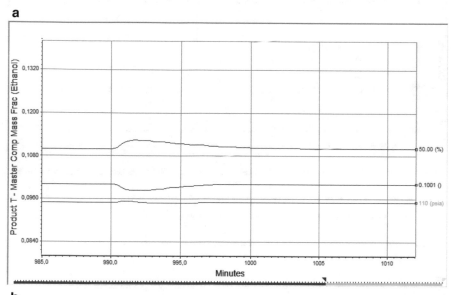

b

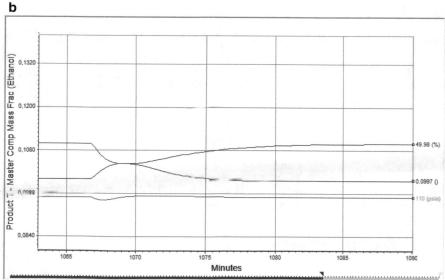

Fig. 8.28 Controller response for a +20 psi (**a**), and −20 psi (**b**) disturbance in the *Gasoline* stream. *Red*: liquid level, *blue*: concentration, *green*: pressure

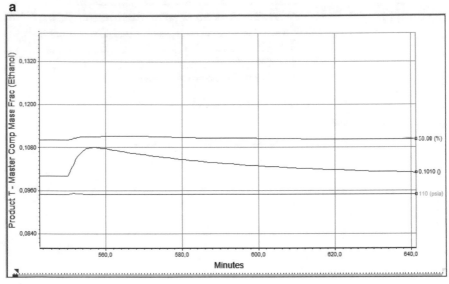

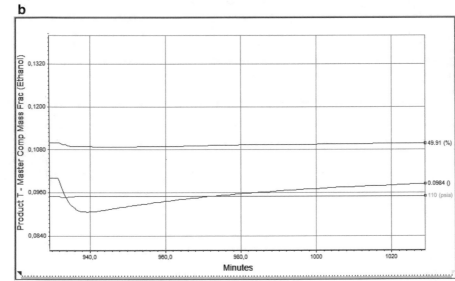

Fig. 8.29 Controller response for a +10 psi (**a**), and −10 psi (**b**) disturbance in the *Ethanol* stream. *Red*: liquid level, *blue*: concentration, *green*: pressure

8.4.4   Recommendations

When dealing with a dynamic simulation the results are more sensitive than with a steady state model. To obtain coherent results the following recommendations are advised.

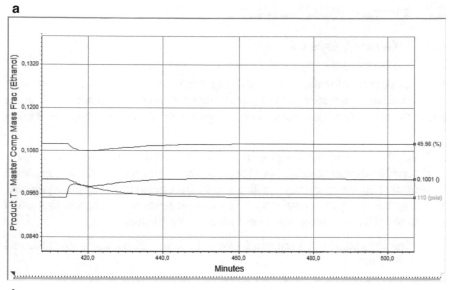

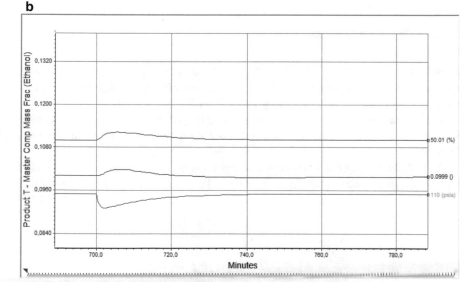

Fig. 8.30 Controller response for a +5 lbmol/h (**a**), and −5 lbmol/h (**b**) disturbance in the *Inert* stream. *Red*: liquid level, *blue*: concentration, *green*: pressure

- Perform adequate selection and sizing of the control valves before jumping to steady state simulation.
- Check the consistency of the control loops, specially the controlled-manipulated variable couples.
- When making changes in existing controllers make small changes so that the transition is smooth.

8.5 Pressure Relief Valves

8.5.1 General Aspects

With the purpose of avoiding explosions or equipment damage in case that over-pressure occurs in a system, relief valves are used, these allow the release of excess gas to the atmosphere or another piece of equipment. There are two types of relief valves:

- Relief valve: Its opening is proportional to the pressure increase. Excess gas is diverted to another vessel to keep the pressure within operation limits. Operation pressure remains constant yet with a small offset from the steady state pressure.
- Safety valve: Opens completely when the set pressure is reached. Operation pressure fluctuates and is far from the steady state value.

Further information is available in the API 520 recommended practice, and related literature.

8.5.2 Application Example

In order to explain the use of relief valves in Aspen HYSYS®, let us continue with the gasoline blending simulation. In this flowsheet disconnect the stream *Vent T* from *V-3*, and install the object *relief valve* from the palette (Fig. 8.31).

Part of the relief valve setting involves the sizing of the orifice that would allow for the gas release in case of a valve failure or a sudden increase in the inert flow rate.

The simulator requires information of the release flow rate, as well as the inlet and discharge pressure to size the relief valve. To do this, divert the entire flow rate

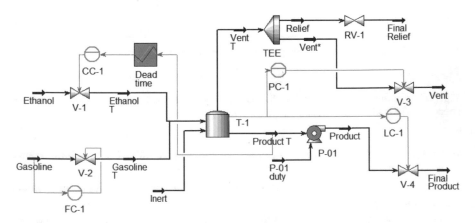

Fig. 8.31 Simulation flowsheet including the relief valve

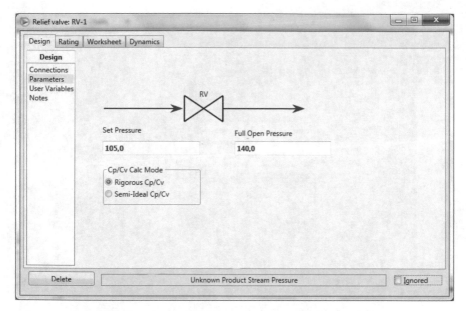

Fig. 8.32 Information required by relief valve

in the Tee (*TEE*) to the *Relief* stream. With this setting, the valve can be properly sized after providing the information required by the module, Fig. 8.32.

To achieve an adequate valve sizing, the following condition must be met.

- The relief valve must be open. A pressure below the operation pressure must be set in the *Set Pressure* field.
- The release flow must be sent to the relief valve.
- The discharge pressure of the relief valve must be provided, this information depends on the release destination (ambient, pressurized tank, etc.).

When defining the *Set Pressure* and *Full Open Pressure* properly, the simulator prompts a warning, this is due to a conflict of the valve being designed to operate fully closed, and the requirement of full opening for dimensioning.

In the *worksheet* tab specify a pressure of 29 psia to the *Final Relief* stream. Once the three conditions are met, the sizing of the device can be checked in the *Rating* tab, as shown in Fig. 8.33.

The held *Orifice Area* shows the calculated orifice diameter that meets the relief requirement. The next step, is to compare the obtained result with the API 526 standard to determine the letter associated with this orifice size. In this specific case, the size is *E*.

Now that the sizing procedure is completed, assign the correct *Set Pressure* to the valve, this pressure is 130 psi. Also, revert the diversion of the *Vent T* stream to the relief valve, by adjusting the setting in the Tee (*TEE*).

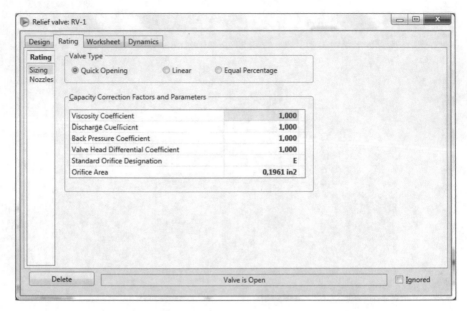

Fig. 8.33 Sizing of the relief valve in the *Rating tab* in Aspen HYSYS®

In the relief valve, select the *E* letter in the *Standard Orifice Designation* of the *Rating* Tab. The simulation is ready to be carried out in the dynamic mode. The dynamic assistant should not generate warnings at this time.

8.5.3   Dynamic State Simulation

In order to demonstrate the performance of the relief valve, a failure in the tank pressure control valve is simulated. To cause it, go to the *Dynamics* Tab of the *V-3* valve. In this window, click on the actuator section; the options shown in Fig. 8.34 are shown.

In this section, information regarding the valve actuator can be entered. For the example purpose, select the *Fail Shut* position in the *Positions* section. When the actuator fails, this simulates the closing of the valve and hence the tank pressure increases, triggering the relief valve.

Aspen HYSYS® carries several options to simulate different valve failure scenarios:

- None: No fail.
- Fail Open: When failure occurs the valve is fully opened.
- Fail Shut: When failure occurs the valve is fully closed.
- Fail Hold: When failure occurs the valve stays at the same position it had just before the failure.

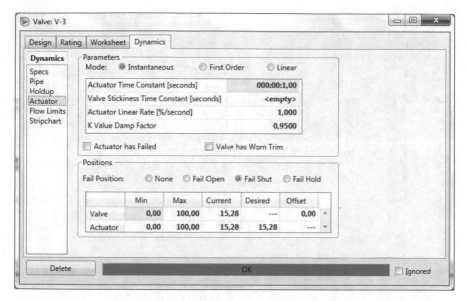

Fig. 8.34 Valve *V-3 Dynamics* window

With these options, several scenarios can be approached and the performance of the control strategy used can be also assessed.

The next task includes generating trend graphs for the pressure of the *Relief* stream and for *RV-1* and *V-3*. To do this, select the *Stripchart* option in the *Dynamics* tab of the valves. The window shown in Fig. 8.35 is displayed for the *RV-1* case.

In this window, select *Small Steady State* from the *Variable Set* drop-down menu. This option only brings the essential variables. However, only *Feed Pressure* is required. Click on *Create Stripchart*. A new window appears for the administration of the stripchart *RV-1-DL1* (See Fig. 8.36). Here you can delete the other two variables with the delete key. A graph with the *Relief* stream pressure will be displayed. However, it is required to add a couple of additional variables to this graph. Click on the *Add* button. In this way the *Variable Navigator* window is accessed, include the additional variables.

The required variables are:

- Relief valve *RV-1* opening percentage
- Control valve *V-3* opening percentage

Then, click on the *Display* button to view the graph (named by default *RV-1-DL1*).

At this point start the dynamic simulation by activating the integrator. Once the system is stable, a failure is caused in *V-3* activating the *Actuator has failed* option in the *Dynamics > Actuator* tab. The obtained response should resemble Fig. 8.37.

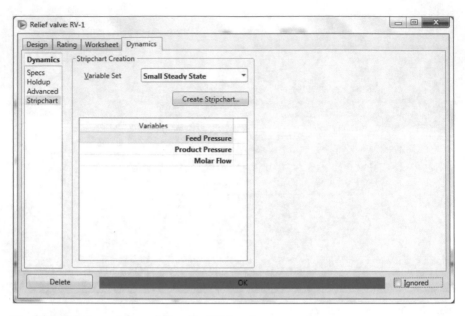

Fig. 8.35 Stripchart window of the valve *RV-1*

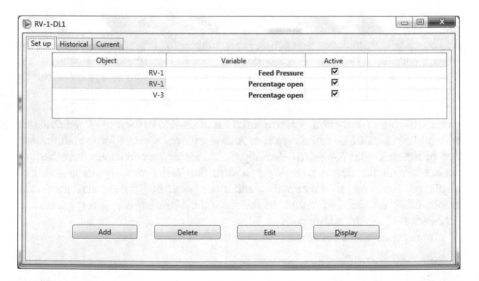

Fig. 8.36 *RV-1-DL1* window in Aspen HYSYS®

In Fig. 8.37, the pressure increase caused by the valve failure can be observed. Pressure rises from 110 to 130.9 psia. When the pressure reaches 130 psi *RV-1* opens up to around 30 % in order to avoid overpressure.

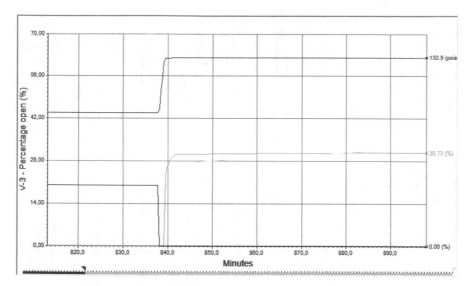

Fig. 8.37 Pressure relief system response to a failure in valve *V*-3. T-1 pressure (*red*), *RV*-1 valve opening (*green*), *V*-3 opening (*blue*)

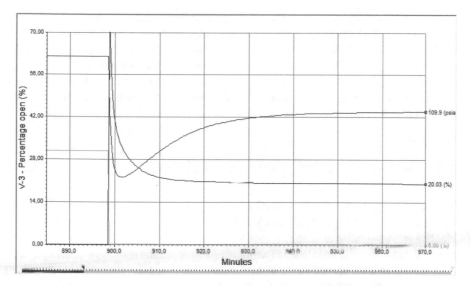

Fig. 8.38 System response to the recovery of *V*-3. T-1 pressure (*red*), *RV*-1 valve opening (*blue*), *V*-3 opening (*green*)

Subsequently, when clicking on the *Actuator has failed* option again, the valve *V-3* recovers its normal function. Figure 8.38 shows the response of the system when *V-3* recovers its function, a typical pressure control performance.

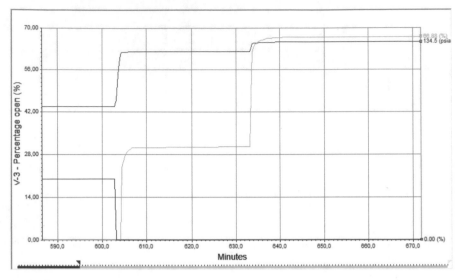

Fig. 8.39 Pressure relief system response to a failure in valve *V*-3, followed by an increase in the inert flowrate. T-1 pressure (*red*), *RV*-1 valve opening (*green*), *V*-3 opening (*blue*)

Now, perform an increase in the inert flowrate from 25 to 50 lbmol/h while the valve *V-3* is malfunctioning. The system response is shown in Fig. 8.39.

When this later disturbance occurs, the relief valve has to increase its opening percentage to about 68 %. This is an indicator that the relief valve was sized properly, since it can respond to both a flow surge and a defective control valve. The tank (*T-1*) pressure increased to about 135 psia, a value far from 150 psia which is the pressure rating of the tank.

8.6   Control of the Propylene Glycol Reactor

This section shows the dynamic performance of the reactor simulated in Chap. 5. This reactor showed multiple steady states due to the reaction of second-order kinetics and its exothermic nature. In this case the Aspen Dynamics® tool is used; this tool translates the steady state simulation into a differential equations system. This system is then solved by means of numerical methods, and using a simultaneous instead of a sequential solving approach.

The control strategy of the reactor with adiabatic operation suggests 3 degrees of freedom. This is translated into the three control valves specified. Valve *V-1* controls the propylene oxide flow rate. Valve *V-2* controls the temperature of the reactor by supplying water that can absorb some of the heat produced, moreover, water flow alters the concentration which in turn affects the reaction rate (heat release rate). Finally *V-3* controls the level in the rector fir safety reason and given that the level is not a self-regulated variable (Fig. 8.40).

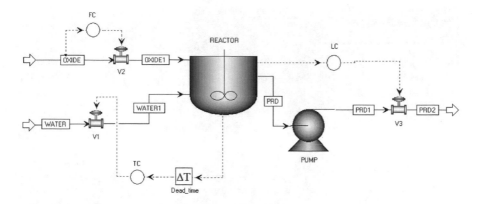

Fig. 8.40 Feedback control loops for the propylene glycol production reactor

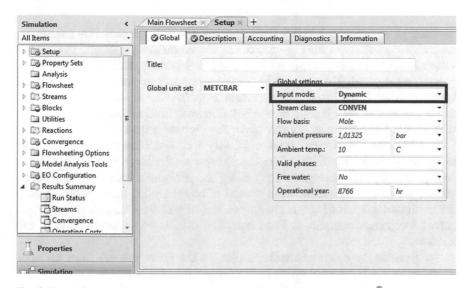

Fig. 8.41 Shift to dynamic mode from a steady state simulation in Aspen Plus®

The transition of a steady state simulation developed in Aspen Plus® to a dynamic state simulation in Aspen Dynamics® begins in the steady state simulator. First, changing the information input mode to dynamic is required; this is done in the path *Setup > Input mode > Dynamic* (Fig. 8.41) and immediately the *Dynamic* option of the *REACTOR* module becomes active. In this section, information about equipment size and geometry can be entered, Fig. 8.42. This information is fundamental for the dynamic model to evaluate the system response with time. In this case, given that the volume has been provided, only the length is required. One meter (1 m) length, elliptical head, and vertical disposition are specified.

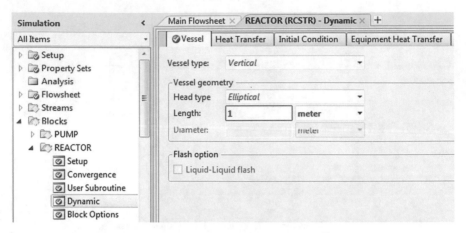

Fig. 8.42 Equipment size and geometry information in Aspen Plus®

Same as with Aspen HYSYS Dynamics, in Aspen Dynamics, it is required to ensure that all flow-pressure specifications are consistently defined. The control valves in this simulation were previously sized in the steady state simulation using a characteristic curve. This information is important to calculate and regulate each stream flow rate through the valves pressure drop. In order to verify the consistency and fulfillment of the dynamic state simulation specifications, Aspen Plus® counts with a tool called *Pressure Checker* under the *Dynamics* tab in the main menu.

The *Pressure Checker* tool is automatically activated when attempting the shift to dynamic mode. By clicking on the *Pressure Checker* button, a window is prompted with some observations regarding the performed analysis. These observations are mainly related to: name change of objects, valve locations to establish the minimum amount of control loop, and use of performance and characteristic curves for pumps and valves. Read carefully the displayed information in this window.

To generate the dynamic simulation file, in the *Dynamics* tab from the main menu. Press the *Pressure Driven* button to select it as file type. Click on *Save*, this creates the Aspen Dynamics® file. Now, open the created file to migrate to the Aspen Dynamics® environment. The main window is displayed in; it is divided into three main sections: On the lower zone of the screen, the browser contains options for equipment, process streams, control signals, and control and dynamic analysis model libraries. On the upper right part is the process flow diagram imported from Aspen Plus® (Fig. 8.43).

Initially, the previously defined control loops are configured. As an example, the configuration of the temperature controller, which includes a dead time function is developed. Go to *Controls > Dead Time* on the lower side section. Drag the *Dead Time* icon to the flowsheet. On the menu, find the model *PIDIncr* that represents a PID controller and drag it to the flowsheet as well. Now, the connections are established through streams that represent control signals. Go to *Streams > Control*

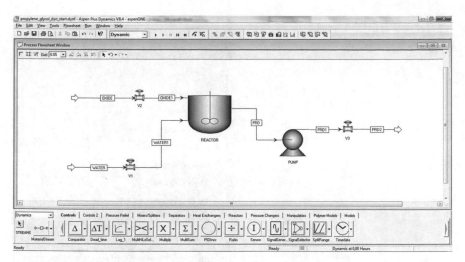

Fig. 8.43 Main window for the dynamic state simulation in Aspen Dynamics®

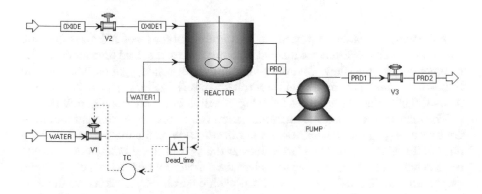

Fig. 8.44 Reactor temperature control loop setup

Signal and drag it to the flowsheet. The connection is made in the following way: First, take an output signal from the reactor (Vessel Temperature), connect this signal to the *Dead_time* module, and then, add a second signal stream from *Dead_time* to *PIDIncr* as *Process Variable* signal. Then, the *PIDIncr* controller output is connected to the control valve that regulated the water flow (Fig. 8.44).

After connections are complete, the controller and the dead time module are configured. For the dead time case, right click on the icon and select the option *Forms > Configure* and enter a dead time of 0 min (this is modified later to check its effect on the controller tuning). Double click on the controller icon, the *Face Plate* window appears, click on *Configure*, select direct action, allow the default tuning parameters, and click on *Initialize Values* to load the steady state parameters (Fig. 8.45). Note that the set point value must be changed to 100 °C, and in the *Ranges* tab, a range of 70–140 °C is entered.

Fig. 8.45 Temperature
control setup of the
propylene glycol reactor

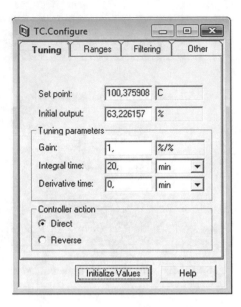

Following an analogous procedure, the controllers for level and mass flow are configured, with the difference that these do not require the dead time module. The parameters for the flow controller are: Reverse action, gain = 0.5, integral time = 0.3 min; for the level controller: direct action, gain = 2, integral time = 20,000 (this is set to void the integral action in the control algorithm).

The next step is to perform an initialization run of the whole simulation; this is done by changing the *Dynamic* option to *Initialization* on the toolbar, and clicking on *Run*. In this way, the boundary values at the start of the simulation are set and consistency is checked. Once again, select the *Dynamic* mode. Also, pull the trend graphs for PV, SP, and opening %, from each controller by clicking on the *Plot* button on the *Face Plate*. Finally, rearrange the windows as suggested in Fig. 8.46 and click on *Run* to start the dynamic state simulation.

The system starts as expected, with the steady state values constant with time. To test the controllability, disturbances must be made both in the process conditions and the controller set points. As a first task, tune the temperature controller. Even though this controller carries the tuning parameters set by default it is important to tune it so that its response is optimal. Aspen Dynamics® uses both open and closed-loop automatic tuning methods with different tuning rules. To tune a controller, click on the *tune* button on the *Face plate*. For this example, run the closed-loop tuning first, and then the open-loop to compare the results.

Figure 8.47 shows the results from the closed-loop tuning with no dead time; an oscillatory response of low amplitude is observed, this leads to an ultimate gain Ku of 33.799 and an ultimate period Pu of 3.0 min. When the simulation is restarted, and a dead time of 3 min is entered before rerunning the tuning, tuning parameters change (Ku of 1.32 and an ultimate period Pu of 10.2 min). Furthermore, a small

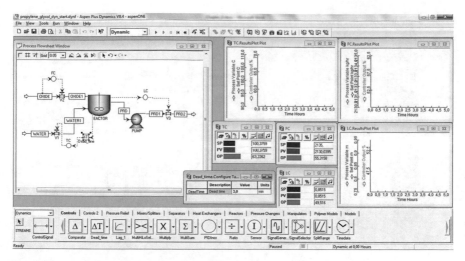

Fig. 8.46 Initialization run of the complete simulation

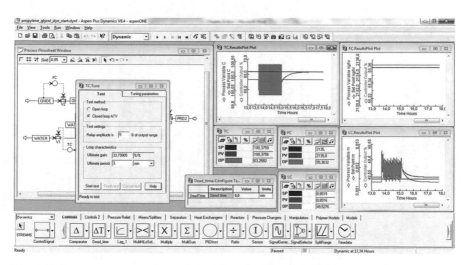

Fig. 8.47 Temperature controller tuning without dead time (closed-loop)

sustained oscillation is observed in the control loop (Fig. 8.47). It can be checked that the oscillation amplitude increased with the dead time specification of the control loop (Fig. 8.48).

Figure 8.49 shows a detailed snapshot of the closed-loop tuning performed with the tuning *Auto Tuning Variation (ATV)* technique. Click on the *Tuning Parameters* tab of the *TC. Tune* window, select the Ziegler-Nichols method for a PI controller and hit *Calculate*. The obtained tuning parameters are shown in Fig. 8.50; these values may be copied to the controller configuration tab by clicking on *Update*

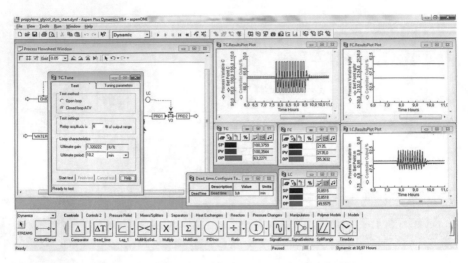

Fig. 8.48 Temperature controller tuning with dead time = 3 min (closed-loop)

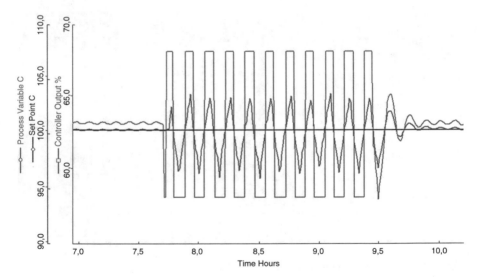

Fig. 8.49 Temperature controller tuning using the *Auto Tuning Variation* (*ATV*) method

Controller. Finally, continue to carry out the simulation, the oscillation induced by the dead time disappears as a consequence on the new tuning parameters.

Now, proceed to reset the simulation to the starting point and set a 3 min delay. Start the dynamic run, after stabilization is achieved, start an open-loop tuning by clicking on *Tuning > Open loop*. An overdamped response resembling a first-order system is obtained. Figure 8.51 depicts the results obtained from the open-loop tuning. The controller graph shows the test start at a time of 10 h, with the step type

Fig. 8.50 Tuning
parameters in the TC. Tune
window in Aspen Plus®

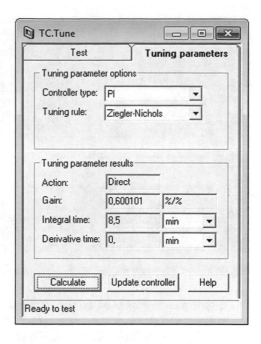

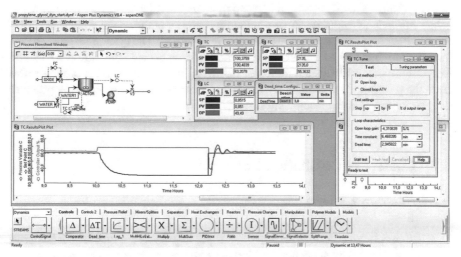

Fig. 8.51 Open-loop tuning results for the temperature controller with a 3 min dead time

disturbance on the valve opening and the first-order response of the temperature. It
also shows the stabilization of the system after the test is completed with some
slight oscillation. Finally, the calculated values for the system are shown in the *TC.
Tune* window; the tuning parameters calculated with the open-loop Ziegler-Nichols
method are comparable to what was obtained from the closed-loop method.

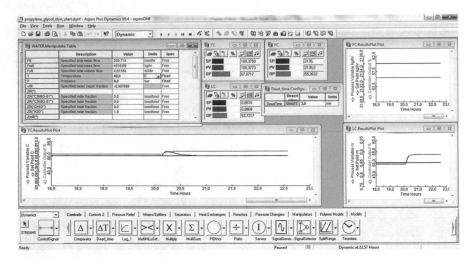

Fig. 8.52 Control system response to a disturbance on the *WATER* stream temperature

Disturbances are induced on the inlet streams to the process. For this reactor, temperature and pressure variations on any of the inlet streams, as well as oxide composition, are possible disturbances. Initially, a disturbance on the water temperature from 25 to 40 °C is made. To do this, double click the *WATER* stream and select *Forms > Manipulate*. A new window opens containing the stream information; in bold font are the variables that can be modified. The system response to the temperature increase is shown in Fig. 8.52. As expected, the liquid level and temperature control loops are affected. However, the control system rapidly stabilizes the process. To conclude, try different disturbances; for example, a pressure drop or increase on the reactor inlets.

8.7   Control of Distillation Columns

8.7.1   General Aspects

Distillation column control is an important topic in process engineering; it has been studied for decades by academic and industry engineers. It is considered as one of the toughest operations to control due to its operational complexity and the dependence on phase equilibrium, which in turn is sensitive to temperature and pressure.

The selection of an adequate process control strategy for distillation involves the installation of the process controllers; then, the problem becomes the selection of controlled-manipulated variable couples that permit the composition control along the column. There are many methods to select control strategies; these make use of criteria based on dynamic and steady state models (Luyben 2002; Luyben 2006; Shinskey 1977; Shinskey 1996).

8.7.2   Distillation Column Example

Distillation processes account for a large percentage of the separation processes in the oil and gas and chemical industry. Additionally, this operation has a significant impact in the energy requirements of the processes and is usually set as a final product purification step, increasing their market value. (e.g., distilled petroleum products, anhydrous ethanol, etc.).

Anhydrous ethanol is widely used in the chemical industry for the synthesis of esters and ethers, as solvent in the paint industry, cosmetics, aerosols, perfumes, medicine, and food products, among others. Moreover, over the last years ethanol has been used as an additive for gasoline, reducing emissions and increasing its octane number.

To begin the study of distillation columns control, let us go back to the extractive distillation of ethanol–water using glycerol as extraction agent developed in Chap. 6.

The control strategy development requires the conversion of the steady state model to dynamic state, in order to assess the effect of disturbances on the extractive distillation performance (Ross et al.2001). The Aspen Plus® model built in Chap. 6 is exported to Aspen Dynamics® as a pressure driven simulation, which calculates the stream flow rates as a function of the pressure differences through the flowsheet. Nevertheless, before shifting the simulation to dynamic, several adjustments are required. The first task is to size the column packing, this is done by using the *Packing Sizing* tool in Aspen Plus®. The liquid accumulators are also sized, defining a residence time of 5 min and a liquid level of 50 %.

Valves and Pumps must be specified with adequate pressure differences to facilitate handling flow rate changes. The Aspen Plus® simulation should undergo a pressure check to ensure flow rate consistency to all the flowsheet operations.

Now, proceed to include control valves that generate pressure drop through the system, ensuring material flowrate in the dynamic mode; as well as pumps to compensate for such pressure drop (Fig. 8.53).

Valves conditions are: pressure drop of 4 bar and design mode; all four pumps are identical with a discharge pressure of 7 bar and 75 % efficiency.

Then, simulation is changed to dynamic mode in the *Setup* section and a pressure check is performed. Export the simulation to Aspen Dynamics®.

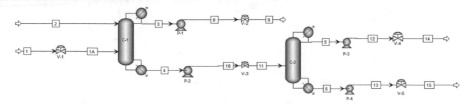

Fig. 8.53 Process flow diagram of extractive distillation with glycerol

Figures 8.54, 8.55, 8.56, and 8.57 show the temperature and composition profiles for both distillation column in the process, these steady state profiles guarantee that the product purity specification is met.

To begin with, a basic control scheme using few independent control loops is established. Details are shown next:

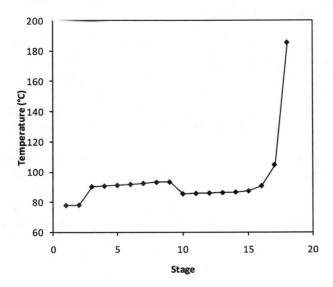

Fig. 8.54 Temperature profile of the extractive distillation column

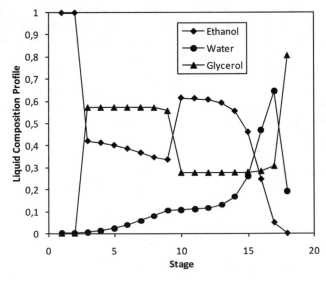

Fig. 8.55 Composition profile of the extractive distillation column

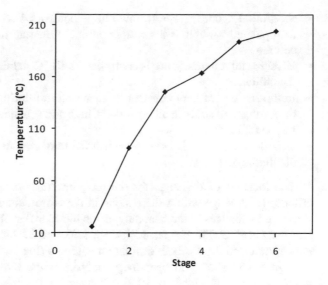

Fig. 8.56 Temperature profile of the solvent recovery column

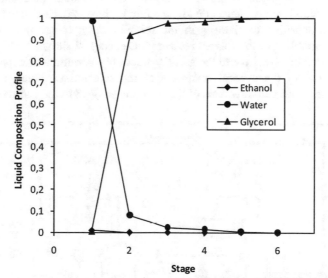

Fig. 8.57 Composition profile of the solvent recovery column

- Reflux tanks level is controlled by manipulating the distillate stream valves *D1* and *D2*.
- Feed flow rate must be controlled to ensure constant flow.
- Top pressure on both columns is controlled by the heat duty of the condenser.
- Bottom level of the extractive column is controlled by the bottom product flow.
- Bottom level of the solvent recovery column is controlled with the makeup flow rate, as per Grassi (1993) and Luyben (2008) recommendations for other extractive distillation systems (Grassi 1993; Luyben 2008).

- Separation agent (glycerol) flow rate is controlled with a ratio controller; the manipulated variable is the bottom product flow rate from the solvent recovery column.
- Glycerol inlet temperature is controlled at 80 °C manipulating the heat duty of the chiller.
- Reflux ratios are kept constant in each column during disturbances. This has been subject of study in other works (Chien and Fruehauf1990; Chien et al.1999; Luyben 2008).
- Reboilers heat duty is used to control the temperature in specific stages of the distillation columns.

The location of the stage for the temperature control is selected with these criteria: (a) A stage with a sharp slope in the temperature profile, and (b) a stage sensitive to changes in the heat duty of the reboiler (Fruehauf and Mahoney 1993; Hurowitz et al. 2003). Figures 8.54, 8.55, 8.56, and 8.57 show the composition and temperature profiles for both distillation columns (Fig. 8.58).

Figure 8.59 shows an open-loop analysis of the temperature profile of both columns when subjected to ±5 % changes on the reboiler heat duty. For the extractive column case, stage 17 shows the highest slope, and according to Fig. 8.59 is sensitive to changes in the reboiler duty; hence it is selected as controlled variable. For the solvent recovery case, stage 2 shows the highest slope yet stage 5 is more sensitive to the reboiler heat duty variations. Keeping in mind the importance of the dynamic response of the selected variables, temperature control of the rectification sections of the columns (stage 8 in the extractive column

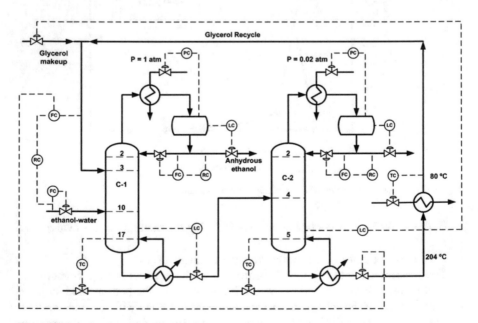

Fig. 8.58 Process flowsheet for the proposed control strategy

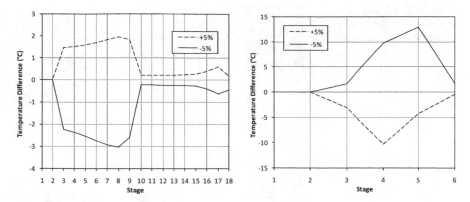

Fig. 8.59 Temperature profile variations on both columns due to ±5 % changes on the reboiler heat duty

and stage 2 in the recovery column) was not selected due to the extended dead time and to avoid a poor performance of the controllers. Stage 5 temperature is selected as manipulated variable for the solvent recovery column control.

Now, the control loops are installed according to the empirical rules for their initialization in Aspen Dynamics®. Level control loops for the reflux drums are proportional with a gain $K_c = 2$ as recommended in (Luyben2002) and $K_c = 10$ for other level controllers with a faster dynamic behavior.

Pressure controllers are Proportional–Integral with $K_c = 20$ and $\tau_I = 12$ min. All flow controllers are Proportional–Integral with $K_c = 0.5$ and $\tau_I = 0.3$ min with a filter $t_f = 0.1$ min. Both column temperature loops are tuned using closed–loop methods to determine ultimate periods and gains which were used in the Tyreus and Luyben (1992) tuning method. The chiller temperature controller (Proportional–Integral) was tuned using the open–loop method and the IMC-PI tuning rule (Chien et al.1999). The results and final parameters for these controllers are displayed in Table 8.13.

The specific configuration of the control loops in Aspen Dynamics® was previously described in this chapter.

From the analysis of the results obtained in Figs. 8.60 and 8.61 the following observations are made:

- The temperature controls respond properly to concentration disturbances (Fig. 8.60), temperature is rapidly driven back to the setpoint value; the temperature variation is not higher than 3 °C in the extractive column or 5 °C in the recovery column.
- When a disturbance on the flow rate occurs, the effect on the temperature is higher; reaching deviations of about 10 and 12 °C for the extractive and recovery columns respectively. This is due to the strong effect of the feed flow rate in the heat duties required for separation.

Table 8.13 Calculated parameters for the process temperature control loops

Variable	Value
TC—Column C-1	
Ultimate gain	1.568
Ultimate period	4.2 min
K_c	0.4902
τ_I	9.24 min
TC—Column C-2	
Ultimate gain	2.727
Ultimate period	4.8 min
Kc	0.8523
τ_I	10.56 min
TC—Chiller	
Open–loop gain	6.77
Time constant	0.59 min
Dead time	0.6 min
K_c	0.13
τ_I	0.899 min

• Product purity is kept within specifications and it is more heavily affected by the feed composition than by the flow rate. However, the ethanol product quality is not negatively affected by the disturbances.
• Inventory control loops are quickly stabilized. Particularly, when feed flow rate is increased, the cascade controller increases the glycerol flow rate to the column making the level on the recovery column to drop. Then, as a result of the increase in the solvent and feed flow rates, the mass balance adjusts the feed flow to the recovery column, causing it to increase again.

8.8   Summary

Developing process dynamic models allows increasing the spectrum of process design enhancements during the conceptual stage. The combination of conceptual and detailed design with the establishment of the conditions that makes a process controllable and dynamically sound, becomes an alternative to generate more robust process flowsheets and unit operations.

The dynamic analysis tools available in commercial process simulators make possible to accurately represent the behavior of systems in which typical dynamic situations occur, such as: dead time, disturbances, valve failure, changes in the feed characteristics, among others. These types of situations are of interest for a process designer or operator, since it allows adjusting the control mechanisms and foreseeing emergency situations in a plant.

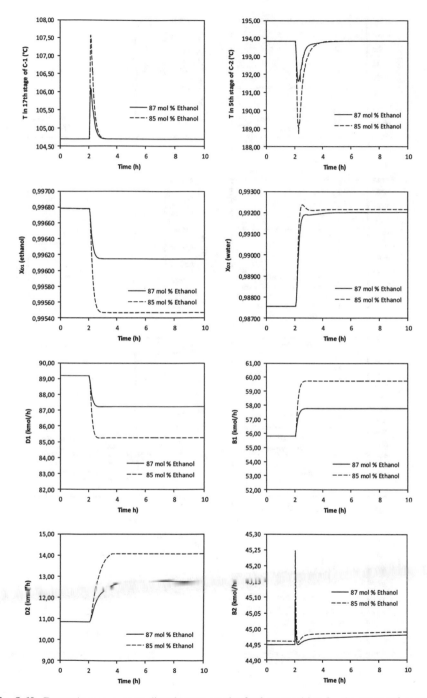

Fig. 8.60 Dynamic responses to disturbances on the feed composition for the proposed control strategy

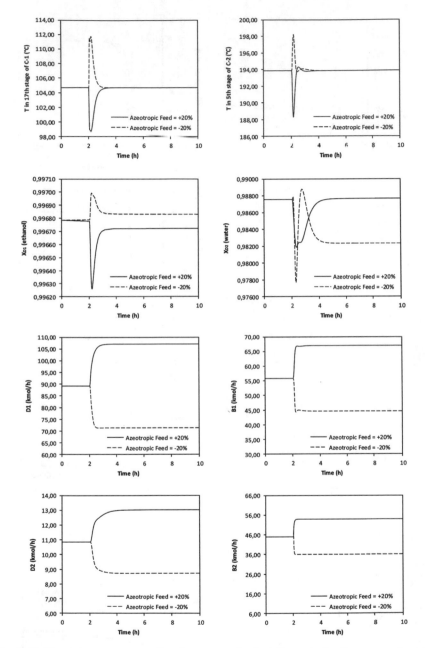

Fig. 8.61 Dynamic responses to disturbances on the feed flow rate for the proposed control strategy

8.9 Problems

P8.1 What is the importance of volume specification in a simulation when converted into dynamic mode? What are the volume effects on the dynamic performance?

P8.2 What is a typical system response when a system is set to manual mode and a disturbance is made? What happens if the set point is changed?

P8.3 The phase separators example (introductory example) was developed using two control loops, one for pressure and one for level. There are still 2 degrees of freedom represented by valves *V-1* and *V-4*. Install control loops for the level and feed flow rate to *LTS*. Set initialization values for the tuning parameters and perform disturbances to assess the control loops performance. What additional control loop can be added to the system?

P8.4 What additional disturbances can occur in real life operation for the gasoline blending example? Perform any and check the control loops performance.

P8.5 Size the relief valve of the gasoline blending example using the D specification of the API 526 standard. What would you expect from its dynamic behavior?

P8.6 If the *Full Open Pressure* of the *RV-1* valve is reduced to 135 psia. Does any change occur regarding the opening percentage during disturbance? What causes this behavior?

P8.7 Using the control strategy portrayed in Fig. P8.1, suggests the tuning parameter settings and observe the dynamic performance when the same disturbances shown in the distillation column control example are applied.

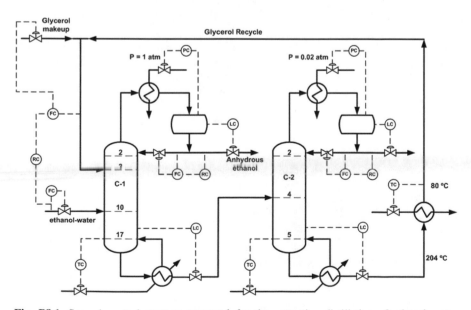

Fig. P8.1 Second control strategy proposed for the extractive distillation of ethanol–water mixture using glycerol as entrainer

References

Aspen Technology, Inc. (2006) Aspen HYSYS 2006: tutorials & applications. Aspen Technology, Cambridge

Chien IL, Fruehauf PS (1990) Consider IMC tuning to improve controller performance. Chem Eng Prog 86:33–41

Chien IL, Wang CJ, Wong DSH (1999) Dynamics and control of a heterogeneous azeotropic distillation column: conventional control approach. Ind Eng Chem Res 38:468–478

Fruehauf P, Mahoney D (1993) Distillation column control design using steady state models: usefulness and limitations. ISA Trans 32(2):157–175

Grassi VG (1993) Process design and control of extractive distillation, Practical distillation control. Van Nostrand Reinhold, New York, pp 370–404

Hurowitz S, Anderson J, Duvall M, Riggs J (2003) Distillation control configuration selection. J Process Control 13:357–362

Luyben WL (2002) Plantwide dynamic simulators in chemical processing and control, Chemical industries series. Marcel Dekker, New York

Luyben WL (2006) Evaluation of criteria for selecting temperature control trays in distillation columns. J Process Control 16:115–134

Luyben WL (2008) Comparison of extractive distillation and pressure-swing distillation for acetone-methanol separation. Ind Eng Chem Res 47(8):2696–2707

Perry R (1992) Chemical engineer's handbook. McGraw-Hill, New York

Ross R, Perkins E, Pistikopoulos E, Koot G, van Schijndel J (2001) Optimal design and control of a high-purity industrial distillation system. Comput Chem Eng 25:141–150

Shinskey F (1977) Distillation control. McGraw-Hill, New York

Shinskey F (1996) Process control systems. McGraw Hill, New York

Tyreus BD, Luyben WL (1992) Tuning of PI controllers for integrator dead time processes. Ind Eng Chem Res 31:2625–2628

Chapter 9
Solids Operations in Process Simulators

9.1 Introduction

Solids operations demand special attention when simulating a process because more experimental information is required. These operations are not usually covered in undergraduate level courses or even in industry technical courses considering that usually oil and gas simulations are more common and in those cases solids presence is neglected.

The introduction of solids in process simulation affects heat and mass balances and can predict more accurately unit operations when solids are present. Remember that a simulation requires a solid component compatibly property models to avoid introducing an excessive error in the results. Solid particles representation must be carefully specified in order to provide information for particle size distribution accurate approach. However, know-how to simulate them is essential to describe solid–solid, solid–liquid, and solid–vapor interactions. Industrially, filtration, drying, crushing, coal combustion, and extraction are vital operations to consider using a process simulation to predict the behavior. In this chapter we will introduce the available modules and how to specify them in order to obtain the more proximate result considering the data provided in each example. It is important to notice that the description of the calculation modules presented here corresponds to the more recent versions of Aspen Plus® (version 8 or higher) in which important improvements and capabilities have been included.

9.2 General Aspects

Before introducing the modules available in process simulators, it is necessary to introduce the required technical background to understand these operations. Main solids operations are explained below.

© Springer International Publishing Switzerland 2016
I.D.G. Chaves et al., *Process Analysis and Simulation in Chemical Engineering*,
DOI 10.1007/978-3-319-14812-0_9

9.2.1   Separation or Classification

Classification is the primary step in solid unit operations, because often there are several particle sizes in a sample and maybe different treatments are implemented by ranges.

9.2.1.1   Hydrocyclones

Hydrocyclones are considered as sedimentation-type clarifiers with no mechanical parts. They are static equipment that use centrifugal force to separate heavy and light particles in a process stream. They are used in many mineral processes due to the following advantages: Fixed and very simple to use, compact and have short residence time for the process, and mostly their low capital cost is the main benefit from the economic point of view.

Hydrocyclones have a cylindrical section closed from one side and an overflow pipe which is fitted axially. Feed is introduced tangent to the hydrocyclone. The bottom of the hydrocyclone is conical, for separation to occur, suspension is pumped from the feed opening. After the feed is sent to hydrocyclone with liquid, heavy particles move outward and collect at the bottom of the vessel whereas bulk liquid and light particles move towards the axis of the hydrocyclone. So, they move towards the upper outlet at the top of the hydrocyclone (Infar 2011). See Fig. 9.1.

Even though, hydrocyclones are cheaper compared to centrifuges, and settling is faster in hydrocyclones than in centrifuges; it is fact that they both have specific applications and each of them has its own importance due to certain features.

9.2.1.2   Cyclones (Infar 2011)

A cyclone separator is almost similar to hydrocyclone. They are almost similar in their operation, construction, and working principle. But the only key difference is that cyclone works for solids suspended in gases whereas hydrocyclone is used for solid–liquid suspension (Gupta 2003).

9.2.1.3   Centrifuges (Infar 2011)

Centrifugal classifiers are of primary importance due to their centrifugal settling method for the division of particles. This division is due to the movement of particles in fluid. When slurry is passed through a centrifuge, larger particles are separated by throwing them out from the liquid while, very fine or light particles might not be able to settle during this time and can be withdrawn with the liquid. See Fig. 9.2.

Fig. 9.1 Schematic representation of a hydrocyclone

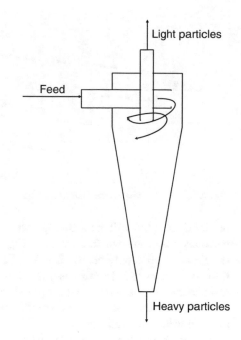

Fig. 9.2 Schematic representation of a centrifuge

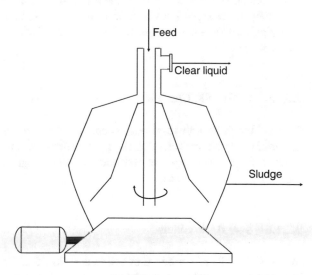

9.2.1.4 Screens (Infar 2011)

Screening is a simple process used for the separation of particles based on size. It is a mechanical process and like other separation processes. It is quite impossible to obtain a complete separation.

For industrial screening, the solids are forced to fall onto or thrown by force against a screen. The bigger particles or tails will stay on screen while undersize

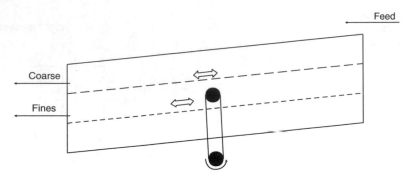

Fig. 9.3 Schematic representation of a screen

particles or fines will pass through the screen. A single screen can divide the particles only into two fractions, see Fig. 9.3. These fractions are called unsized fractions because only one limit is known i.e., upper or lower limit and the other limit is unidentified. In this case, fractions of both limits are known. Sometimes, wet screening is used but the most preferably and commonly dry screening is used. When it is required to achieve a specific particle size, the process stream will pass through different sized screens.

Screening is normally used for the separation of coarse particles. The efficiency of screens for fine particles is poor with normal screens and fine screens are very costly as well for fineness. So, the particle size for the separation should be more than 250 μm.

9.2.1.5 Hydraulic Classifiers

A classifier is an industrial equipment in which particles are sorted by specific gravity in a stream of water that rises at a controlled rate; heavier particles gravitate down and are discharged at the bottom, while lighter ones are carried up and out (Infar 2011) (Fig. 9.4).

9.2.1.6 Spiral Classifiers

In spiral Classifiers a mixture of solids and liquid is fed to separate solid particles into fractions according to particle size or density by methods other than screening. The products resulting are a partially drained fraction containing the coarse material (known as underflow) and a fine fraction along with the remaining portion of the liquid medium (known as overflow) (Infar 2011). See Fig. 9.5.

The classifying operation is carried out in a pool of fluid confined in a tank arranged to allow the coarse solids to settle out, where they are removed by gravity, mechanical means, or induced pressure. Solids which do not settle report as overflow.

Fig. 9.4 Hydraulic
classifiers

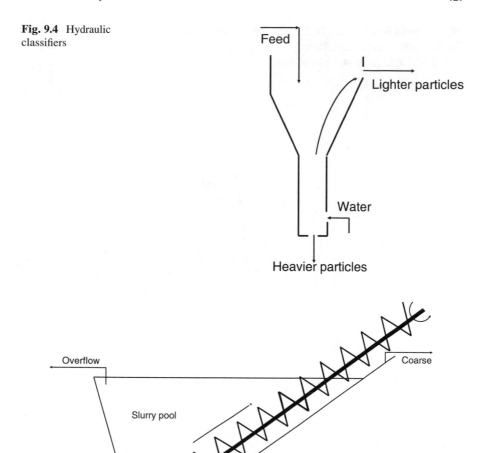

Fig. 9.5 Spiral classifier (Adapted from Infar 2011)

9.2.2 *Comminution*

Reducing the size of a product to specific requirements for a further use is important
in any kind of industry, for example, to increase reactivity or increase specific
contact area for solubility purposes.

These operations can be divided into two large sections: Crushing and Grinding.
Both operations are required for achieving a specific size in the product. Crushing is
mainly used for the reduction of big particles into smaller sizes for further reduction. On the other hand, grinding refers to the reduction of smaller sized particles
produced by crushing, into fine powder. During crushing, heat losses are lower than
grinding. Compression is used for crushing while grinding is normally done by

using impact technique. Crushing is done in the initial stages while grinding will be
the final step in comminution.

Typical crushing equipment are:

- Gyratory crushers

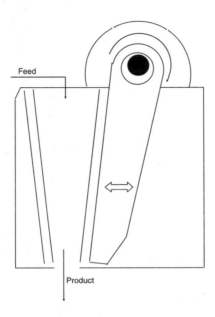

- Cone crushers
- Jaw crushers

    Typical grinding equipment are:

- Impact breakers
- Rod mills
- Ball mills

- Hammer mills
- Jet mills

Table 9.1 Typical energy requirements for crushing and grinding operations (Rosenqvist 2004)

Operation	Energy requirements (kWh/t)
Coarse crushing	0.2–0.5
Fine crushing	0.5–2
Coarse grinding	1–10
Fine grinding	2–25
Micronizing	100

Table 9.2 Crushing and grinding equipment with its characteristics (Couper et al. 2010)

Equipment	Feed size (mm)	Product size (mm)	Reduction ratio	Capacity (tons/h)	Power consumption (kW)
Gyratory crushers	200–2000	25–250	8	100–500	100–700
Jaw crushers	100–1000	25–100	8	10–1000	5–200
Cone crushers	50–300	5–50	8	10–1000	20–250
Impact breakers	50–300	1–10	40	10–1000	100–2000
Rod mills	5–20	0.5–2	10	20–500	100–4000
Ball mills	1–10	0.01–0.1	100	10–300	50–5000
Hammer mills	5–30	0.01–0.1	400	0.1–5	1–100
Jet mills	1–10	0.003–0.05	300	0.1–2	2–100

Information reported in Table 9.1 indicates that smaller particles require more power. This phenomenon can be reduced using water or chemical agents to reduce friction and lower the energy requirements.

To select proper technology for reducing particle size is important to consider the nature of raw material. Table 9.2 shows the typical ranges for different equipment available.

9.2.3   Filtration

Filtration is a separation technique of suspended particles contained in a fluid by using a filter as medium. This separation is done by passing a fluid through a porous medium. The solid particles are retained on the surface of medium whereas, the fluid i.e., filtrate, passes through the pore or voids of a membrane (Cheremisinoff 1998).

Two main different types of filtration are practiced in the industry:

- Cake filtration
- Deep-bed filtration

Different types of filters are used for these filtration processes. Cake filtration is done by surface filters or granular filters whereas deep-bed filters are used in deep-bed filtration.

In cake filtration, the initial pressure drop for the medium is relatively low and particles of same size or greater than the orifices of medium are trapped or stay at the medium surface. By this way, the orifices of the medium are closed and produce small ways which can remove the small particles from the fluid. In this way, a filter cake is obtained which works as a medium for filtration. In order to avoid clogging of the medium, filter aids are helpful in precoating that forms an initial layer on the medium (Svarovsky 2000).

9.2.4 Crystallization

Crystallization is the process of formation of solid crystals precipitating from a solution, melt or more rarely deposited directly from a gas.

Industrial crystallization process can be defined as:

- Cooling crystallization
- Evaporating crystallization

9.2.5 Particle Size Distribution Meshes

In order to represent the particle size distribution in a material stream, it is recommended to realize a screening test in a specialized lab.

9.3 Modules in Aspen Plus®

Aspen Plus® has different modules for these fluid handling operations, all in the *Solids* and *Solids Separations* tabs. Different modules are summarized in Table 9.3.

9.4 Modules in Aspen HYSYS®

See Table 9.4.

Table 9.3 Models available in Aspen Plus®

Icon	Name	Description
	Crusher	This module simulates different types of crushers: Gyratory, Jaw, Cone, Hammer, Ball, and so on. It can calculate the PSD from de outlet or calculate equipment power and dimensions
	Granulator	This module considers particle growth in four ways: granulation, agglomeration, specifying a distribution function or specifying outlet PSD
	Crystallizer	Simulates crystal formation of solids entering crystallization kinetic and solubility information
	Centrifuge	Simulates the solid separation due to centrifugal force considering different equipment configurations
	Cyclone	This module simulates the solid particle separation from a gas stream. An example is solid carry-over in an air stream
	Hydrocyclone	This module simulates the solid particle separation from a liquid stream. An example is solid separation from water recollection stream
	Dryer	Simulates convective and spray dryers entering a drying curve from experimental data or an equation which describe the phenomenon
	Screen	This module simulates the particle separation driven by particle and screen size. This module includes single and multideck screens

Note: Depending of your Aspen Plus® version, solid unit operations will include new modules. For more information search "Chapter 8: Solids" on Aspen Plus *Help* menu, *Help topics* submenu

Table 9.4 Models available in Aspen HYSYS® (Aspen Technology, Inc. 2012)

Icon	Name	Description
	Baghouse filter	This module is based on empirical equations relating separation efficiency with particle size. This unit operation is not available for dynamic simulation
	Cyclone	It is used to separate solid particles above 5 μm from gas streams. The separation is achieved due to the centrifugal force which moves the particles toward the wall of the equipment. This unit operation is not available for dynamic simulation
	Hydrocyclone	It is similar to the cyclone module but separates solids from a liquid stream. This unit operation is not available for dynamic simulation
	Rotary vacuum filter	This module calculates the retention of solvent in a particle cake. This operation estimates the behavior regarding the particle diameter and sphericity provided. This unit operation is not available for dynamic simulation
	Simple solid separator	This module performs nonequilibrium separation calculation considering solids. It does not perform energy balance, only calculates carry-over in liquid and gas streams. This unit operation is not available for dynamic simulation

9.5 Crusher Introductory Example

To illustrate the use of solid unit operations, a simulation of a crusher is presented including how to specify solids in an Aspen Plus® simulation.

9.5.1 General Aspects

Crushers are a useful solid operation in industry, because some raw materials or chemical substances are used in bulk. Reducing particle size is essential to improve the chemical reaction capability, solubility or just for packaging purposes.

Calcite is a mineral crystal form of calcium carbonate ($CaCO_3$) which can be crushed to small particles for further processes. Currently, there are several calcite crusher plants in South Africa, Saudi Arabia, India, Colombia, and Chile.

Calcium carbonate has plenty of applications in different fields, like pharmacology, medicine, paper industry and construction, among others. Calcite minerals are used as raw materials for cement production, in paint manufacturing and as formula in PVC and other polymer production.

Aspen Plus® has a wide database of solid components, however cannot have different crystal configurations for them. Depending of the process can be specified, the hardness of the specific raw material used. Remember that simulation involving solid materials requires some level of knowledge of the properties and behavior of the solids.

9.5.2 Simulation in Aspen Plus®

This simulation seeks to illustrate the specification of solid components in Aspen Plus® and how to generate solid PSD (Particle Size Distribution) for a stream. This procedure is applicable to every solid simulation carried out in Aspen Plus®.

This simulation is composed of a crusher module as shown in Fig. 9.6.

For this example a simulation with English units (Solids with Metric Units) and the run type Flowsheet are selected. After selecting the appropriate options, it must be clicked on Accept and, in this way, the simulation environment can be accessed.

In the *Main Flowsheet* section install the *Crusher* module located in the *Solids* tab in the *Model Palette* shown in Fig. 9.7.

Click the *Next* button and the *Components* tab is shown. Click in the *Find* button and search for *CaCO₃* compound. Select the component from the *Solids* databank and add to the list clicking in the *Add selected compounds* button as shown in Fig. 9.8.

Once added this component, in the *Type* row, select *Solid* to enable the inlet of solid data to specify streams (Fig. 9.9).

Fig. 9.6 Simulation flow diagram for calcite crusher example

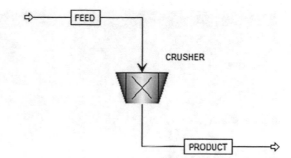

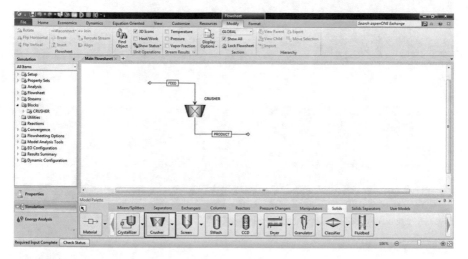

Fig. 9.7 Installing the crusher module in Aspen Plus®

In the *Methods* menu, select the *Ideal* method. This method is usually used to represent solid components when there is not equilibrium calculations required.

In *the Setup > Specification* menu, select the *METSOLID* Global unit set.

The next step is to define the inlet data to be introduced to define solids present in streams and how models perform this calculation accordingly. In *the Setup > Solids* menu, define the stream class according to the next options available:

- *CONVEN*: This stream class is designed to handle mixed streams (when no solids are present).
- *MIXNC*: This stream class can handle nonconventional solids but without particle size distribution.
- *MIXCISLD*: This stream class handles conventional solids but without particle size distribution.
- *MIXNCPSD*: This stream class allows working with nonconventional solids and providing particle size distribution.
- *MIXCIPSD*: This stream class allows working with conventional solids and providing particle size distribution.

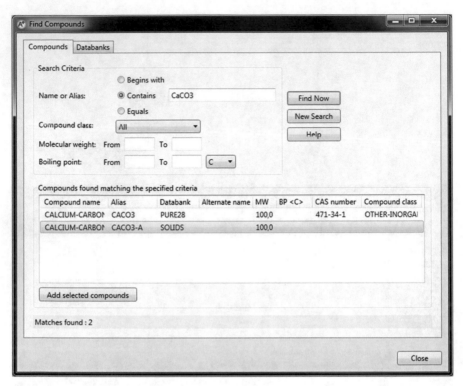

Fig. 9.8 Specifying solid components in Aspen Plus®

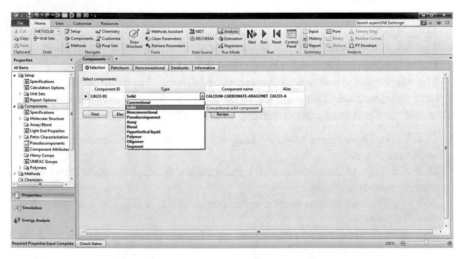

Fig. 9.9 Specifying solid components in Aspen Plus®

- *MIXCINC*: This stream class handles conventional and nonconventional solids but without particle size distribution.
- *MCINCPSD*: This stream class handles conventional and nonconventional solids providing particle size distribution.

For this simulation purpose, stream class *MIXCIPSD* is selected. In the *PSD Mesh* tab, click the *New* button to define the particle size distribution (PSD) for this simulation input data. There are four PSD mesh types to be selected:

- *Equidistant*: The differences between intervals are constant. Divide the total range into the number of intervals defined.
- Geometric: The grid will be filled in with intervals sized so that the volume ratio between consecutive sizes is 2 (since diameter is shown, the ratio of diameters will be the cube root of 2). The upper limit will be enforced strictly; the number of intervals will be adjusted so that the lower limit appears within the first interval (Aspen Plus®Help).
- Logarithmic: The grid will be filled in with intervals sized so that ratio between consecutive sizes is constant (Aspen Plus® Help).
- User: The grid will be provided by the user using experimental data of particle size.

For this simulation, an equidistant grid with ten intervals between 0.1 and 10 cm will be introduced as shown in Fig. 9.10. Click the *Generate PSD Mesh* button.

The next step is to define the inlet stream called *FEED* with the information reported in Table 9.5. In this case because it is a solid stream, these values must be specified in *CI Solid* tab in the *Streams > FEED* menu as shown in Fig. 9.11.

The next step is to specify the weight fraction of the particle size distribution created before, there are two possible ways:

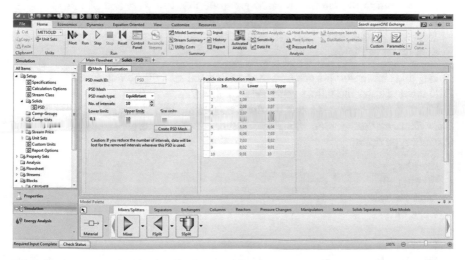

Fig. 9.10 Specifying particle size distribution (PSD) in Aspen Plus®

Table 9.5 Inlet conditions
for FEED stream

Feed	
Temperature (°C)	25
Pressure (bar)	1
Mass Flow (kg/h)	5000
Mass composition	
CaCO$_3$	1

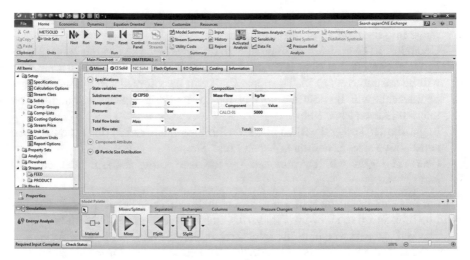

Fig. 9.11 Specifying solid streams in Aspen Plus$^®$

- User-specified values: Allows entering specific values from experimental data. Remember to configure the PSD according to the available data.
- A distribution function: Allows to generate the data with one of the following models: GGS (Gates-Gaudin-Schuhmann), RRSB (Rosin-Ramler-Sperling-Bennet), Normal and Log Normal. This feature is used when experimental data is not available.

For this simulation purpose, user-specified values will be entered. Click in the PSD tab and enter the values reported in Table 9.6.

Then, it is necessary to specify the Crusher module. In the *Blocks > CRUSH-ER > Specifications* menu, there are different types of crushers to select in the Crusher type option:

- Gyratory
- Single roll
- Multiple roll
- Cage mill
- Jaw
- Cone
- Impact mill

Table 9.6 PSD specification
for FEED stream

Segment	Weight fraction
1	0.0
2	0.0
3	0.0
4	0.39
5	0.24
6	0.16
7	0.09
8	0.07
9	0.03
10	0.02

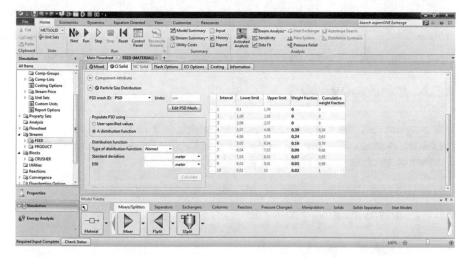

Fig. 9.12 Specifying particle solid distributions in Aspen Plus®

- Hammer mill
- Ball mill
- Rod mill

Leave in the default option, Gyratory. In the *Breakage function parameters* section, enter the *Maximum particle diameter* of 2 cm (Figs. 9.12 and 9.13).

Now enter in the *Grindability* tab in the *Blocks > CRUSHER > Specifications* menu, here it is necessary to specify a variable call Bond Work Index. According to Oates (2008), Bond Work Index for this type of rocks (limes and limestones) is in an interval between 4 and 10 kWh/t depending on the hardness of raw material. Considering the Mohs scale, limestone is one of the softest rocks, so the value for Bond Work Index to be used will be 6 kWh/t. In the *Communition Law* options select *Rittinger's Law*.

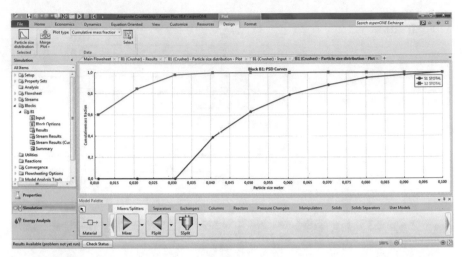

Fig. 9.13 PSD from inlet and outlet streams for aragonite crusher example in Aspen Plus®

9.5.3 Results Analysis

The program calculates a power of 3.19 hp which is a reasonable value for this operation. Additionally, can be displayed graphically the inlet and outlet solids distribution to observe the change in particle size.

9.6 Solids Handling Example

9.6.1 General Aspects

Natural gas treatment includes unit operation involving solids. An example of these operations is adsorption; sometimes natural gas must achieve cryogenic process conditions (hydrocarbon and water dew point below -180 °F) in order to obtain high purity petrochemical products as methane, ethane, propane, LPG, etc.

Adsorption processes involve a solid which removes water and hydrocarbons from gas stream but sometimes small solid particles are carried over with gas. These particles must be separated and different equipment can be used.

In this example several equipment calculations to remove these particles from gas stream and how to enter the required information in Aspen Plus® are presented.

9.6.2 Simulation in Aspen Plus®

This simulation seeks to illustrate the specification of solid model units in Aspen Plus®. This procedure is applicable to every solid simulation carried out in Aspen Plus®.

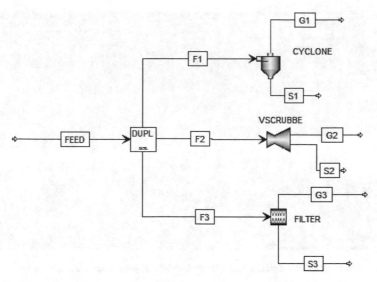

Fig. 9.14 Simulation flowsheet for natural gas treatment example in Aspen Plus®

Table 9.7 Components for natural gas treatment example

Component	Type
CO_2	Conventional
N_2	Conventional
Methane	Conventional
Ethane	Conventional
Propane	Conventional
Butane	Conventional
i-butane	Conventional
Pentane	Conventional
i-pentane	Conventional
Hexane	Conventional
H_2O	Conventional
Dust	Nonconventional

This simulation flow diagram is presented in Fig. 9.14.

Start an Aspen Plus® simulation using the *MIXNCPSD* stream class and enter the following components (Table 9.7).

The property method to be used is *IDEAL*. The next step is to set the property method for the nonconventional component. In *Methods > NC Props* route is shown by the different methods available for this component. Select the *HCOALGEN* method for enthalpy calculations and *DCOALIGT* method for density calculations. In the *Required component attributes* section, select *PROXANAL*, *ULTANAL*, and SULF*ANAL* options (Fig. 9.15).

The next step is to define the *FEED* stream. Enter the information reported in Table 9.8.

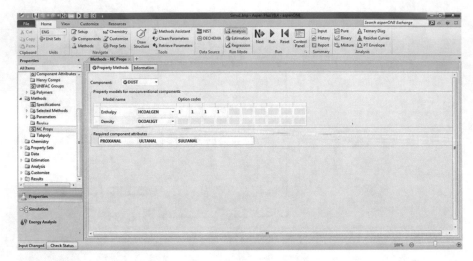

Fig. 9.15 Property method selections for nonconventional components in Aspen Plus®

Table 9.8 Inlet conditions for FEED stream

Feed	
Temperature (°F)	80
Pressure (psig)	1500
Std Gas Flow (MMSCFD)	50
Molar composition	
CO_2	0.01
N_2	0.02
Methane	0.8
Ethane	0.07
Propane	0.08
Butane	0.008
i-butane	0.006
Pentane	0.003
i-pentane	0.002
Hexane	0.001
H_2O	0.000

Install a DUPL module and generate three new streams called *F1*, *F2*, and *F3* leaving the new module. Then, it is required to install three solids separator modules: A cyclone called *CYCLONE*, a *VScrub* module called *VSCRUBBE* and a *FabFl* module called *FILTER*.

For *Cyclone* module, define the outlet streams *G1* for gas outlet and *S1* for solids outlet. For *Scrubber* module, define the outlet streams *G2* for gas outlet, an inlet liquid stream, *LIQ* and *S2* for solids outlet. For *LIQ* stream specify the same conditions of *FEED* stream (pressure and temperature) and a flow of water of 1000 kg/h.

Table 9.9 Calculation information for cyclone

Cell	Value
Model	Cyclone
Mode	Design
Calculation method	Leith-Licht
Type	Stairmand-HE
Maximum no. of cyclones	200

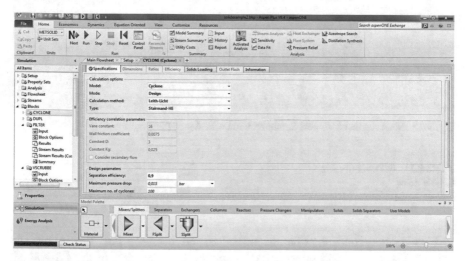

Fig. 9.16 Main screen of cyclone module in Aspen Plus®

Table 9.10 Calculation information for Venturi Scrubber

Cell	Value
Model	Venturi Scrubber
Mode	Design
Calculation method	Yung
Separation efficiency	0.26

And for *Filter* module, define the outlet streams *G3* for gas outlet and *S3* for solids outlet.

The next step is to specify the different module characteristics. For cyclone module, in the *Calculation options* section select the following options (Table 9.9).

Finally, in *Design parameters* section, enter a value of 0.9 in the *Separation* efficiency cell (Fig. 9.16).

For *VScrubber* module, enter the information reported in Table 9.10.

For Filter module, in the *Fraction of solids to solids outlet* option enter a value of 0.99. In *Solid load of vapor outlet* option enter a value of 0.1. In the *Outlet flash* tab, enter a pressure of 0.1 bar to specify pressure drop (Fig. 9.17).

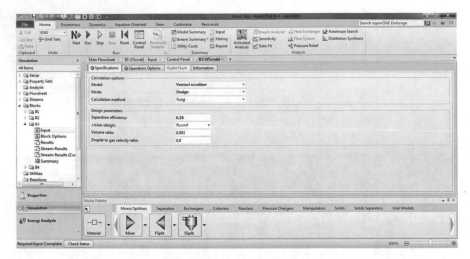

Fig. 9.17 Main screen of Venturi Scrubber module in Aspen Plus®

Table 9.11 PSD information
to be entered

Cell	Value
PSD mesh type	User
No. of intervals	10

Finally, the solid distribution needs to be entered. For this enter the *Setup > Solids* route and select water as moisture component. In the *PSD Mesh* tab, enter a new PSD with the information reported in Table 9.11.

And then, click the *Create PSD Mesh* button. In the *FEED* stream, enter the NC Solid tab and enter the same conditions for fluid stream (pressure and temperature). A mass flow of 100 kg/h of *DUST* component.

In the *Component Attribute* section enter the following information:

- PROXANAL: Moisture 10 %, Ash 90 %. (Enter the values in every option)
- ULTRANAL: Carbon 90 %, Ash 10 %
- SULFANAL: 0 for the three options available.
- GENANAL: 0 for the elements available.

In the Particle Size Distribution section, enter the information reported in Table 9.12.

Now the simulation is ready to run the calculations.

9.6.3 Results Analysis

In Table 9.13 the main results are reported.

Solids entered in this example could be produced from degradation of adsorbent material that is the reason the coal property methods are used.

Table 9.12 PSD information to be entered

Interval	Lower limit	Upper limit	Weight fraction
1	0	20	0.2
2	20	40	0.1
3	40	60	0.1
4	60	80	0.05
5	80	100	0.05
6	100	120	0.05
7	120	140	0.05
8	140	160	0.1
9	160	180	0.1
10	180	200	0.2

Table 9.13 Simulation results

Cyclone	
Variable	Value
Number of cyclones	136
Diameter of cylinder (ft)	4.92
Efficiency	0.9
Length of vortex (ft)	12.19
Length of cylinder (ft)	7.38
Length of cone (ft)	12.3
Venturi Scrubber	
Throat diameter (ft)	0.49
Throat length (ft)	1.00
Filter	
Number of cells	38
Number of bags per cell	78
Filtering time (h)	8.96

Solids modules in Aspen Plus® can design different equipment alternatives for this separation. However, by using a Venturi Scrubber it is not possible to achieve adequate separation efficiency whereas the other modules are able to.

9.7 Summary

Aspen Plus® in its version 8 and higher has powerful solid models to represent several industrial processes. It is worth to notice that solids simulation can be used to represent particle size distributions and to describe in a more specific way solid–solid, solid–liquid, and solid–vapor mixtures. Many industrial processes have solids operations like filtration, drying, crushing, coal combustion, and extraction that can be calculated and estimated by means of solids calculation modules discussed in this chapter.

References

Aspen Technology, Inc. (2012) Aspen HYSYS operations guide. Aspen Technology, Cambridge

Aspen Technology, Inc. Aspen Plus® help menu

Cheremisinoff NP (1998) Liquid filtration (2nd edn). USA: Blueworth

Couper JR, Penney WR, Fair JR, and Walas SM (2010) Chemical Process Equipment - Selection and Design 3rd Edn. Elsevier

Oates JA (2008) Lime and limestone: chemistry and technology, production and uses, 1st edn. Wiley, New York

Svarovsky L (2000) Solid-liquid separation 4th Edn. UK: Butterworth-Heinimann

Gupta CK (2003) Chemical metallurgy: principles and practice. Mumbai, India: Wiley-vch Verlag GmbH & Co.KGaA

Infar HM (2011) Simulations of Solid Processes by Aspen Plus. Lappeeranta University of Technology

Rosenqvist T (2004) Principles of extractive metallurgy. Trondheim: Tapir Academic Press

Chapter 10
Case Studies

10.1 Introduction

The case studies are examples of slightly more complex systems, where many of the topics reviewed in the previous chapters are integrated. Here we attempt to address cases involving the topics studied in this text as well as address additional tools available in the Aspen Engineering Suite Tech® that solve problems holistically.

In these examples, sensitivity analyses and/or comparisons between differing provisions are made to take advantage of computational tools incorporated into simulators that help decrease the time to get answers and be able to use this time to analyze situations.

One of the main drawbacks when performing process simulations is the selection of property packages, since the goal of the simulation is to achieve values that are very close to actual operation conditions. Added to this, the difficulty in establishing the recycles for processes with many interconnected unit operations, where many of them are highly nonlinear, justifies the need to develop cases of complete processes to illustrate some strategies to address these problems.

10.2 Simulation of Nylon 6,6 Resin Reactor

10.2.1 Problem Description

To study the polymerization of adipic acid with hexamethylenediamine to produce nylon 6,6 resin uses the Aspen Polymer Plus® available from the engineering suite of AspenTech®. The reaction involved is as follows:

The production process of these resins can be performed either in a continuous process (Giudici and Nascimiento 1999; Giudici 2006), or in batch (Kumar and Gupta 2003); along this case study both possibilities are studied. The main reasons

© Springer International Publishing Switzerland 2016
I.D.G. Chaves et al., *Process Analysis and Simulation in Chemical Engineering*,
DOI 10.1007/978-3-319-14812-0_10

to use a batch process are: the production, the initial investment and the possible implementation as a new product.

A large number of studies on the operating conditions of this reaction are available in the literature. This is because both resins such as Nylon 6,6 fibers have wide application and polymer production is emerging as one of the most profitable businesses nowadays.

To study this reaction what is called the "nylon salt" is employed which is a mixture of the reactants (Adipic acid and Hexamethylenediamine) in aqueous solution. As the reaction advances, water is produced; in the same way that the initially fed water is evaporated.

The polymers module, Aspen Polymer Plus, should be installed additionally to the basic Aspen Plus package and includes databases, kinetic models, properties, and thermodynamic models for polymer processes as well as other compounds commonly used in the polymer industry. In the case of chain growth reactions (such as the polymerization of Nylon 6,6) a few components, named segments, are required by Aspen Polymer Plus for computations. These correspond both to the end and the repeated fragments within a polymer molecule and determine its size.

The reactions used are of two types: condensation and arrangement. The general form of these chemical reactions is shown next:

$$P_n + P_m \leftrightarrow P_{n+m} + H_2O \tag{10.1}$$

$$P_n + P_m \leftrightarrow P_{n+m-q} + P_q \tag{10.2}$$

Aspen Polymer Plus denotes direct condensation reactions as CONDENSATION, reverse reactions as REV-CONDENS. While reactions of arrangement, direct or inverse, are cataloged as POLYMERIZAT.

There is a differentiation in the simulator between end and repetitive segments, which are explained with the polymer of study in this problem.

In Fig. 10.2 one can see the repetitive fragment of the molecule Nylon 6,6. The red square segment is to be derived from adipic acid; similarly, the blue segment is from hexamethylenediamine (HMDA). Therefore, it is required to enter a repeating segment of adipic acid as well as of hexamethylenediamine (HMDA) (Odian 2004) (Fig. 10.2).

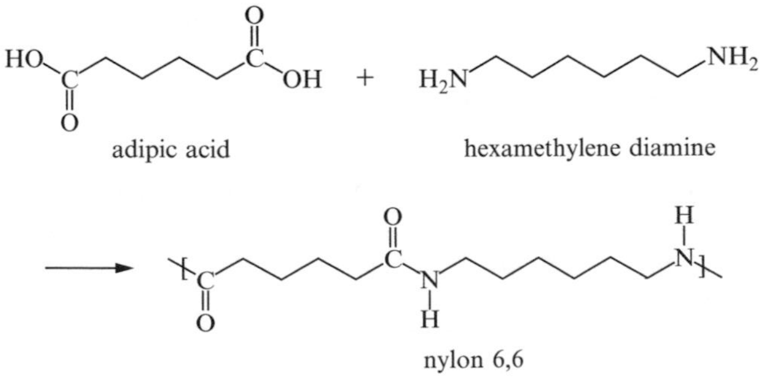

adipic acid   hexamethylene diamine

nylon 6,6

Fig. 10.1 Polymerization reaction for Nylon 6,6 production

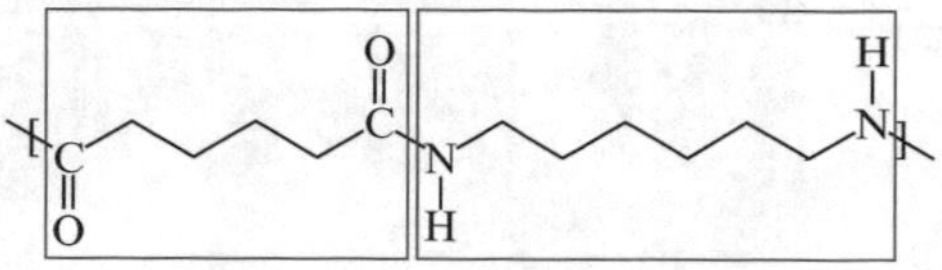

Fig. 10.2 Molecule
segments for Nylon 6,6

As the end segments correspond to the ends of the molecule; end segments of each reagent for each side of the molecule should be added. The reaction kinetics of polymerization varies depending on the mechanism of polymerization in which the reaction is carried out. There are several polymerization mechanisms that are not explained in this text but can be found in the bibliographical sources. In this case, the reaction is carried out by the chain growth mechanism, and the following data are reported in the literature (Seavey and Liu 2008):

$$r_1 = 7 \times [\text{NH}_2][\text{COOH}] \times e^{\frac{-2.5 \times 10^6 \text{J/kmol}}{RT}} \tag{10.3}$$

$$r_2 = 0.014 \times [\text{H}_2\text{O}] \times e^{\frac{-2.5 \times 10^6 \text{J/kmol}}{RT}} \tag{10.4}$$

These data can be used for the simulation; however, the kinetics input and how to observe the results, especially the specific properties of the polymer is shown first.

10.2.2 Polymerization Reaction Kinetics

First, start a simulation in Aspen Plus using *Polymers with metric units* option in the main window. In Flowsheet Run Type leave the default option *Flowsheet*, considering that equipment information will be entered (Fig. 10.3).

Now, proceed to enter the components as is typically done. The components to be added are shown in Table 10.1.

To change the component type click the box in the *Type* column and select the drop-down list. The components list as shown in Fig. 10.4 should appear.

Once the components are selected and after selecting the component NYLON-66 as Polymer type, the simulator requests information about the polymer in the route Components > Polymers. Click on this folder and a window that asks for information required for calculation, depending mainly on the role of the polymer in the simulation and, in this case, the polymerization mechanism by which the reaction takes place.

Click the Components > Polymers route; the Segments tab where you specify which segments are terminal or repetitive appears. Use the information reported in Table 10.2.

Note that segments with the "E" or "R" termination are named to differentiate segments at any time during the simulation (Fig. 10.5).

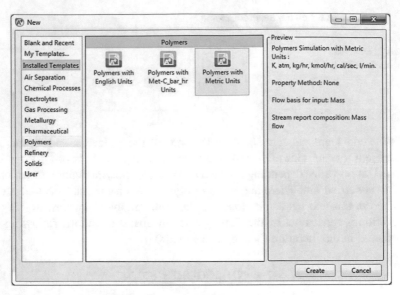

Fig. 10.3 New window for polymers in Aspen Plus®

Table 10.1 Components to be entered for this simulation in Aspen Polymer Plus®

Component	Type	Name	Formula
HMDA	Conventional	Hexamethylene-diamine	$C_6H_{16}N_2$
ADA	Conventional	Adipic acid	$C_6H_{10}O_4-D_1$
H_2O	Conventional	Water	H_2O
HMDA-E	Segment	Hexamethylene-diamine-E	$C_6H_{15}N_2-E$
ADA-E	Segment	Adipic acid-E	$C_6H_9O_3-E$
HMDA-R	Segment	Hexamethylene-diamine-R	$C_6H_{14}N_2-R$
ADA-R	Segment	Adipic acid-R	$C_6H_8O_2-R$
NYLON66	Polymer	Nylon-6,6	NYLON6,6

Fig. 10.4 Entered components in Aspen Polymer Plus®

Table 10.2 Segments input
in Aspen Polymer Plus®

Segment	Type
HDMA-E	End
ADA-E	End
HDMA-R	Repeat
ADA-R	Repeat

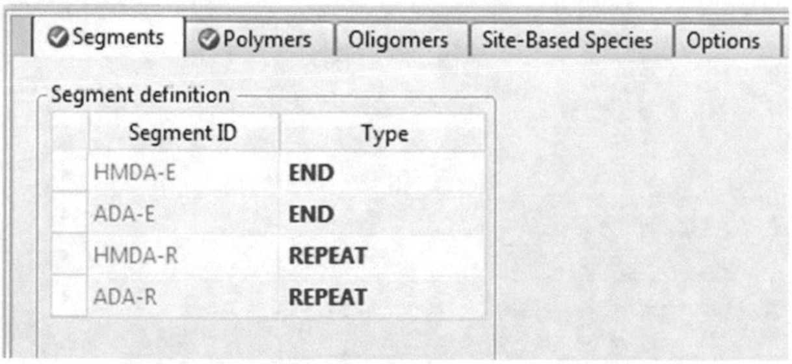

☑ Segments	☑ Polymers	Oligomers	Site-Based Species	Options

Segment definition

Segment ID	Type
HMDA-E	**END**
ADA-E	**END**
HMDA-R	**REPEAT**
ADA-R	**REPEAT**

Fig. 10.5 *Segments* window in Aspen Polymer Plus®

Now click on the *Polymers* tab where information about polymer characterization that allows observing the properties of importance is specified. This is used to monitor the reaction progress or results of operations involving the polymer.

In this window it can be observed that the Polymer-ID checkbox is selected for polymer *NYLON-66* because it is the only polymer declared in the Components window. Now, select the *Step-Growth* option in the *Built-in attribute group* to specify the appropriate set of attributes for the reaction mechanism for the production of this polymer. On the *Edit* button add or remove properties to adequately characterize the polymer depending on its nature and its role within the simulation. In this case, properties are not edited (Fig. 10.6).

With the information entered, the properties and nature of each of the components previously entered is defined. Now, the right property package that accurately represents the system is selected.

It is important to mention that the properties of the polymer and selected segments are calculated by a group contribution method developed by Van Krevelen (2009) Aspen Polymer Plus® has this method as default. However this can be edited by entering the groups that make up each molecule using the respective lists for the selected method (For more information see Aspen Technology, Inc. 2001).

In the path *Methods > Specifications* enter the model properties as usual, with the difference that in the *Process Type* box POLYMER is selected by default. Select the POLYNRTL model. This model is reported in the literature (Seavey and Liu 2008) as accurate to represent the system. However remember that thorough application the simulation results should be compared with experimental data to verify the correct selection of the thermodynamic model (see Chap. 2).

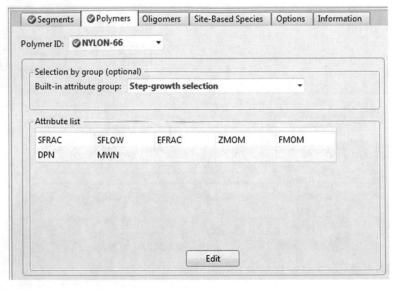

Fig. 10.6 Polymers window in Aspen Polymer Plus®

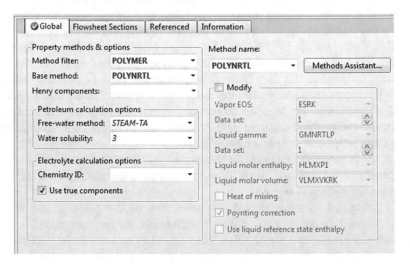

Fig. 10.7 Properties Package selection window in Aspen Polymer Plus®

Now the polymer-forming reaction is entered; select the following path *Simulation > Reactions* and click *New*. A window is displayed where the type of reaction is selected; select the option *STEP-GROWTH* (Fig. 10.7).

After selecting the STEP-GROWTH reaction type, corresponding to the chain growth mechanism of polymerization, a tab called *Species* which specifies the polymer to be produced, and the species (components, segments and groups)

Fig. 10.8 Reaction type selection window in Aspen Polymer Plus®

Table 10.3 Species entered in the reaction kinetics in Aspen Polymer Plus®

Species	Type	Observation
T-NH$_2$	E-GRP	Electrophilic group
T-COOH	N-GRP	Nucleophilic group
ADA-R	EE-GRP	Two-sided electrophilic group
HDMA-R	NN-GRP	Two-sided nucleophilic group

involved are displayed. Select the option *NYLON-66 Polymer* in the *Polymer* box (Fig. 10.8).

Reactions should be entered below; substances are represented with the corresponding simulation names (Wakabayashi 2008):

$$\text{HMDA} \rightarrow 2[T - NH_2] + [\text{HMDA} - R] \qquad (10.5)$$

$$\text{ADA} \rightarrow 2[T - COOH] + [\text{ADA} - R] \qquad (10.6)$$

$$[\text{HMDA} - E] \rightarrow [T - NH_2] + [\text{HMDA} - R] \qquad (10.7)$$

$$[\text{ADA} - E] \rightarrow [T - COOH] + [\text{ADA} - R] \qquad (10.8)$$

$$[\text{ADA} - R] \rightarrow [\text{ADA} - R] \qquad (10.9)$$

$$[\text{HMDA} - R] \rightarrow [\text{HMDA} - R] \qquad (10.10)$$

$$H_2O \leftrightarrow [T - COOH] + [T - NH_2] \qquad (10.11)$$

The information is entered in matrix form in the section called *Reacting species structure*. In the *Species* column enter the components declared according to the equations presented above. In the *Group* row input four groups with the characteristics summarized in Table 10.3.

In Fig. 10.9 the information is displayed considering the equations presented above:

Now on the Reactions tab, click the *Generate Reactions* button. With the information provided in matrix form, the simulator generates reactions taking place during the simulation from the reactions database (Fig. 10.10).

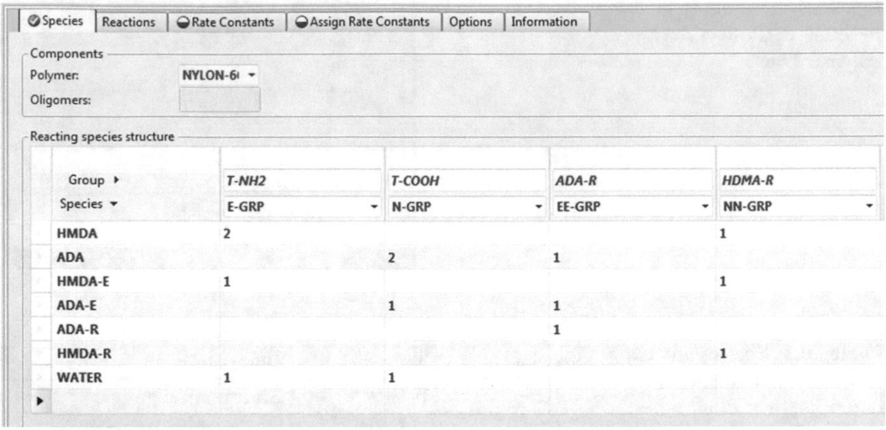

Fig. 10.9 *Species* window in Aspen Polymer Plus®

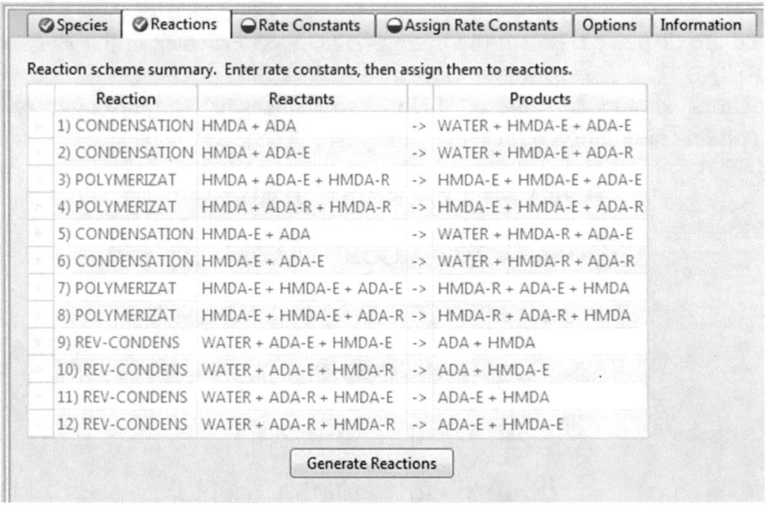

Fig. 10.10 *Reactions* window in Aspen Polymer Plus®

After the simulator generates the reactions, their kinetic parameters (pre-exponential factor and activation energy) are entered, only for the controlling steps: direct and reverse condensation; in this case the information reported in (10.3) and (10.4). Verify that the units of the pre-exponential factor are in sec^{-1} and of the activation energy in J/kmol.

Finally, the groups participating in each reaction (1 and 2 in the simulator) are assigned. From (10.3) and (10.4) it is observed that in the first reaction and the acid and amide groups are involved; and in the second reaction the term corresponding to the water concentration appears (Fig. 10.11).

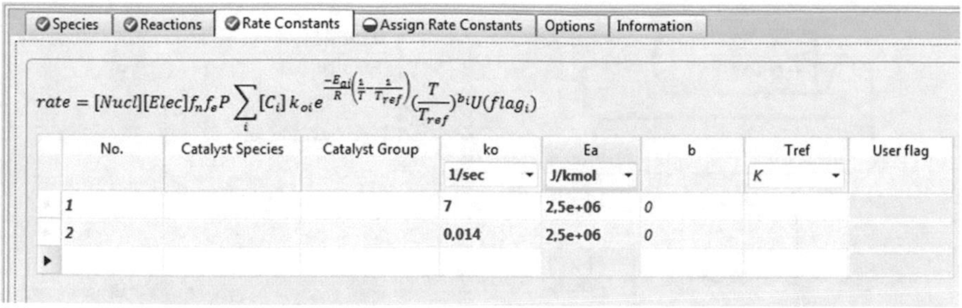

| | Species | | Reactions | | Rate Constants | | Assign Rate Constants | Options | Information |

$$rate = [Nucl][Elec]f_nf_eP\sum_i[C_i]k_{oi}e^{\frac{-E_{ai}}{R}\left(\frac{1}{T}-\frac{1}{T_{ref}}\right)}\left(\frac{T}{T_{ref}}\right)^{b_i}U(flag_i)$$

No.	Catalyst Species	Catalyst Group	ko	Ea	b	Tref	User flag
			1/sec ▾	J/kmol ▾		K ▾	
1			7	2,5e+06	0		
2			0,014	2,5e+06	0		
▸							

Fig. 10.11 *Rate Constants* window in Aspen Polymer Plus®

| | Species | | Reactions | | Rate Constants | | Assign Rate Constants | Options | Information |

Reaction set assignment
◉ Global   ○ Individual

No.	Attacking reactant			Victim reactant			Rate constant sets
	Nucleophilic Species	Electrophilic Group	Electrophilic Species	Nucleophilic Group	Nucleophilic Species	Electrophilic Group	
1		T-NH2		T-COOH			1
2	WATER						2
▸							

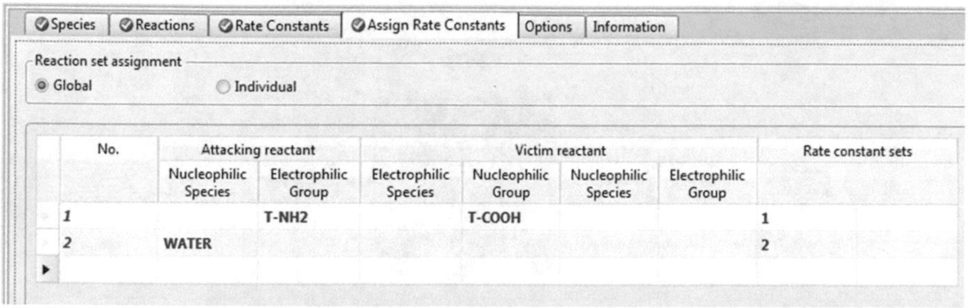

Fig. 10.12 *Assign Rate Constants* window in Aspen Polymer Plus®

Go to the *Assign Rate Constants*, select *Global* in the *Reaction set assignment* section, ensuring that the information entered is used for the global reaction set and not for individual reactions. In the lower part of the window, enter the information available above equations. The entered information is displayed as seen in Fig. 10.12.

Note that in the first reaction, the amide group (electrophilic group) attacks the acid group (nucleophilic group) and that in the *Constant Rate Sets* column the kinetic information of reaction 1 was assigned in the second reaction water is formed and the kinetic set number 2 is assigned.

With the above the reaction for the production of Nylon 6,6 was completely defined. Now the two production possibilities, continuous or batch, are studied. For this purpose information on this process is available in the literature.

10.2.3   Continuous Production

The continuous production process is widely found in literature. A diagram of typical process employed for the production of Nylon 6,6 is shown next.

In Fig. 10.13 the main operations for the industrial production of Nylon 6,6 industrially is shown. First, nylon salt is pumped to the evaporator where it is taken

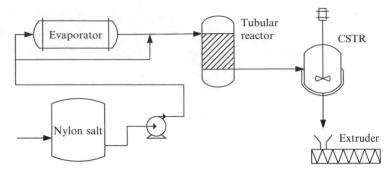

Fig. 10.13 Process flow diagram for continuous production of Nylon 6,6 resins (Giudici 2006)

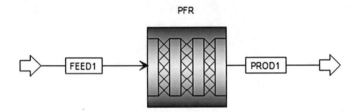

Fig. 10.14 Simulation flow diagram for production of Nylon 6,6 resins

to the appropriate concentration to enter the reactor. Storage takes place in a tank and the nylon salt is diluted to prevent extensive corrosion.

After evaporation, the salt enters the tubular reactor of about 1000 m in length with a bank of helical tubes ranging in diameter along the reactor to compensate for viscous effects prevailing during the reaction of the polymer chain growth.

Subsequently, a second reaction step is carried out in a CSTR reactor in which water is separated and the chain growth process continues. Finally, the resulting viscous mixture, falls by gravity into an extruder that shapes the product while excess water that was not removed previously is evaporated.

For purposes of this text the analysis is limited to the reactor and the analysis is made around the polymerization behavior. To do that a PFR reactor model is employed (Fig. 10.14).

The operating conditions of the process must be optimized beforehand for best results. Sensitivity analysis was performed, besides some experimental data reported in the literature was used. In Table 10.4 the results for the operating conditions are reported.

Aspen Polymer Plus is very robust for the analysis of the degree of polymerization and chain length distribution of the reaction product. For this analysis, the chain length behavior along the reactor is shown since it is an indicator of the polymer growth.

It can be seen in this graph that the chain grows along the reactor length, reaching a value sufficient to be considered a resin. If the reactor had a greater

Table 10.4 Experimental data for production of Nylon 6,6 reaction

Variable	Value	Source
Temperature	280 °C	(Van Krevelen 2009)
Pressure	5 atm	(Kumar and Gupta 2003)
Feed composition (% Mass)		
Adipic acid	20	(Kumar and Gupta 2003)
HMDA	20	
Water	60	

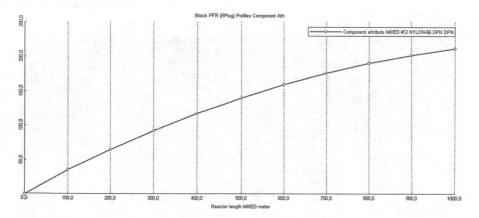

Fig. 10.15 Growth behavior for Nylon 6,6 chain across the reactor

residence time (which translates into a greater length) the chain would grow until it is large enough to be a limiting factor of its own growth. This behavior is observed in the change in slope of the curve presented in Fig. 10.15.

10.2.4 Batch Production

In order to study a batch process operation, a reaction time of 12 h was selected to reach a high enough growth of the polymer chain. The flow diagram for the simulation is shown in Fig. 10.16. Install an *RBatch* model which calculates a batch reactor, having as output a stream of water vapor *VENT*, and the other with the product of the polymerization *PROD1*.

The *FEED2* stream has the same composition and conditions as the inlet stream as the PFR reactor, since the two alternatives shall be evaluated to determine which is apparently more promising (Fig. 10.17).

On the *Reactions* tab enter the information corresponding to the reactions. Activate the *Reactive system* option and select the reaction set *R-1* which was introduced in the previous case. The window should be as shown in Fig. 10.18.

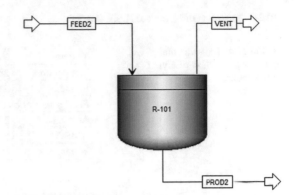

Fig. 10.16 Simulation flow diagram for batch simulation

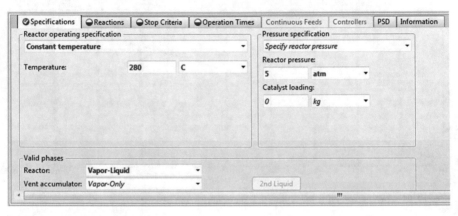

Fig. 10.17 Main screen of batch reactor configuration in Aspen Polymer Plus®

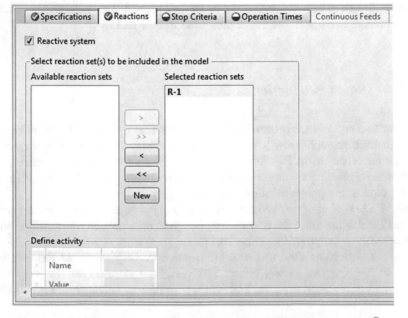

Fig. 10.18 *Reactions* window in batch reactor configuration in Aspen Polymers Plus®

Table 10.5 Inlet information for stop criteria for batch reactor

Cell	Value
Criterion	1
Location	Reactor
Variable type	Time
Stop value	12

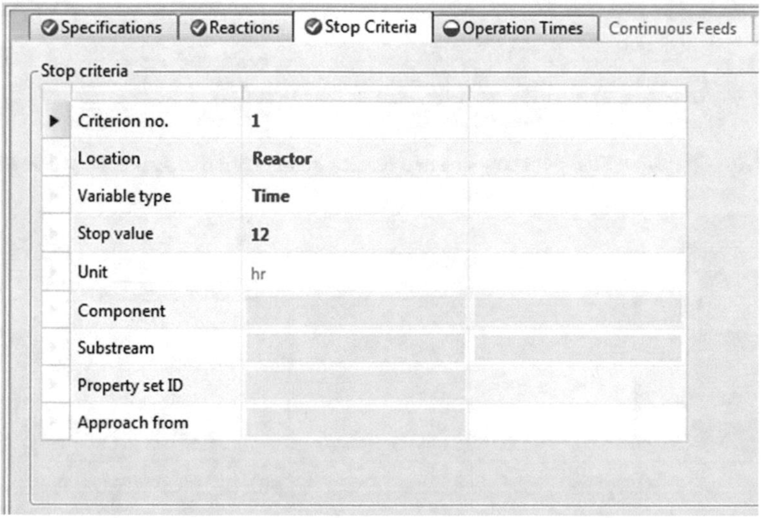

Fig. 10.19 *Stop Criteria* window in batch reactor configuration in Aspen Polymers Plus®

Table 10.6 Inlet information for batch reactor operation

Cell	Value
Batch feed time	1 h
Down time	1 h
Maximum calculation time	22 h
Time interval between profile points	0.1 h
Maximum number of profile points	222

Now in the *Stop Criteria* tab, the criteria considered for the operation to stop are entered. For this case study, time is considered the stop criterion; input the information in Table 10.5. Consider that another stop criterion can be implemented such as pressure, any substance concentration in case the reaction time required to achieve a given conversion is to be estimated, etc. (Fig. 10.19).

Finally, in the *Operation Times* tab, enter the time information for batch reactor; this is feeding time, cycle time and information for the mathematical calculation. The information to be entered is reported in Table 10.6.

With this information batch reactor module is completely defined. In Fig. 10.20 the *Operation Times* window completed is shown.

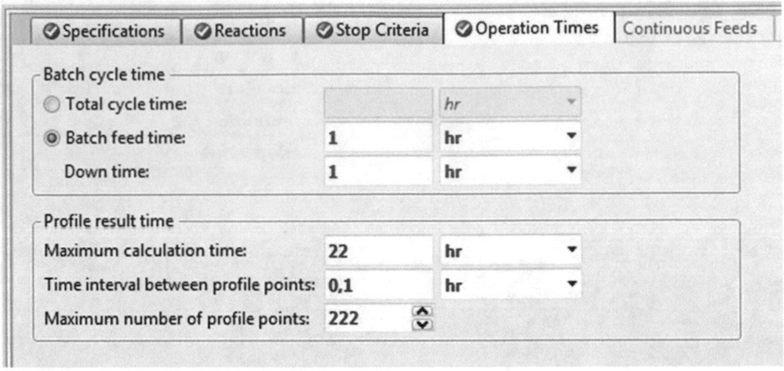

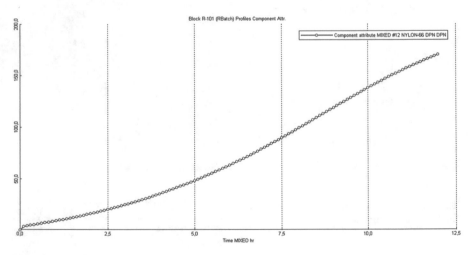

Fig. 10.20 *Operation Times* window in batch reactor configuration in Aspen Polymers Plus®

Fig. 10.21 Nylon 6,6 polymer chain growth behavior as function of batch time

After specifying the batch reactor the corresponding calculation is performed and the polymer chain growth is obtained as function of batch time is shown in Fig. 10.21.

10.2.5   Results Comparison

In order to make an objective comparison among results a graph must be generated showing polymer chain growth in terms of PFR reactor residence time. This can be compared in a single graph such compiling the results of the two reactors (Fig. 10.22). This graph can be obtained by taking the data tabulated in the *Profiles* tab of the two reactors and selecting the *DPN NYLON-66* component attribute.

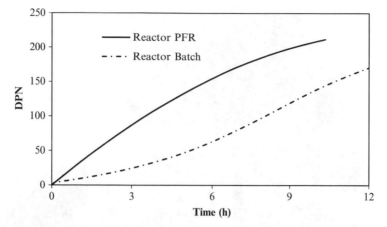

Fig. 10.22 Polymer chain growth comparison for the batch and continuous reactors

This graph can corroborate what is observed when the fundamentals of chemical reactors are studied; continuous reactors achieve greater conversion for most reactions. The analysis can be extended to polymers and in this case, can take the polymer chain growth as an indicative parameter of reaction yield.

However, when production quantities are small (<500 t/year) it is more favorable to make production in a batch system. For this system it is observed that the difference between the chain lengths after 10 h did not exceed 25 %. This may be acceptable depending on the Nylon 6,6 application for which the product is intended.

Additionally, batch production allows a much better fixed cost and large enough manufacturing flexibility when working with small markets and productions. This is the reason why many industries prefer batch production schemes.

With the results observed in Fig. 10.22 it can be concluded that it is possible to perform the manufacture of Nylon 6,6 using batch reactors. Even if that implies slower growth rate in the polymer chain, due to reduced capital investment requirements.

Proper production scheduling, parallel operation, and reaction time plan depending on customer needs can allow batch production to be much more beneficial from an economic point of view for a company.

In Fig. 10.23, a basic 3D model of a batch reactor designed for the production of Nylon 6,6 resin is shown.

Note that operational information for reactor design was entirely a result of Aspen Polymer Plus® simulation. It is evident that simulation can be a powerful tool for process design in the chemical industry.

However, additional information is required about the substances involved in the material selection, process control fundamentals and complemented with some heuristics for decision making which can lead to a proper process equipment design. However, in early stages the design assistance provided by computational tools is useful.

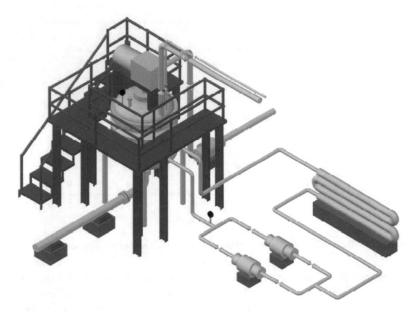

Fig. 10.23 Basic 3D model for Nylon 6,6 resin batch reactor

10.3   Azeotropic Distillation of Water–Ethanol Mixture Using Cyclohexane as Entrainer

10.3.1   General Aspects

It is desired to produce 300,000 L/day of anhydrous alcohol for use as fuel. For this, the technology employed azeotropic distillation using cyclohexane as a stripping agent is studied, which modifies the relative volatility of the components of the mixture, thus overcoming the barrier created by the azeotropic distillation. The azeotropic distillation is characterized by the formation of new azeotropes between components, which in this case are two minimum-boiling points: A binary one (water and cyclohexane) and a ternary one; the formation of these new azeotropes define the distillation zones, in each of them, pure components may be obtained. That is the reason why the simulation of azeotropic distillation shows convergence problems, which makes it necessary a preliminary study on the possible operating conditions to obtain the desired products.

The ethanol dehydration process using cyclohexane as entrainer is performed in two distillation columns and a decanter. The first distillation column operates without a condenser, and azeotropic mixture is fed with the solvent. In the bottom, ethanol is obtained anhydrous, whereas on the top stream a ternary mixture is obtained near the ternary azeotrope, which is then cooled down to be sent to a decanter, where, by liquid–liquid heterogeneous equilibrium effects, ethanol is

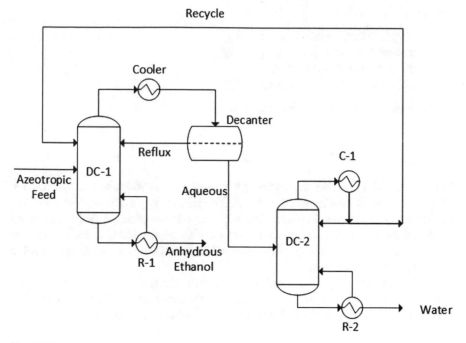

Fig. 10.24 Azeotropic distillation process flow diagram

separated from water. The organic phase is returned to the top of the azeotropic column as reflux, containing cyclohexane serving as solvent. Furthermore, the aqueous phase is fed into the second distillation column, in which cyclohexane is recovered as top product, which is recycled to the first column, thereby completing the amount required to perform the dehydration; as bottom product water is obtained. Figure 10.24 shows the process flow diagram.

Azeotropic distillation simulation presents convergence problems when closing recycles. The problems are associated to the proper estimation of flow rates and compositions for reflux and recycle streams, corresponding to the organic phase and the distillate of the second column, respectively. The following explains in detail procedure for closing recycles in the simulation.

10.3.2 Process Simulation

It must initially establish the solvent and azeotropic ethanol flow entering the dehydration column to meet the established production goal. To determine the azeotropic mixture amount use the following equation:

$$F = \frac{q}{x_{\text{EtOH}}} \times \overline{\rho_{\text{Mixture}}} \times \frac{1 \text{ day}}{24 \text{ h}} \tag{10.12}$$

where:

F: Azeotropic ethanol mole flow (kmol/h).
q: Volumetric flow to be produced (m³/day).
x_{EtOH}: Ethanol feed mole fraction.
$\overline{\rho_{Mixture}}$: Mixture molar density (kmol/m³).

Substituting the known values into (10.12) is:

$$F = \frac{300 \text{ m}^3/\text{day}}{0.885} \times 17.32 \frac{\text{kmol}}{\text{m}^3} \times \frac{1 \text{ day}}{24 \text{ h}} = 244.64 \frac{\text{kmol}}{\text{h}} \tag{10.13}$$

The solvent which assists the separation is recycled into the column: the organic phase from the decanter, which acts as the dehydration column reflux and the distillate from the second column. Initially, the organic phase flow (reflux) is set to 489.28 kmol/h, which corresponds to twice the azeotropic feed, and the amount of distillate from the second column (recycle) is set to 244.64 kmol/h, equivalent to the azeotropic mixture feed flow.

To estimate the recycle and reflux compositions, a residue curves map (RCM) system is used; this was generated using the ternary map tool from Aspen Plus. Figure 10.25 presents RCM at 2 atm and 40 °C where three distillation zones are

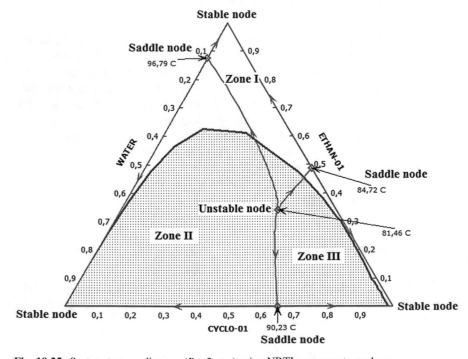

Fig. 10.25 System ternary diagram ($P = 2$ atm) using NRTL as property package

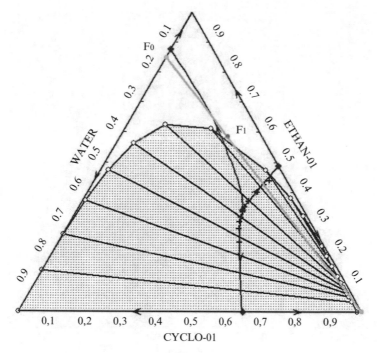

Fig. 10.26 Steps 1 and 2 represented in ternary diagram

observed, in all of them the unstable node is the ternary azeotrope, binary azeotropes are the saddle nodes, and the vertices of each pure component correspond to
the stable nodes. This indicates that ethanol and water can be obtained as bottom
products if mixtures located in zones I and II of the map are distilled, and as a result
a ternary mixture with a near-azeotrope composition as top product, its composition
may vary depending on the balance line to be established in each column. To set
proper balance lines in each column follow the steps described below (see
Fig. 6.28):

1. Azeotropic feed (F0) is located in the map.
2. As purity of cyclohexane stream entering the column is unknown, F0 is initially
   mixed with pure cyclohexane, obtaining F1 mixture, which should be located in
   zone I of the MCR (Fig. 10.26).
3. Subsequently, the mixture F1 is distilled to obtain as bottom product anhydrous
   ethanol (B1), whilst on top product a mixture with a near the ternary azeotrope
   (D1) (Fig. 10.27).
4. As D1 is located in the partial miscibility area, the mixture is decanted. The
   products of this operation are shown in the β and α points corresponding to reflux
   and feed to the recovery column, respectively.

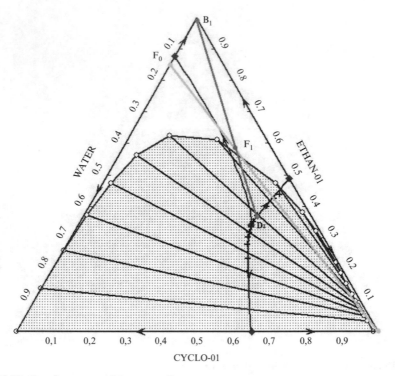

Fig. 10.27 Step 3 represented in ternary diagram

5. The aqueous phase (α) enters the recovery column to be distilled, thereby obtaining pure water as bottom product (B2), and while as the top product a ternary mixture (D2).

6. Considering that what actually enters to the dehydration column is a mixture of F0, D2, and β, proceed to make such a mixture, resulting in a new feed stream (F1–II) to the column.

7. Repeat as necessary to achieve convergence of values the compositions of each stream.

8. Calculate the flows of each stream using the lever rule for each balance line (Fig. 10.28).

By this method the results for the different process streams were obtained. The results of the iterations are reported in Table 10.7.

Once the amounts and compositions of reflux and recycle streams are set, define the inlet pressure and temperature. It is known that the temperature of the ternary azeotrope at 1 atm pressure (operating condition to the decanter and the recovery) is 62.4 ° C, and considering that the reflux and recycle streams correspond to ternary mixtures, it is established that the inlet temperature of such streams into the first column is 60 ° C. Regarding pressure, it is necessary

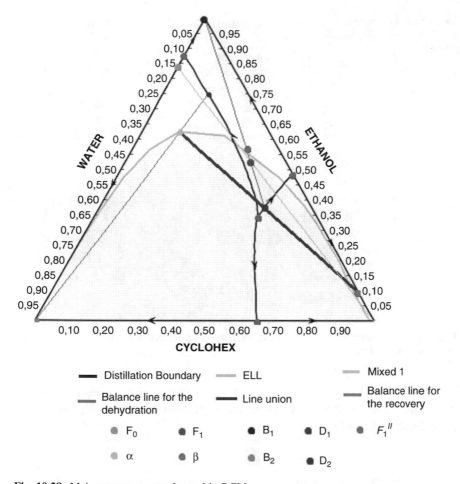

Fig. 10.28 Main process streams located in RCM

Table 10.7 Iteration results for β y D_2 stream compositions

		Mole fraction		
Iteration	Stream	Ethanol	Water	Cyclohexane
1	B	0.1	0.01	0.89
	D_2	0.65	0.1	0.25
2	B	0.11	0.01	0.88
	D_2	0.72	0.1	0.18
3	B	0.12	0.01	0.87
	D_2	0.81	0.1	0.09
4	B	0.07	0.01	0.92
	D_2	0.8	0.1	0.1
5	B	0.06	0.01	0.93
	D_2	0.8	0.1	0.1
Average	B	0.092	0.01	0.898
	D_2	0.756	0.1	0.144

Table 10.8 Inlet stream conditions to the dehydration column

Parameter	F_0	Reflux	Recycle
Temperature (°C)	78.2	60	60
Pressure (atm)	2	2	2
Flow (kmol/h)	244.64	489.28	244.64
Mole fraction			
Ethanol	0.885	0.092	0.756
Water	0.115	0.01	0.1
Cyclohexane	0	0.898	0.144

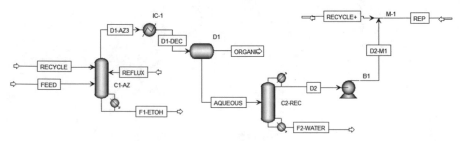

Fig. 10.29 Simulation flow diagram of ethanol–water azeotropic distillation using with open recycles in Aspen Plus®

that such streams enter at a minimum of 2 atm, which is the operating pressure of the column. Table 10.8 summarizes the main characteristics of the feed streams to the dehydration column.

10.3.2.1   Process Equipment

The flowsheet in Aspen Plus® is going to be implemented at first is shown in Fig. 10.29. As a first approach to the problem, recycles are not closed since it is still necessary to adjust the conditions of the inlet streams. Also because the system is very sensitive and some of its operating points are very close to the limits of distillation.

To set the initial assumption of the flow and composition of reflux and recycle streams entering the dehydration column, it is necessary to simulate the entire process, and after that to vary the reflux flow until a match between the quantity and composition of the *Organic* stream with *Reflux* and *Recycle* with *Recycle* + streams (Fig. 10.29). Then the operating conditions of each process equipment are specified.

The first equipment corresponding to the C1-AZ column is simulated as a type RADFRAC column without condenser, with 31 ideal stages, and operating at 2 atm, because at this pressure zone I is wider, which helps obtaining of anhydrous

Table 10.9 Operating conditions for C1-AZ column

Cell	Value
No. of stages	31
Feed stages	
Feed	15
Reflux	1
Recycle	10
Condenser type	None
Bottoms mole flow (kmol/h)	216.51
Stage 1 pressure (atm)	2
Convergence option	Azeotropic

ethanol. The feed enters the stage 15 and the recycle stream enters at stage 10. Since this column has no condenser, it is necessary to supply the reflux stream in the first stage, ensuring that the column has a reflux. In addition, it specifies that the convergence is azeotropic. These data are summarized in Table 10.9.

Then the *HE-1* cooler must be specified. The vapor stream coming out as distillate from the dehydration column cools down in this heat exchanger. It is simulated as a *Heater* module and operates with a pressure drop of 0.1 atm with an outlet temperature of 40 °C, at which there is partial miscibility, producing two liquid streams.

Then, vessel D-1 is installed, which is simulated as *Decanter* type separator. The organic (light phase) corresponds to the *First Liquid* outlet stream in the decanter, while the aqueous phase (heavy phase) is specified as a *Second Liquid*. The decanter operates at 2 atm, because as mentioned above, at this pressure the distillation zone is bigger and contains a minor part of the immiscibility region, this aid in the convergence of the simulation. Set it as an adiabatic decanter.

Now the regenerating column C2-Rec is installed. This also is simulated as a RadFrac type column with 22 ideal stages and total condenser operating at a pressure of 1 atm. The feed, which corresponds to the aqueous phase leaving the decanter, enters the column at stage 11. Set the convergence as azeotropic, which makes the column module run quickly and easily when there are changes in the inlet stream.

This column recovers most of the water entering the separation system, specify that;

$$F_{\text{bottoms},2} = F_{\text{Feed}} \times x_{\text{Water,Feed}} = 244.64 \, \frac{\text{kmol}}{\text{h}} \times 0.115 = 28.13 \, \frac{\text{kmol}}{\text{h}} \quad (10.14)$$

The mixture to be separated in this column consists mainly of water, which makes the separation easily achieved, and a small amount of reflux is required, a mole reflux ratio of 0.3 is specified. Table 10.10 summarizes data for *C2-REC* column.

The distillate of the second column is recycled to the dehydration column, and it is necessary that its pressure is greater than the pressure entering the stage; as the dehydration column is operated at constant pressure, the pressure of the stream

Table 10.10 Operating
conditions for C2-REC
column

Cell	Value
No. of stages	22
Feed stages	
Aqueous	11
Reflux ratio	0.3
Condenser type	Total
Bottoms mole flow (kmol/h)	28.13
Stage 1 pressure (atm)	1
Convergence option	Azeotropic

should be greater than the operating pressure of such column. To achieve this, a pump is used, which specifies a discharge pressure of 2 atm.

Finally you must install the mixer M-1. To this block enters the outlet stream from the pump to mix with the cyclohexane makeup stream, which enters at 2 atm and 60 °C. The flow rate is specified as 1 kmol/h, as will subsequently calculating the flow. The unit is simulated as a mixer and operation conditions set by default is allowed.

10.3.2.2 Convergence and Specifications for Recycles

When making a first calculation of simulation it is indicated that the bottoms flow of the distillation columns are not producing ethanol and water with the desired purity. The reason being that the amount of cyclohexane entering the dehydration column is high, and making the operation distillation zone II instead of I, that is, as bottom product mostly cyclohexane is obtained, and as top product a mixture close to the azeotrope between ethanol and cyclohexane, plus some water. This makes the compositions of the aqueous and organic phase leaving the decanter differ from the expected values, and by distilling the aqueous phase, which is now part of distillation zone I rather than II, most of cyclohexane is recovered by top, while the water recovery is very poor. Table 10.11 shows the simulation results when the *Reflux* stream flow is 489.29 kmol/h.

The first precise modification is on the compositions and flow of the recycle and reflux streams. This should correct the result described above decreasing the reflux stream flow. As the flow of this stream decreases, the purity of ethanol and water increases in bottoms in the dehydration and regenerating columns respectively. The initial value of mole reflux stream flow was reduced to obtain a correlation between the composition of the Reflux and Organic streams, as well as Recycle and Recycle +streams. In Table 10.12 the results of the simulation are shown when the reflux stream flow is 269.1 kmol/h, which is the value that produces best results, which corresponds to 55 % less than the assumed value.

By comparing the simulation results (Table 10.11) with assumed values for composition of Reflux and Recycle streams (Table 10.12), it may show that the values are very close together, so take the results of composition of such streams as

Table 10.11 Simulation results for reflux flow rate of 489.29 kmol/h

Stream	Feed	Recycle	Reflux	D1-AZ3	F1-Etoh	Organic	Aqueous	D2	F2-Wat
Temperature (°C)	78.2	52	52	83.3	84.8	41.9	41.9	63.3	78.8
Pressure (atm)	2.027	2.027	2.027	2.027	2.027	1.013	1.013	1.013	1.013
Flow (kmol/h)	244.64	244.64	489.28	762.05	216.5	334.44	427.61	399.48	28.13
Mole fraction									
Ethanol	0.885	0.755	0.092	0.421	0.58	0.096	0.675	0.676	0.657
Water	0.115	0.1	0.01	0.075	0	0.003	0.132	0.117	0.343
Cyclohexane	0	0.145	0.898	0.504	0.42	0.902	0.193	0.206	0

Table 10.12 Simulation results for reflux flow rate of 269.1 kmol/h

Stream	Feed	Recycle	Reflux	D1-AZ3	F1-Etoh	Organic	Aqueous	D2	F2-Wat
Temperature (°C)	78.2	52	52	82.6	92.7	42.8	42.8	64	80.4
Pressure (atm)	2.027	2.027	2.027	2.027	2.027	1.013	1.013	1.013	1.013
Flow (kmol/h)	244.64	244.64	269.1	541.87	216.5	252.59	289.28	261.1	28.13
Mole fraction									
Ethanol	0.885	0.755	0.092	0.404	0.957	0.087	0.68	0.709	0.407
Water	0.115	0.1	0.01	0.102	0	0.003	0.189	0.145	0.593
Cyclohexane	0	0.145	0.898	0.494	0.043	0.91	0.131	0.146	0

Table 10.13 Iteration result of *Reflux* and *Recycle* streams compositions

Iteration	I_0	I_1	I_2
Recycle flow (kmol/h)	244.64	244.64	244.64
Reflux flow (kmol/h)	498.28	498.28	498.28
Organic mole fraction			
Ethanol	0.096	0.087	0.083
Water	0.003	0.003	0.003
Cyclohexane	0.902	0.910	0.914
D_2 mole flow (kmol/h)	399.48	261.15	254.13
D_2mole fraction			
Ethanol	0.676	0.709	0.704
Water	0.117	0.145	0.179
Cyclohexane	0.206	0.146	0.117

the new group of input data to the simulation. When making this change, the values of streams compositions iteration closer to the second assumption, and the value of the mole flow of the distillate 2 is closer to the assumed value. Table 10.13 shows the results of the iterations, wherein the third column corresponds to the input data to the simulation are summarized.

The results above demonstrate that the simulation is very sensitive to the amount of cyclohexane entering into the dehydration column in the *Reflux* stream. Through a trial and error process a mole flow value for this stream was found with which it is possible to obtain most ethanol as bottoms product from the dehydration column, most of the water is recovered as bottoms product of the regeneration column, besides obtaining values close to the assumed composition in the iteration streams. However, there is still no certainty about the value of the Reflux stream flow value, and anhydrous ethanol is not being produced.

Using a first design specification in the dehydration column is possible to obtain the value of the mole flow of *Reflux* stream where anhydrous ethanol is obtained: it is desired that the cyclohexane mole recovery in bottom product to be 0.003, by varying the mole flow of *Reflux* stream. If the specification is done with the ethanol mole recovery and no cyclohexane, most ethanol comes out as bottom product. This causes the amount that is obtained as top product to be less than that required to meet the established balance line (which involves obtaining a product with a composition close to the ternary azeotrope as top product) and this would result in convergence problems. Also, when specifying such a small value of cyclohexane as bottom product and knowing that when operating in distillation zone I, the amount of water cannot exceed 1 %, ensures that ethanol losses are minimal.

The value of mole flow for *Reflux* stream resulting from the first design specification value is taken as input to the simulation. Once results are obtained with this new value, a change is made in the design specification: it is desired that the mole cyclohexane recovery reaches 0.003 in the bottoms, but now varying bottoms product flow rate instead of the *Reflux* stream because once recycles are closed, such stream disappears.

Table 10.14 Design specification data for distillation columns

Parameter	C1-AZ (a)	C1-AZ (b)	C2-Rec
Specification	Mole recovery	Mole recovery	Mole recovery
Component	Cyclohexane	Cyclohexane	Ethanol
Value	0.003	0.003	0.0001
Stream	B1	B1	B2
Varying	Reflux flow	Bottoms product flow	Bottoms product flow
Limits	220–280 kmol/h	200–230 kmol/h	15–35 kmol/h

Table 10.15 Design specification data for mixer

Parameter	Value
Variable	*MKUP*
Type	Mole flow
Stream	*Recycle+*
Component	Cyclohexane
Specification	*MKUP*
Value	23.324
Tolerance	0.001
Varying	*MKUP* mole flow
Component	Cyclohexane
Limits	0.05–5 kmol/h

Furthermore, in order to obtain pure water on the bottoms from the second column, it is specified that mole ethanol recovery of the column to be 0.0001 varying the flow of bottoms product. Again, ethanol recovery instead of water is specified to ensure compliance with established balance line, and water losses at top product to be minimal.

Despite willing to produce pure ethanol and water as bottoms product in the columns, always small fractions of cyclohexane are carried over with the products, i.e., solvent lost in the separation progresses. To adjust the amount of pure cyclohexane makeup necessary to enter the system, a design specification in the mixer is performed, where the makeup stream is mixed with the distillate 2; this specification should be entered in the *Flowsheeting Options* menu navigation tree. The specification is defined as the mole flow of cyclohexane in the stream exiting the mixer (*Recy*) corresponds to the flow in *Recycle* stream which is being replaced to close the recycle varying the flow of cyclohexane in the makeup stream.

In Tables 10.14 and 10.15 the details of the design specifications are summarized. Keep in mind that the values shown in these tables apply only to this simulation. Due to the possibility of multiple steady states, the values may differ, especially in the design specification for the mixer (Table 10.15).

Before closing the recycles it is necessary to adjust the values of pressure and temperature on the initialization streams (Recycle and Reflux, Table 10.8) to the values obtained after running the simulation (*Organic* and *D2*, respectively). These adjustments are made to facilitate convergence of the simulation, since once the

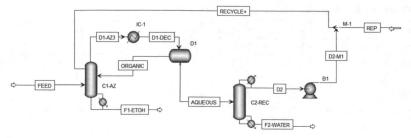

Fig. 10.30 Final simulation flow diagram for extractive distillation process

recycles are closed, the simulator iterates on the pressure and the enthalpy of the tear stream(s). In the case of reflux stream to the dehydration column, the temperature changed from 60 to 44.1 °C; also, the temperature of the recycle stream is manipulated from 60 to 65.4 °C. It is also important to adjust the temperature of the makeup stream, which changes from 60 to 65.4 °C.

To close recycles *Reflux* and *Recycle* streams should be deleted, *Recycle +* and *Organic* streams are connected to the dehydration column: *Organic* stream enters to the first stage of the column, while the *Recycle +* stream enters to the stage 10. It is important to close the two recycles simultaneously so that the simulation converges. In Fig. 10.30 the final flow diagram is shown in Table 10.16 and the main results of the simulation are reported after closing the recycles.

The results reported in Table 10.16 shows that the products of interest are obtained with the desired purity and quantity, which will have lower losses than 5 %. If one keeps track of the results of makeup mole flow (MKUP) these show that when closing the recycles the amount of solvent required is much lower, because the amount of cyclohexane leaving the system is much smaller, and the streams have a greater amount of solvent being recirculated. The results are important since they give reliable foundation for future studies in dynamic state and economic studies that consider the true solvent amount required for the operation; here lies the importance of closing the two recycles in the steady-state simulation.

10.3.3   Convergence Recommendations

While constructing the simulation some errors and warnings may occur; the following are some recommendations that may be useful to solve the most common problems:

- When iterating on the flow of Reflux stream, it is observed that as it lowers, the similarity between the corresponding streams is higher. Then, it is natural to continue lowering that value. If by doing so the dehydration column presents a convergence problem due to not fulfill the mass balance, it is necessary to increase the flow; since having a very low flow can cause some stages to run dry.

Table 10.16 Simulation results after closing recycles

Stream	Aqueous	F1-Etoh	F2-Wat	D1-AZ3	D1-DEC	D2
Temperature (°C)	44.1	96.6	99.8	81.5	40	65.3
Pressure (bar)	2.027	2.027	1.013	2.027	2.027	1.013
Vapor fraction	0	0	0	1	0	0
Mole flow (kmol/h)	311.46	217.24	28.12	596.75	596.75	283.34
Enthalpy (MMkcal/h)	−19.97	−13.85	−1.88	−25.32	−31.06	−17.88
Mole fraction						
Ethanol	0.646	0.997	0.001	0.374	0.374	0.71
Water	0.282	0	0.999	0.148	0.148	0.21
Cyclohexane	0.072	0.003	0	0.478	0.478	0.08
Stream	D2-M1	FEED	MKUP	ORGANIC	RECYCLE+	
Temperature (°C)	65.4	78.2	65.4	44.1	65.3	
Pressure (bar)	2.027	2.027	2.027	2.027	2.027	
Vapor fraction	0	0	0	0	0	
Mole flow (kmol/h)	283.34	244.64	0.71	285.29	284.05	
Enthalpy (MMkcal/h)	−17.87	−15.87	−0.03	−11.08	−17.9	
Mole fraction						
Ethanol	0.71	0.885	0	0.077	0.708	
Water	0.21	0.115	0	0.003	0.21	
Cyclohexane	0.08	0	1	0.92	0.082	

- It is very common for distillation columns to not converge within the number of iterations established by default before and after closing the recycles. If column convergence (*Configuration* tab) is azeotropic, change to *strongly non ideal liquid* option; then run the simulation, and after obtaining results, rerun with azeotropic convergence. Generally the azeotropic solution facilitates convergence of the column and has fewer errors when the simulation runs with recycles closed.
- If after changing the mode of convergence error still occurs, increase the number of iterations to *100* in the *Convergence* folder of the column module.
- When the convergence error involves that design specifications were not met because the manipulated values (what varies in each specification) are off limits, it is necessary to change them. In order to know which of the two limits to vary, it is necessary to know the result of the specification, which is observed in the *Results* tab of the *Vary* folder on the corresponding equipment, or the *Results* tab of the design specification in the *Flowsheeting Options* folder, when the specification is carried out on the equipment, as in the case of the mixer. The purpose of doing this is to extend the iteration interval, and then depending on the outcome the lower or upper limit varies accordingly.
- When running the simulation for the first time after closing the recycles, do not restart the simulation, since the values of the previous run are very close to the final solution. This also applies after running the simulation with closed recycles.

- The simulator selects a default tear stream. If after making a topological analysis concludes that the best tear stream is different than it is by default, it can be changed from the *Convergence* folder in the navigation tree, in the *Tear* option.
- Finally, it is recommended that the simulation progresses keep backup copies with different names; for example, save once each equipment converges, the complete simulation without design specifications, then the specifications before closing recycles, and finally, the simulation with closed recycles. Doing this facilitates comparison of results in order to improve the final result.

10.4 Ethylene Oxide Production

Ethylene oxide is a gas with a pleasant odor, colorless, flammable, and miscible in water and most organic solvents. It is a compound of very high reactive power and can lead to explosions when in contact with certain metals such as copper, silver, mercury, and magnesium. Furthermore, it is an industrial product which is obtained with high purity regardless of the method used for their production.

Ethylene oxide is an important product used as raw material in the glycols, polyglycols, and polyol production, which itself are used from the fiber industry, to foams and refrigerants, additives to detergents. Additionally, it can be used as a disinfectant, sterilant or fumigant when mixed with nitrogen, carbon dioxide, or dichlorofluoromethane, in nonexplosive proportions.

10.4.1 Process Description

The production of ethylene oxide is important in the chemical industry for both the polymer industry and for the production of ethylene glycol, the main component in automotive antifreeze. For the production of ethylene oxide, a partial oxidation of ethylene with oxygen is performed. Taking into account the explosive range of ethylene; it must be handled in excess of ethylene, which is recycled later.

Besides the reaction of interest (10.15), two undesirable reactions occur: The first, total oxidation ethylene (10.16) and the other is the oxidation product (10.17). However, the third is avoided with ethylene excess and the oxygen as limit reactant. For purposes of this exercise were taken into account only the first two reactions.

$$C_2H_4 + 0,5\ O_2 \rightarrow C_2H_4O \tag{10.15}$$

$$C_2H_4 + 3\ O_2 \rightarrow 2\ CO_2 + 2\ H_2O \tag{10.16}$$

$$C_2H_4O + 2,5\ O_2 \rightarrow 2\ CO_2 + 2\ H_2O \tag{10.17}$$

The production reaction is generally a low conversion one, which is carried out in the vapor phase in the PFR reactor (Plug Flow Reactor) operating at 180 °C and 2.1 MPa. A solid catalyst of silver supported on silica is used. The kinetic equations of the two main reactions are shown below:

$$-r_1 = \frac{3.1574 \times 10^4 \times \exp(-69.49/RT) \times P_{ET} \times P_O}{1 + 3.8414 \times 10^{-10} \times \exp(79.92/RT) \times P_{CO_2} + 9.5916 \times 10^{-16} \times \exp(134.10/RT) \times P_{O_2}^{0.5} \times P_{H_2O}}$$

$$(10.18)$$

$$-r_2 = \frac{1.6294 \times 10^8 \times \exp(-81.77/RT) \times P_{ET} \times P_{O_2}^{0.75}}{1 + 3.8414 \times 10^{-10} \times \exp(79.92/RT) \times P_{CO_2} + 9.5916 \times 10^{-16} \times \exp(134.10/RT) \times P_{O_2}^{0.5} \times P_{H_2O}}$$

$$(10.19)$$

Where:

   r_i = Reaction rate (kmol/s kg cat)
   P_i = Component i partial pressure (MPa)
   Ea_i = Activation energy (kJ/mol)
   ΔH_i = Heat adsorption (kJ/mol)

   Both rates are referred to ethylene production.
   Commercially, the ethylene oxide has been produced by two methods. The first involves the reaction of ethylene with hypochlorous acid followed by dehydrochlorination of the resultant chlorohydrin and ethylene oxide obtaining calcium chloride. However, currently the entire world production of ethylene oxide is carried out by the process of direct oxidation of ethylene in the presence of a silver catalyst (Fig. 10.31).
   In this case study, the gas compression from its source to the reaction conditions required is considered. Oxygen is considered to come from a cryogenic air distillation, so the initial conditions were set at 1 bar and 25 °C. Ethylene can come from a previous process or storage; for this reason, the conditions were set at 50 bar and 25 °C. Methane was considered coming from a pipeline at 10 °C and 8 bar.
   Thus, oxygen compression is performed in three steps with intermediate cooling to mitigate the thermal effects associated with compression. In a first compression is carried from 1 bar to 350 kPa, which leads to a temperature of 189 °C, then cooled to 35 °C and compressed to 700 kPa and then another cooling is the reaction pressure (2650 kPa). Finally it is heated to 230 °C.
   Then, entering the first reactor reaches a conversion of ethylene oxide of about 4 %; in order to perform the subsequent absorption using water must be compressed since it operates at 3000 kPa. Therefore, the thermal effect must be considered whereby the stream is cooled prior to compression and then again afterwards. This avoids an operation using refrigerant replacing it with two more conventional operations.

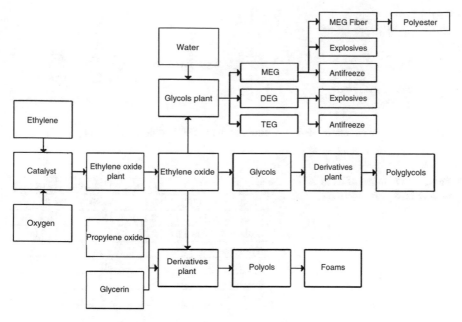

Fig. 10.31 Production line of products derived from ethylene oxide

The gas obtained from this operation is throttled and fed to the second reactor, because it has even high amounts of oxygen and ethylene, which achieves a conversion of about 4 % and the effluent is led to a second absorption using water.

10.4.2 Aspen HYSYS Simulation

In Fig. 10.32 the complete flow diagram from the simulation of the ethylene oxide (EO) production process is observed.

To start the simulation, enter the required components, the properties method, and the chemical reactions involved. To do this, enter the following components: Ethylene, Ethylene Oxide, Water, CO_2, Oxygen, and Methane. The properties method most suitable for the system according to the literature is NRTL-RK.

As explained above each of the gases involved in this process that are considered comes from previous processes therefore you must enter the information in Table 10.17.

10.4.2.1 Gas Compression

Since the reaction is carried out at high pressure, it is required to raise each carry streams installed to the reaction pressure. Because the flows are large enough

Fig. 10.32 Simulation flow diagram of ethylene oxide (EO) production process

Table 10.17 Information to be entered for raw materials

Name	Oxygen	Ethylene	Methane
Temperature (°C)	25	180	10
Pressure (MPa)	0.101	5	0.8
Mole flow (kmol/h)	2100	7840	18,060
Mole fraction			
Oxygen	1	0	0
Ethylene	0	1	0
Methane	0	0	1

Table 10.18 Information to be entered for compression in ethylene oxide process

Equipment	K-1	K-2	K-3	K-4
Inlet stream	Methane	Oxygen	Oxygen1*	Oxygen2*
Outlet stream	Methane1	Oxygen1	Oxygen2	Oxygen3
Energy stream	W-K1	W-K2	W-K3	W-K4
Outlet pressure (psia)	304.6	50.76	101.5	304.6

oxygen compression is performed in several steps since it is known that the compression efficiency is compromised and lots of energy is required to compress all in one step. Compression of methane is not excessive and can be considered in a single step. Instead the ethylene stream is being throttled to reduce its pressure is higher than as required for the reaction.

A compressor for the methane stream, a valve for the ethylene stream and three compressors for the oxygen stream are installed with the information shown in Table 10.18.

For ethylene case must be installed a throttle valve with the following characteristics (Table 10.19).

To counteract the effects of heating caused by the compression of the gases shall be installed intercoolers as can be seen in Fig. 10.32. In the case of methane heating is performed to take it to reaction conditions because it is not sufficiently heated. Information for coolers are reported in Table 10.20.

Finally a mixer is installed to join all gaseous streams for subsequent entry into the reactor. The equipment will be called *M-1* and the output stream *To Reactor*.

Table 10.19 Data to be entered in valve *V-1*

Module	V-1
Inlet stream	Ethylene
Outlet stream	Ethylene1
Outlet pressure (psia)	304.6

Table 10.20 Information for intercoolers

Equipment	E-1	E-2	E-3	E-4
Inlet stream	Methane1	Ethylene1	Oxygen1	Oxygen2
Outlet stream	Methane1*	Ethylene1*	Oxygen1*	Oxygen2*
Energy stream	Q-E1	Q-E2	Q-E3	Q-E4
Outlet temperature (°C)	180.6	180	50	40
Pressure drop (psia)	0	0	0	0

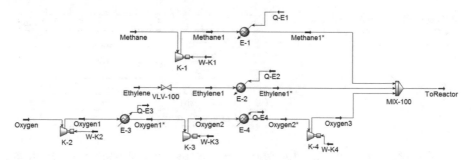

Fig. 10.33 Compression system flow diagram for ethylene oxide production

So far the simulation has the compression system and a flow diagram as shown in Fig. 10.33 is obtained.

10.4.2.2 Chemical Reaction

Prior to calculating the reaction conditions a sensitivity analysis on the conversion and selectivity should be performed since for this type of reaction a greater conversion causes a decrease in selectivity and vice versa. The ethylene–oxygen ratio was varied to find an optimal and a better selectivity. In Fig. 10.34 the results of the sensitivity analysis are reported.

The chemical reaction, when present, is always the most important stage of any process. In this case the reaction is carried out in a PFR reactor at 180 °C and 2.1 MPa (304.6 psia) in the gas phase. The reactor is specified as discussed in Chap. 5 with the example of methanol reforming reactor. The additional information for calculating the reactor is reported in Table 10.21.

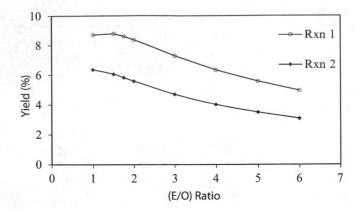

Fig. 10.34 Sensitivity analysis on reactor feed to improve reaction selectivity

Table 10.21 Geometric data for PFR reactor of ethylene oxide production

Parameter	Value
Tube diameter (mm)	31.3
Number of tubes	8821
Packing density (kg/m^3)	590
Void fraction	0.5
Packing length (m)	8.2
Coolant temperature (°C)	224

With the corresponding information the composition and temperature profiles along the reactor can be constructed as shown in Figs. 10.35 and 10.36. The reactor has an output stream called *To Sep*, the module is called *PFR-100* and the outlet temperature is 230 °C.

10.4.2.3   Separation

Similar to the sensitivity analysis for the reaction stage, another analysis for separating ethylene oxide from the gas stream must be performed. The process most commonly used is the absorption using water. For this operation, it is vital to properly determine the L/G ratio involving gas and liquid flows; this variable determines the diameter of the absorption column and the degree of separation that can be achieved. The number of stages of the column is important, as it generates more or less contact between the phases, though in a lesser degree.

For this analysis the L/G ratio was varied, that is fundamentally altering the water flow (solvent) since the amount of gas is already fixed by the process. The behavior of ethylene oxide fraction in the liquid phase recovered after absorption using the following formula was observed.

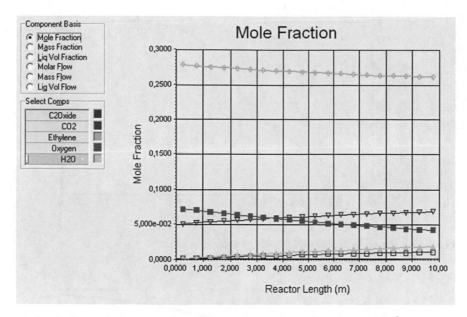

Fig. 10.35 Composition profiles along the PFR reactor obtained in Aspen HYSYS®

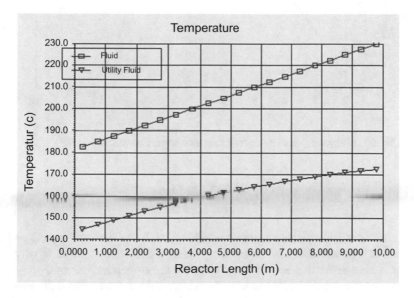

Fig. 10.36 Temperature profiles along the PFR reactor obtained in Aspen HYSYS®

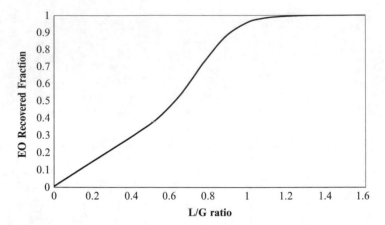

Fig. 10.37   Sensitivity analysis of L/G ratio on separation process

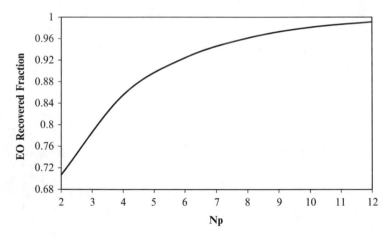

Fig. 10.38   Sensitivity analysis of number of stages on separation process

$$\text{R.F.} = \frac{\text{EO Liq Flow}}{\text{EO feed flow}} \tag{10.20}$$

Results from such analysis are reported in Fig. 10.37. It can be seen that with an L/G ratio one can obtain about 96 % recovery of ethylene oxide, which is why that amount is entered for the flow of water entering the absorption column.

Subsequently, an additional sensitivity analysis is performed to determine the number of stages required for the separation. The above analysis was carried out using a number of steps equal to 10, however it is possible that many stages are not required. The results of this analysis are reported in Fig. 10.38.

Table 10.22 Data to be
entered for absorption column

Parameter	Value
Name	C-1
Number of stages	8
Stage 1 pressure (psia)	431
Stage 8 pressure (psia)	435
Gas inlet stream	*Gas to Abs*
Gas outlet stream	*Gas*
Liquid inlet stream	Water
Liquid outlet stream	Wat + EO

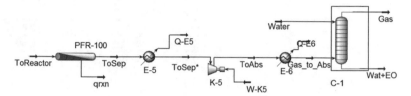

Fig. 10.39 Reaction and separation flow diagram for ethylene oxide production

It can be seen that with a L/G ratio equal to 1, 8 stages can be installed to achieve a recovery of 96 % of the ethylene oxide present in the reactor outlet. This is the information that enters the module in the absorption column simulation in Aspen Plus®.

A similar analysis can be performed for each stage of the simulator to provide the best information on the various operations.

The absorption takes place at a pressure of 435 psia thus is compressed *To Sep* stream leaving the reactor again. However it is known that compressing a gas heats it and it is also known to enter a cooler gas to the absorption stage, the separation is better; reason why is first cooled *To Sep* stream using a cooler (*E-5*) to a temperature of 75 °C.

Then, the mixture is compressed in compressor K-5 to a pressure of 435 psia and again using a cooler (E-6) the mixture is brought to a temperature of 64 °C. It can be considered that the two coolers have no pressure drop.

Now enter a Column Absorber module with the information found in Table 10.22.

With this information it can perform the calculation of the absorption column C-1. Please check the recovery with the selected operating conditions is reached after the calculation performed. In Fig. 10.39 the flow diagram shows from the reaction step to the absorption column.

The stream *Gas* then goes to a CO$_2$ removal process consisting of absorption with chemical reaction using a solution of K$_2$CO$_3$ in water. For the purposes of this exercise, this process is omitted. The second separation operation is to separate the ethylene oxide from the stream of water, to this end, two distillation columns were installed and will be placed a valve (*V-2*) that leads the *Water + EO* stream to a

Table 10.23 Data to be entered for distillation column *DC-1*

Parameter	Value
Name	DC-1
Number of stages	15
Top pressure	290
Bottom pressure	290
Feed	To Dist
Feed stage	7
Condensator	Partial
Gas outlet stream	Gases
Distillate outlet stream	EO*
Bottom outlet stream	Rec Wat 1
Specifications	
Reflux ratio	0.4579
Gas mole flow (kmol/h)	0.39
Distillate mole flow (kmol/h)	13.23

Table 10.24 Data to be entered for distillation column *DC-2*

Parameter	Value
Name	DC-2
Number of stages	10
Top pressure	290
Bottom pressure	290
Feed	EO*
Feed stage	5
Condensator	Partial
Gas outlet stream	Gases2
Distillate outlet stream	Ethylene oxide
Bottom outlet stream	Rec Water2
Specifications	
Reflux ratio	1.0
Gas mole flow (kmol/h)	0.1
EO mole fraction on top	0.9476

pressure of 290 psia. The outlet stream of this valve is to be called *Water + EO**. Thereafter the stream is heated to about 95 °C to enter the distillation column *DC-1* whose function is to remove small amounts of gases carried over in the absorption; reason why the pressure is decreased. And take ethylene oxide stream to about 50 % by weight. The column specifications are shown in Table 10.23.

Finally the *EO** stream is taken to a second distillation column to obtain 95 % pure ethylene oxide as top product and water as bottoms product. The specifications for this column are reported in Table 10.24.

With this information, the ethylene oxide process is complete. In Fig. 10.40 the distillation section of the process is shown.

To conclude, in this exercise one of the applications that process simulation has in process design is illustrated. Initially only known information about the kinetics

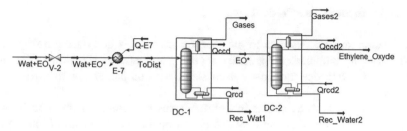

Fig. 10.40 Distillation section flow diagram for ethylene oxide production

of the process and certain conditions such as pressure were given. Doing sensitivity analyses and ensuring that the thermodynamic model is adequate, the other conditions can be estimated to perform a complete simulation of a process as it was carried out in this exercise.

10.5   Economic Evaluation Using Aspen Icarus® (Guevara 2010)

The economic evaluation is perhaps one of the most challenging tasks for chemical engineers, and is in turn one of the most important tasks to be done to implement a design of a process, a modification of the existing process or decision about purchase of equipment.

For this analysis, Aspen Tech Engineering Suite® has a powerful tool: Aspen Icarus Process Evaluator®. This program is linked with the process simulators used in the present text, Aspen Plus® and Aspen HYSYS® for simulation information for sizing, estimating equipment costs, and ultimately economic evaluation.

10.5.1   General Aspects

For proper economic evaluation it should be performed prior in-depth analysis of some economic indicators that influence the behavior of investment in time such as wages, wage growth, inflation, price forecasts for raw materials and products, percentage of income tax, utility pricing, etc. This analysis should be done with official data and referenced to reliable sources.

For the following example information related to Colombia and any information relevant to the region of Valle del Cauca in order to perform the evaluation of the construction of the plant in that region is taken. The other parameters were taken from the Asocaña annual balance (Local association), the Ministry of Social Protection, and finally the Ministry of Mines and Energy. The economic indicators are vital for economic evaluation as they vary in time as depending on the location in which the project is built.

10.5.1.1   Income Tax Percentage

The income tax is a payment to the state for income earned during a fiscal year.
In the case of companies are assessed on net profits after purification of the profit
and loss at a rate determined by each state. All companies are required to pay
this tax.

Under the current tax statute in Colombia this rate corresponds to 35 % of the
result of subtracting administrative and operational costs, financial of total operating nonoperating income.

10.5.1.2   Inflation

Inflation is defined as the general increase in the price level of goods and services. It
is also defined as the fall in the market value or purchasing power of a currency in a
particular economy, which differs from the devaluation, since the latter refers to the
fall in the value of the currency of a country relative to another currency traded in
international markets, such as the dollar, the euro, or the yen (Fig. 10.41).

10.5.1.3   Product Price Increase

Product price, fuel alcohol is regulated by the Ministry of Mines and Energy
through decrees. These prices have increased considerably and can be found in
Asocaña, Fedebiocombustibles and Ministry website. In Fig. 10.42 the behavior of

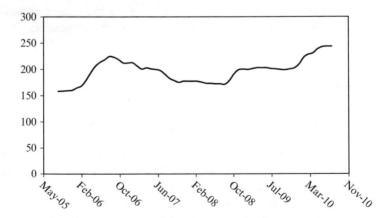

Fig. 10.41 Behavior of the consumer price index (CPI) for the sugar industry in Colombia.
Source: Asocaña annual balance. September 2010

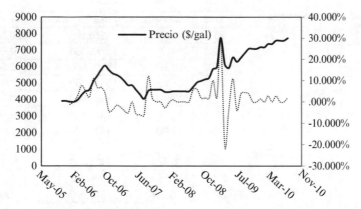

Fig. 10.42 Behavior of the price of fuel alcohol in Colombia. *Source*: Ministry of Mines and Energy of Colombia. September 2010

the price of fuel alcohol in the Colombian market in the last 5 years and the annual price variation, both data of great importance to the economic evaluation, and thus the viability of the project is shown.

10.5.1.4   Raw Material Price Increase

The increase in raw materials is important for the development of economic evaluation because it must build cash flow and it is important to properly estimate the increase of raw materials throughout the project life.

For further analysis only the part of dehydrated alcohol is taken; reason why commodities correspond to glycerol and estimate the cost of azeotropic ethanol. The price of glycerol and azeotropic ethanol was considered in the previous section of this text.

10.5.1.5   Minimum Wage Increase

In Colombia the minimum wage is set in the early days of each year performing a consultation between the representatives of major industries and representatives of the most influential guilds. If no agreement is reached within a reasonable time, the government intervenes and declares a value increase in the minimum wage taking into account the proposals of each of the parties.

For the present study took into account the minimum wage increases in the last 10 years and entered into the interface Aspen ICARUS® to estimate the cash flow of the project (Table 10.25).

Table 10.25 Minimum wage increase for the last 10 years in Colombia

Year	Daily amount	Monthly amount	Increase (%)
2000	$8670	$260,100	10.00
2001	$9533	$286,000	9.96
2002	$10,300	$309,000	8.04
2003	$11,067	$332,000	7.44
2004	$11,933	$358,000	7.83
2005	$12,717	$381,500	6.56
2006	$13,600	$408,000	6.95
2007	$14,457	$433,700	6.30
2008	$15,383	$461,500	6.41
2009	$16,563	$496,900	7.67
2010	$17,167	$515,000	3.64

Source: Ministry of Social Protection. September 2010

Table 10.26 Utilities prices in the Valle del Cauca region

Utility	Price
Electricity	0.2058 U\$/kWh ($380.77/kWh)
Service water	0.848 U\$/m^3 (1569.35 \$/m^3)

10.5.1.6 Wages

Wages must be entered for three existing price ranges within the industry hierarchy and would be found in charge of construction and/or operation: assistant, operator, and supervisor.

Wages are set on a monthly minimum wage (SMMV) for assistants, 2 SMMV for operators and finally 3 SMMV for supervisors considering recommended by the Superintendent of Industry and Commerce values.

10.5.1.7 Utilities Prices

For the economic evaluation of the plant's location in the department of Valle del Cauca is studied. These data were taken from the pages of Emcali and ESSA; are referred to the industrial rate specified for the current period (September 2010). The data is reported in Table 10.26.

10.5.2 Simplifications

It is assumed that the utilities are contracted in some of these companies; however, the values of the other companies are around the same order of magnitude. Is it possible to have a lower water cost of any natural source such as a well, pond, or swamp, this would be reflected in a significant improvement to profitability. Electricity could be generated within the same plant if it is profitable or if the service is already available. The effect on profitability would be similar to the water.

10.5.3   Aspen Icarus® Simulation

To illustrate the use of this program, example of the extractive distillation of the ethanol–water mixture using glycerol as a separation agent is retaken which is present in Chaps. 6 and 8, the simulation is used only in steady state which was built in Chap. 6.

Open the Aspen Plus® simulation and ensure that the simulator perform the calculations again so that Aspen ICARUS have the appropriate information to export the simulation. If you want to perform an evaluation of a simulation carried out in Aspen HYSYS® the process is the same for export.

10.5.3.1   Property Sets

The first step is to create a set of units that will provide sufficient information to Aspen ICARUS to evaluate and size the equipment according to the given simulation flow diagram.

To do this, go to *Properties > Prop-Sets* path and enter a new property set that should be known as *IPE1*. A window where the properties that belong to this set can be entered is displayed; to do this click on the *Physical Properties* column and select the property from the list to enter. For this set, enter the following properties:

- TEMP: Mixture temperature.
- PRES: Mixture pressure.
- MASSFLMX: Mass flow of the mixture.
- VOLFLMX: Volumetric flow of the mixture.
- MWMX: Molecular weight of the mixture.
- MASSSFRA: Solid mass fraction of the mixture.
- MASSFLOW: Components mass flow.
- MASSRHOM: Mass density of the mixture.
- MASSVFRA: Mass vapor fraction of the mixture.

Figure 10.43 shows how the window is defined after entering the properties on the set.

Now define two new sets called *IPE2* and *IPE3* with the information reported in Table 10.27.

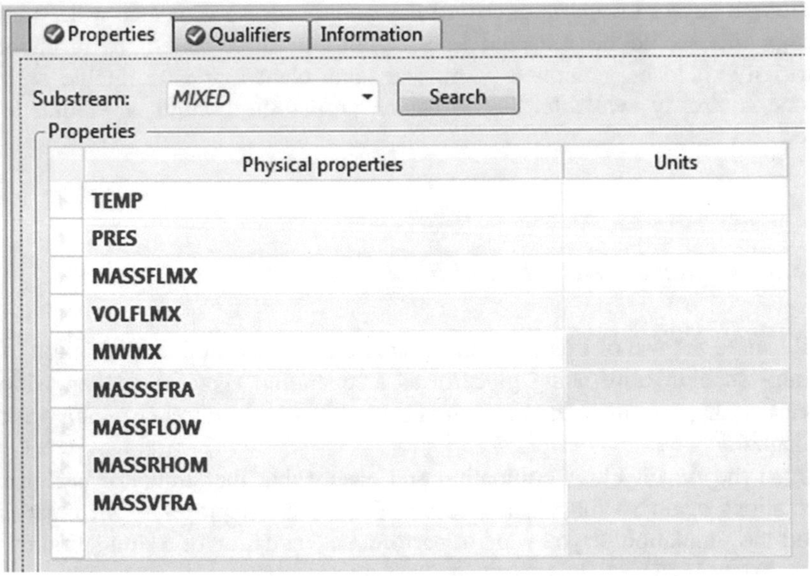

Fig. 10.43 *Properties set* specification window in Aspen Plus®

Table 10.27 Properties to be entered in the new two sets

Property	IPE2	IPE3
CPMX	X	X
MWMX	X	X
MASSFLMX	X	X
KMX	X	X
SIGMAMX	X	
MUMX	X	X
VOLFLMX	X	X
MASSRHOM	X	X
MASSVFRA	X	X

Be sure to select the CPMX property, which corresponds to the specific heat of the mixture is reported in J/mol K. To do this, select such units of the Units column in front of the property in the two sets.

After specifying sets of properties for Aspen Icarus®, they should be added to the calculation routine. Go to Route *Setup* > *Report Options* and enter on the Streams tab. At the bottom right is a button called *Property Sets*, click it. A window as shown in Fig. 10.44 is displayed, then select sets *IPE1*, *IPE2*, and *IPE3*; and add to the list on the right. Do the same for all the sets that appear in the list on the right to provide any information possible Aspen Icarus®.

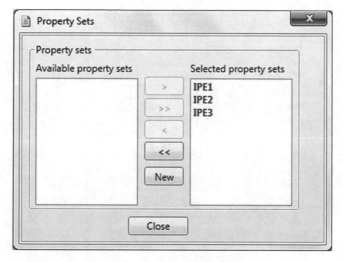

Fig. 10.44 Properties set Selection window in Aspen Plus®

10.5.3.2 Utilities Definition

The next step is to define the utility streams to be used and link the different units of the process to ensure the necessary flows calculation for each utility and be able to estimate the operating costs of the process.

Go to the Utilities option from the navigation tree of the simulator and click the New ... button. This value corresponds to the cooling water so it should be called this utility as WATER. Select the Utility type box, the Water option. Enter Utility costs as 8.48×10^{-4}/kg which corresponds to price reported in Table 10.26 but in mass terms. Leave *Calculation Option* section marked in *Specify the inlet/outlet conditions* option to specify the input and output utility streams. In Fig. 10.45 is observed the specified window.

In the State variables tab can be specified the conditions for entry and leaving the utility stream. For this utility enter the information reported in Fig. 10.46.

Thus, the cooling water outlet is completely specified. For the same operation is performed two additional utilities: Medium pressure steam and coolant whose information is reported in Table 10.28.

Now in each equipment is specified the corresponding process utility. To do this go to the menu for each block and select the *Utility* tab and choose the corresponding utility stream. By example for the distillation column where it should be specified in the *Reboiler* and *Condenser* tabs respectively as shown in Fig. 10.47.

In the bottom of the *Utility Specification* section, the *WATER* utility stream is selected. Finally the calculation of the entire simulation is performed and can be exported to Aspen Icarus®.

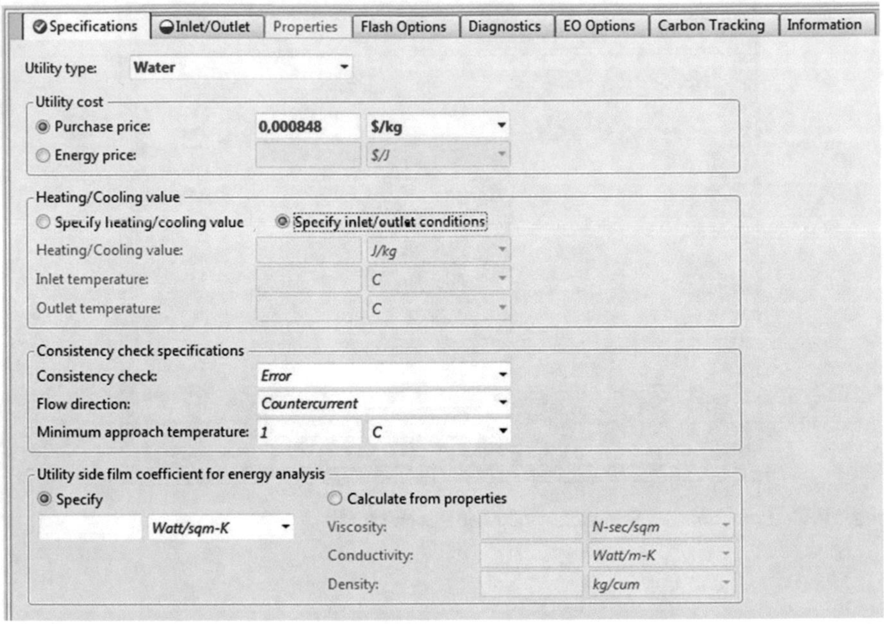

Fig. 10.45 Utility *Specifications* main window in Aspen Plus®

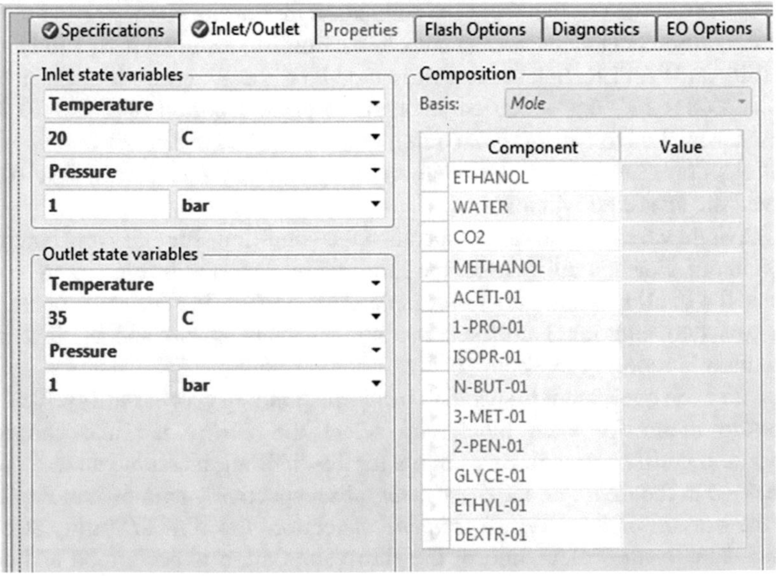

Fig. 10.46 *Inlet/Outlet* tab of utility specification in Aspen Plus®

Table 10.28 Information to be entered for additional utilities

Variable	MPS	Ref
Utility type	Steam	Refrigeration
Price ($/kg)	0	0
Operating conditions		
Inlet	180 °C/sat	0 °C/1 bar
Outlet	180 °C/sat	15 °C/1 bar

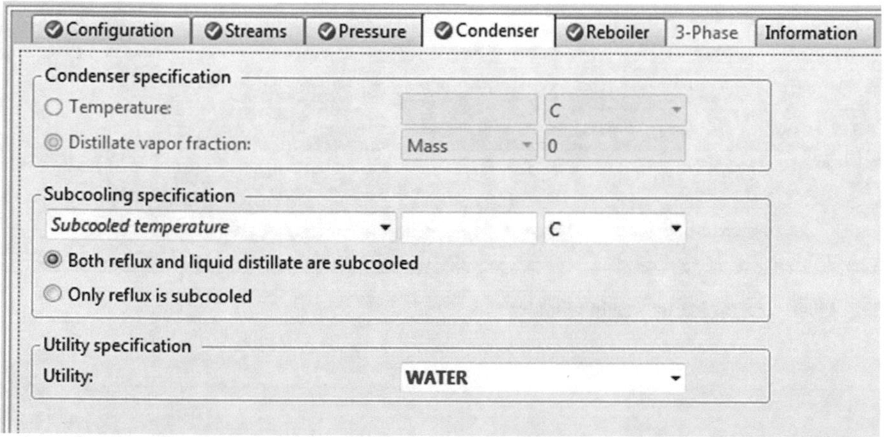

Fig. 10.47 *Condenser* tab of *TD-101* column module in Aspen Plus®

10.5.3.3   Transition to Aspen Icarus®

Now select *File > Send To > Aspen Icarus* option to send all relevant information to Aspen Icarus®. The Aspen Icarus® main window is displayed as shown in Fig. 10.48.

This first window allows you to assign name both the project and the scenario. It should explain that Aspen Icarus® creates for the same project several scenarios to assess different aspects of each of them: location, type of equipment, change in any of the economic indicators, etc. In the *Project Name* box enter: *Ethanol Dehydration Plant* and the *Scenario Name* box, name it as *Scenario1*. Click the *OK* button.

In Fig. 10.49 the process description or information regarding various other scenarios to be created is entered. For the purpose of this exercise is left blank this window, but remember to specify any changes you make to avoid confusion when comparing results. Additionally, the units system to be used can be selected.

Subsequently, a window where you should specify the units of work is displayed. All units are left by default so click on the *Close* button (Fig. 10.50).

In Fig. 10.51 basic information about the costs estimation is entered: project name, date of the estimate, project author, etc. Complete the information and click on the *OK* button.

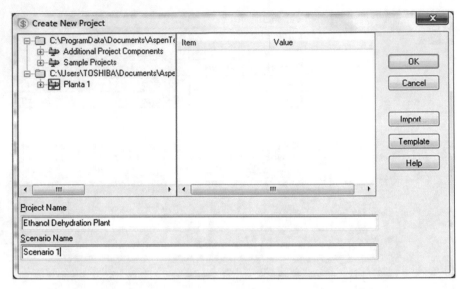

Fig. 10.48 Aspen Icarus® main window

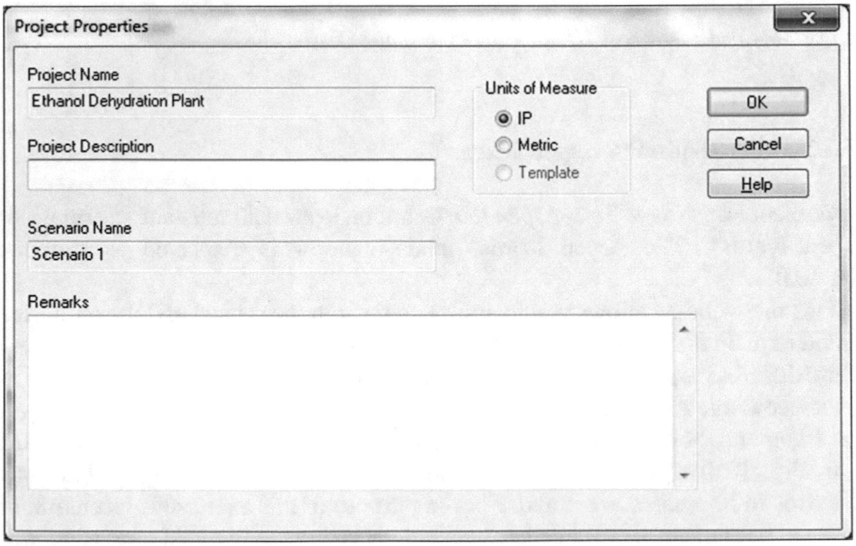

Fig. 10.49 Project Properties window in Aspen Icarus®

Fig. 10.50 Unit specification windows in Aspen Icarus®

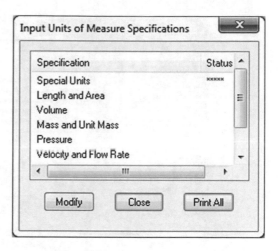

Fig. 10.51 General Project data window in Aspen Icarus®

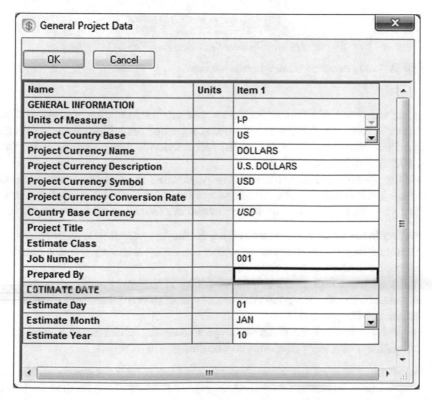

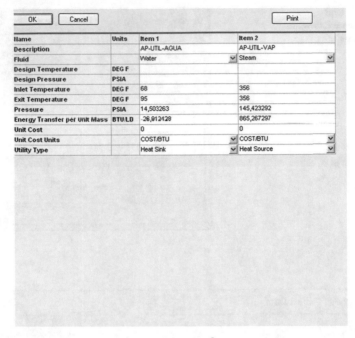

Fig. 10.52 Utility Resources window in Aspen Icarus®

After entering basic information, Aspen Icarus® information requested if the simulator is loaded. Click the Yes button. A report of the utilities entered into the Aspen Plus simulation to verify that the information is adequately specified (see Fig. 10.52) is displayed.

Finally, once the information of utility streams is confirmed, the program interface is displayed (see Fig. 10.53) where there are three main work areas: on the left you can see the navigation pane where the following navigation options:

- *Project Basis View*: Allows observing information regarding sizing parameters, economic and other measures by which the equipment sizing and subsequent economic evaluation is performed.
- *Process View*: Displays the equipment assigned to the simulation data that was exported. You can add additional elements: both equipment and components external to the process.
- *Project View*: Here the different project areas are given when you want to estimate the necessary structures and buildings, for example.

Select the *Project Basis View* window. You must enter economic values and sizing criteria in this section so that Aspen ICARUS has full information for sizing all equipment involved in the process (Fig. 10.54).

In the navigation tree is displayed, select the path *Project Basis > Basis for Capital Costs > General Specs*. In this window, the overview of the project is

Fig. 10.53 Main windows in Aspen Icarus®

Name	Units	Item 1
Construction workforce number		
Number of shifts		1
Productivity adjustment		N
Indirects		
ALL CRAFTS PERCENT OF BASE		
Workforce reference base		
Wage rate percent of base	PERCENT	100
Productivity percent of base	PERCENT	100
ALL CRAFTS FIXED RATES		
Wage rate all crafts	USD	
Productivity all crafts	PERCENT	
WORK WEEK PER SHIFT		
Standard work week	HOURS	40
Overtime	HOURS	0
Overtime rate percent standard	PERCENT	150
GENERAL CRAFT WAGES		
Helper wage rate	USD	
Helper wage percent craft rate	PERCENT	50
Foreman wage rate	USD	
Foreman wage percent craft rate	PERCENT	150

Fig. 10.54 Wages specification window in Aspen Icarus®

specified; the type of project, the level of complexity, the control scheme, location, land features and standards to be met by some of the equipment. Enter the information reported in Fig. 10.55 for this example. For your projects please evaluate each of the aspects taking into account the needs of it.

Name	Item 1	
Process Description	Proven process	▼
Process Complexity	Typical	▼
Process Control	Digital	▼
PROJECT INFORMATION		
Project Location	South America	▼
Project Type	Grass roots/Clear field	▼
Contingency Percent	18	
Estimated Start Day of Basic Engineering	1	
Estimated Start Month of Basic Engineering	JAN	▼
Estimated Start Year of Basic Engineering	11	
Soil Condition Around Site	FIRM CLAY	▼
EQUIPMENT SPECIFICATION		
Pressure Vessel Design Code	ASME	▼
Vessel Diameter Specification	ID	▼
P and I Design Level	FULL	▼

Fig. 10.55   General specification window in Aspen Icarus®

Enter the *Project Basis > Basis for Capital Costs > Construction Workforce* route. In this window corresponding to the workforce, information is specified. In Fig. 10.54 it can be seen the input values to the current estimation.

50 % salary for assistants, and 150 % for foremen was fixed. Subsequently, the minimum wage rate is fixed for all these roles are referenced to that amount. After entering the data, click on the *OK* button shown on the top left of the central panel of the interface; if this button is NOT pressed, any changes made are not saved. The *Apply* button also performs the same function.

In the *Project Basis > Basis for Capital Costs > Indexing* route can be entered information about indexes for painting, civil work, location, etc. For the current year is leave it all by default, however, consider entering appropriate values to properly secure these criteria.

In *Basis Project > Process Design > Design Criteria* route can vary the parameters used for Aspen ICARUS to equipment size. Remember that each team has a characteristic design parameter for the estimation of the cost as the correlations found in the literature. For the current year these criteria are not modified.

At the previous stage of the exercise, corresponding utility currents are defined for supplying the heat transfer process. Now in the *Project Basis > Process Design > Design Criteria > Utility Specification* route makes sure that the values correspond to the entered above; otherwise modify them from the window shown in Fig. 10.56.

The next step in the economic evaluation specifications corresponds to the entry of the economic parameters presented in Sect. 10.5.1; in Fig. 10.56 the information to be entered in the *Project Basis > Investment Analysis > Investment Parameters* route is displayed. The values entered are mainly:

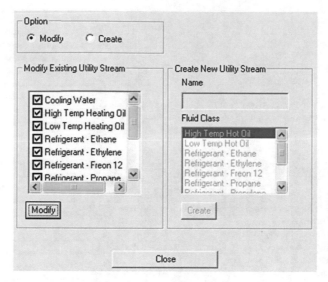

Fig. 10.56 Utility streams specification window in Aspen Icarus®

- *Number of Periods of Analysis*: indicate for how long will be considered the project. Normally for chemical processes, a period between 10 and 20 years is selected.
- *Tax Rate*: Corresponds to the income tax rate according to the Colombian tax statute (35 %).
- *Depreciation Method*: Can be selected here as the method to depreciate the equipment.
- *Project Capital Escalation*: Corresponds to the variation of money over time. It can be taken as the average over the past 5 years of inflation in the economy of the country or sector.
- *Products Escalation*: Corresponds to the change in the products price over time: It can be taken as the average over the past 5 years the annual change in the price.
- *Raw Material Escalation*: Corresponds to the change in the raw materials price over time: It can be taken as the average over the past 5 years the annual change in the price.
- *Maintenance Labor Escalation*: Corresponds to the wages variation over time. It can be taken as the average over the past 5 years the increase in the minimum wage in the country or sector.
- *Utilities Escalation*: Corresponds to the variation in the utilities price over time. It can be taken as the average over the past 5 years the increase in the country or sector.

Finally, at the bottom in the section *Facility Operation Parameters* can enter information related to the type of process and the nature of the fluids used. Additionally to fix plant operating time and estimated starting period (Fig. 10.57).

Name	Units	Item 1
Period Description		Year
Number of Weeks per Period	Weeks/period	52
Number of Periods for Analysis		20
Tax Rate	Percent/period	35
Interest Rate/Desired Rate of Return	Percent/period	20
Economic Life of Project	Period	20
Salvage Value (Percent of Initial Capital Cost)	Percent	20
Depreciation Method		Straight Line ▼
ESCALATION PARAMETERS		
Project Capital Escalation	Percent/period	4,88
Products Escalation	Percent/period	2,5
Raw Material Escalation	Percent/period	-6
Operating and Maintenance Labor Escalation	Percent/period	6,7786
Utilities Escalation	Percent/period	3
PROJECT CAPITAL PARAMETERS		
Working Capital Percentage	Percent/period	5
OPERATING COSTS PARAMETERS		
Operating Supplies	Cost/period	25
Laboratory Charges	Cost/period	25
Operating Charges	Percent/period	25
Plant Overhead	Percent/period	50
G and A Expenses	Percent/period	8
FACILITY OPERATION PARAMETERS		
Facility Type		Chemical Processing Facility ▼
Operating Mode		Continuous Processing - 24 Hou ▼
Length of Start-up Period	Weeks	20
Operating Hours per Period	Hours/period	8.000
Process Fluids		Liquids ▼

Fig. 10.57 Economic parameter specification windows in Aspen Icarus®

Now in *Project Basis > Investment Analysis > Operation Unit Costs* route must be specified the rates of workers and utilities. Using the values reported in Fig. 10.58 corresponding to the data reported in Sect. 10.6.1 is recommended. For the wages of employees used 2 and 3 monthly minimum wages in the appropriate units.

Now you must specify the price of raw materials and products with their respective streams in the process diagram. To do this, go to *Project Basis > Investment Analysis > Raw Material Specification* route where a window like the one shown in Fig. 10.59 is displayed.

With the *Create* option selected, enter as *Azeo* name corresponding to the azeotropic ethanol stream. In the Section *Basis* select the option *Mass* and in *Phase* section, the *Liquid* option. Click the *Create* button. By selecting the *Modify* option and clicking on the button with the same name a window as shown in Fig. 10.60 is displayed. In it, enter a value of 0.767 U/kg and the process stream entering the *DT-101* column.

Name	Units	Item 1
LABOR UNIT COSTS		
Operator	Cost/Operator/H	3,48
Supervisor	Cost/Supervisor/H	5,22
UTILITY UNIT COSTS		
Electricity	Cost/KWH	0,2382
Potable Water	Cost/M3	0,893
Fuel	Cost/MEGAWH	0,007442
Instrument Air	Cost/M3	0

Fig. 10.58 Operating costs specification window in Aspen Icarus®

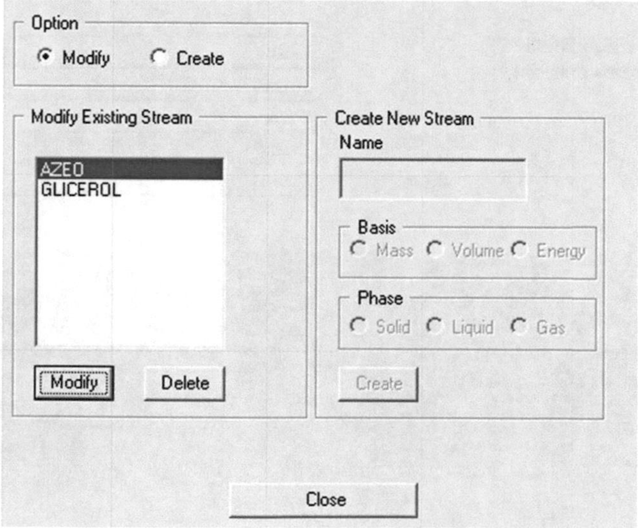

Fig. 10.59 Raw material specification windows in Aspen Icarus®

The information reported in Table 10.29 reports the information to enter for the next raw material, glycerol makeup.

Similarly the product streams are entered, for this case, the anhydrous alcohol in the *Project Basis > Investment Analysis > Product Specification* route. In Fig. 10.61 it can be seen the information entered for the product stream, ETHANOL.

Thus all relevant information on prices and economic parameters is entered.

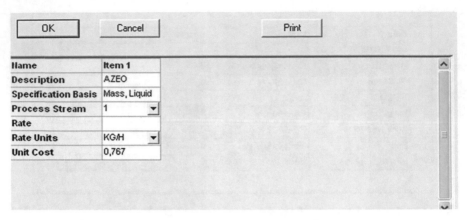

Fig. 10.60 Raw material specification windows in Aspen Icarus®

Table 10.29 Information to be entered for raw material streams

Cell	Value
Name	Glycerol
Basis	Mass
Phase	Liquid
Price	0.819 U$/kg

Fig. 10.61 Product specification window in Aspen Icarus®

10.5.3.4 Equipment Sizing

In order to size the equipment is required to use the information generated in the simulation of Aspen Plus® process done called mapping modules consisting specification to Aspen Icarus® what equipment of its database corresponds to each

Fig. 10.62 *Process View* tab in Aspen Icarus®

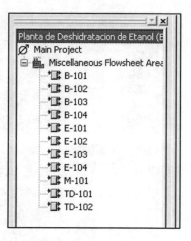

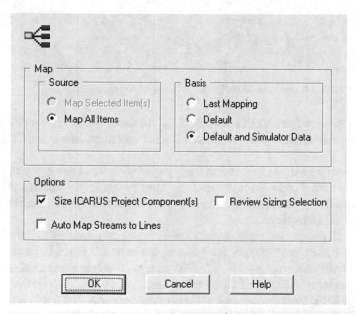

Fig. 10.63 *Map Simulator Items* window in Aspen Icarus®

module installed in the Aspen Plus® diagram. Also this mapping can be done with a simulation of Aspen HYSYS®.

Therefore, in the left pane select the *Process View* tab. A list folds where all the equipment for the diagram entered into Aspen Plus® in yellow as shown in Fig. 10.62 are named.

Now, click the upper right button called *Map Simulator Items* that can be found to the right of the save and print buttons. A window as shown in Fig. 10.63 is displayed.

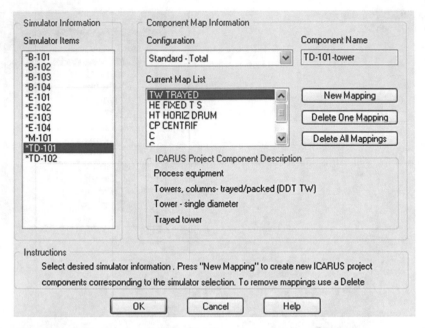

Fig. 10.64 *Project Component Map Preview* window in Aspen Icarus®

In this window, you must select the *Map All Items* in *Source* section and then ensure that in the *Basis* option is selected *Default and Simulator Data*. Click the *OK* button.

A window where can be selected the types of equipment corresponding to each block within the simulation in Aspen Plus® and Aspen Icarus® makes them a first approximation to what should go in each appears. However, you should verify that the selection is adequate.

In Fig. 10.64, the example module *DT-101* that corresponds to the distillation column to obtain pure ethanol for bottoms is observed. For this module, Aspen Icarus® identifies a column, *TW TRAYED* as a plate column, the condenser and a heat exchanger (*HE FIXED TS*), the reflux tank (*HT HORIZ DRUM*), reflux pump (*CP CENTR*IF), the reboiler as a U-tube exchanger (*RB U-TUBE*) and the flow dividers of bottoms and top and two unknown components.

Click the OK button once to check each of the modules are displayed in the left section of the window. The mixer *M-1* is linked to a vertical tank in which the mixture of the two streams takes place. To do this select the appropriate module and click the *Delete One Mapping* window to delete the default associated equipment Aspen Icarus®.

Click the *New Mapping* button where a window like the one shown in Fig. 10.65 where you can add new equipment associated with a module or new equipment that is not part of the simulation is displayed.

Select the component in the following path: *Vessels, pressure, storage > Vessel−Vertical tank > Vertical Vessel Process* to select the above

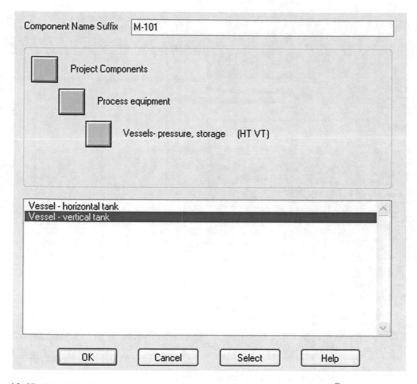

Fig. 10.65 *ICARUS Project Component Selection* window in Aspen Icarus®

mentioned tank. At the top of the window it can be seen the path and you can go to higher levels.

All heat exchangers must be sized using Aspen HTFS +® because it is specialized software to design this type of equipment, leading to more accurate results that are entered in the simulation of Aspen Icarus® as illustrated below. Enter the default equipment in each module and click the *OK* button to complete the mapping of the equipment. For more information about the design in Aspen HTFS +® see Chap. 4 of this text.

A list of the "mapped" items appears and Aspen Icarus® takes information from the simulation to perform the calculations and import it to size them. This process may take a few minutes depending on the computer and the number of equipment related. After this the modules turn green and the list of equipment is observed in the central panel.

To modify information in the modules, select from that list the equipment to be modified, for example select exchanger *E-101*. Right-click and select *Modify Item*. A window as shown in Fig. 10.66 is displayed.

In this window you can see the information extracted from the simulator and which has by default Aspen Icarus® to size each type of equipment. It will enter the information for the design of the exchanger done in Aspen HTFS +® which is

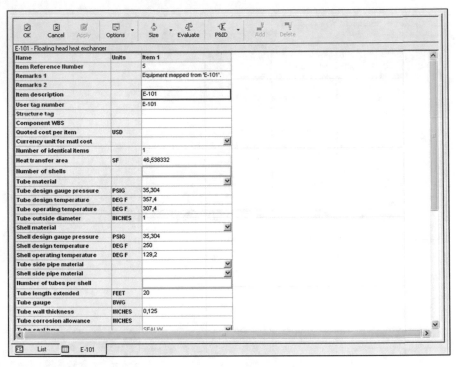

Fig. 10.66 Equipment specification window in Aspen Icarus®

Table 10.30 Information for heat exchangers designed in Aspen HTFS +®

Equipment	E-101	E-102	E-103	E-104
Heat transfer area (m²)	16.9	47.41	57.69	57.69
Equipment cost (USD)	7697	12,085	14,004	14,004

Table 10.31 Cost for heat exchangers designed in Aspen HTFS +®

Equipment	TD-101 cond	TD-101 reb	TD-102 cond	TD-102 reb
Equipment cost (USD)	31,400	25,000	40,000	30,000

reported in Table 10.30. The information is entered to *Heat Transfer Area* and *Quoted cost per item* options respectively.

For auxiliary equipment for distillation columns (condensers and reboilers) must enter the information reported in Table 10.31.

A vacuum pump that reduces pressure within the second column is added. To do this, add new equipment going to the Project View tab, right-clicking on the object and *Miscellaneous Flowsheet Area* selected the *Add Project Component* option. Enter the information reported in Table 10.32.

Table 10.32 Information for vacuum pump to be entered in Aspen Icarus®

Equipment	VP-101
Type	Mechanical oil-sealed vacuum pump
Gas flow (m³/h)	55

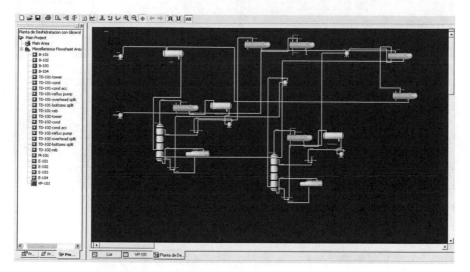

Fig. 10.67 Process flow diagram in Aspen Icarus®

Finally check that all tanks (check those that are associated with the distillation columns), in the *Vacuum design pressure gauge* option, negative values are found, if so delete the value and leave it blank. If these values are left, when estimating the cost are going to fail.

10.5.3.5   Economic Evaluation

Before the economic evaluation, verify that equipment and streams are properly installed in the data imported to Aspen Icarus®. To do this, select from the tool bar the *View > Process Flow Diagram* route. A diagram as shown in Fig. 10.67 is displayed. In this window should be verified that the connections are identical to those used in the simulation in Aspen Plus®. To change any connections, right-click on the stream and select the appropriate option.

Finally, select the *Evaluate Project* button located next to the *Map Simulator Items* button. Click the *OK* button.

10.5.4   Results Analysis

Aspen Icarus® generates reports within the same program and additionally allows you to generate reports in Microsoft Word® and Microsoft Excel® respectively. Figure 10.68 is an example of external reports generated by Aspen Icarus®.

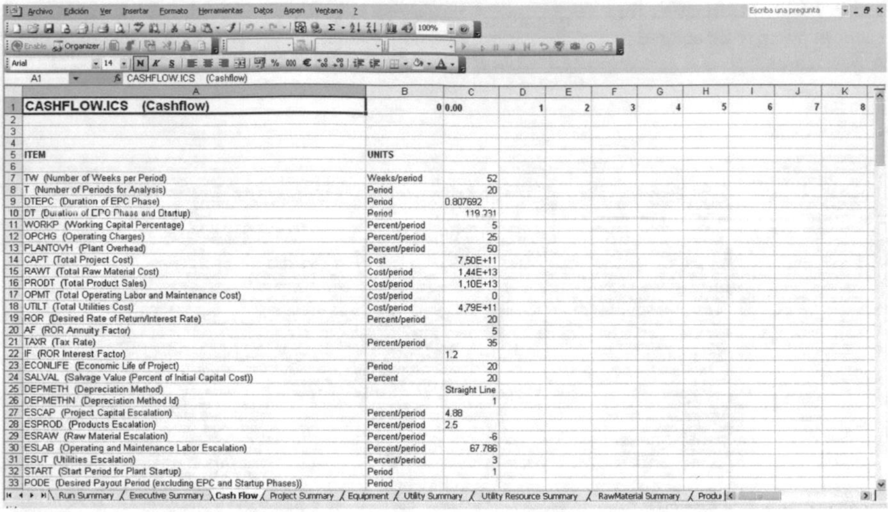

	A	B	C	D	E	F	G	H	I	J	K
1	CASHFLOW.ICS (Cashflow)	0	0.00	1	2	3	4	5	6	7	8
2											
3											
4											
5	ITEM	UNITS									
6											
7	TW (Number of Weeks per Period)	Weeks/period	52								
8	T (Number of Periods for Analysis)	Period	20								
9	DTEPC (Duration of EPC Phase)	Period	0.807692								
10	DT (Duration of EPC Phase and Startup)	Period	119.231								
11	WORKP (Working Capital Percentage)	Percent/period	5								
12	OPCHG (Operating Charges)	Percent/period	25								
13	PLANTOVH (Plant Overhead)	Percent/period	50								
14	CAPT (Total Project Cost)	Cost	7,50E+11								
15	RAWT (Total Raw Material Cost)	Cost/period	1,44E+13								
16	PRODT (Total Product Sales)	Cost/period	1,10E+13								
17	OPMT (Total Operating Labor and Maintenance Cost)	Cost/period	0								
18	UTILT (Total Utilities Cost)	Cost/period	4,79E+11								
19	ROR (Desired Rate of Return/Interest Rate)	Percent/period	20								
20	AF (ROR Annuity Factor)		5								
21	TAXR (Tax Rate)	Percent/period	35								
22	IF (ROR Interest Factor)		1.2								
23	ECONLIFE (Economic Life of Project)	Period	20								
24	SALVAL (Salvage Value (Percent of Initial Capital Cost))	Percent	20								
25	DEPMETH (Depreciation Method)		Straight Line								
26	DEPMETHN (Depreciation Method Id)		1								
27	ESCAP (Project Capital Escalation)	Percent/period	4.88								
28	ESPROD (Products Escalation)	Percent/period	2.5								
29	ESRAW (Raw Material Escalation)	Percent/period	-6								
30	ESLAB (Operating and Maintenance Labor Escalation)	Percent/period	67.786								
31	ESUT (Utilities Escalation)	Percent/period	3								
32	START (Start Period for Plant Startup)	Period	1								
33	PODE (Desired Payout Period (excluding EPC and Startup Phases))	Period									

Fig. 10.68 Microsoft Excel generated report in Aspen Icarus®

10.5.4.1 Equipment Cost

In the report exported to Microsoft Excel you can find information regarding various aspects of the project, including the cost of equipment and direct costs including installation, commissioning, etc.

In Table 10.33 it can be seen a summary of the teams were evaluated and estimated price by Aspen Icarus®.

10.5.4.2 Cash Flow

Cash flow is presented in Fig. 10.69. We can see that in the first 2 years the investment is not recovered because it is operating at 50 % of the production design. Subsequently, from year 3 to year 20 (end of project life) an increase in the value of the project is observed. After 4 years, the money invested in the design and implementation of the project is paid back.

In terms of economic indicators, the internal rate of return is defined as the interest rate that makes the net present value of the investment zero. It indicates as profitable an investment and the higher its value the project economically attractive it is for investors. The net present value to determine whether an investment complies with the basic financial goal: MAXIMIZE investment. The net present value allows determining if the investment can increase or reduce the initial amount to invest. The change in the estimated value can be positive, negative, or remain the same. If positive means that the value of the firm will have an equivalent to the amount of the net present value increase. If it is negative it means that the firm

Table 10.33 Information of cost estimation for equipment calculated using Aspen Icarus®

TAG	Equipment type	Direct cost (USD)	Equipment cost (USD)
B-101	Centrifugal pump	24,500	3500
B-102	Centrifugal pump	20,100	3400
B-103	Centrifugal pump	29,400	4300
B-104	Centrifugal pump	28,900	3400
TD-101-tower	Packed tower	338,000	128,400
TD-101-cond	Shell and tube heat exchanger	65,800	31,400
TD-101-cond acc	Horizontal vessel	91,700	12,100
TD-101-reflux pump	Centrifugal pump	33,700	4500
TD-101-overhead split	Tee	230	150
TD-101-bottoms split	Tee	230	150
TD-101-reb	Shell and tube heat exchanger U-type	42,900	40,000
TD-102-tower	Packed tower	200,700	48,500
TD-102-cond	Shell and tube heat exchanger	59,400	25,000
TD-102-cond acc	Horizontal vessel	86,600	12,700
TD-102-reflux pump	Centrifugal pump	20,600	3900
TD-102-overhead split	Tee	230	150
TD-102-bottoms split	Tee	230	150
TD-102-reb	Shell and tube heat exchanger U-type	32,900	30,000
M-101	Horizontal vessel	81,800	12,600
E-101	Shell and tube heat exchanger	54,200	7697
E-102	Shell and tube heat exchanger	66,400	12,085
E-103	Shell and tube heat exchanger	67,800	14,004
E-104	Shell and tube heat exchanger	67,800	14,004
VP-101	Vacuum pump	23,600	5400
Total		1,437,720	417,490

reduces its wealth in the value yielding the VPN. If the result is zero NPV, the company does not change the amount of its value.

10.5.4.3   Economic Indexes

From the results of calculation of economic indicators reported in Table 10.34 (IRR and NPV) can be analyzed as previously assumed that the effect of the cost of utilities is not significant to change the investment behavior. However, it can be seen a small economic window to build the plant in the department of Valle del Cauca.

Additionally must be considered that ethanol plants are currently in operation in this department and this project can be considered as a modification to them, or to

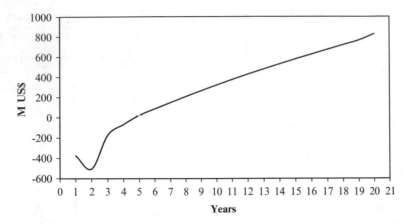

Fig. 10.69 Cash flow for chemical plant investment

Table 10.34 Economic indexes for chemical plant investment

Index	Value
IRR	16.61 %
NPV	$2,907,888,552

implement new ones. The region competitive advantage is that the production of sugar generates the raw material for ethanol production in situ so that the transportation costs are reduced significantly. In the case of Santander should be considered the implementation of the entire plant and in addition to the issue of transportation of raw material considering that the country has no road infrastructure in optimum condition for the continuous transport without complications during any season.

References

Aspen Technology, Inc. (2001) Aspen polymers plus 11.1 user's guide. Aspen Technology, Cambridge

Giudici R, Nascimiento C (1999) Mathematical modeling of an industrial process of nylon-6,6 polymerization in a two-phase flow tubular reactor. Chem Eng Sci 54(15–15):3243–3249

Giudici R (2006) Simulation model of polyamide-6,6 polymerization in a continuous two-phase flow coiled tubular reactor. Ind Eng Chem Res 45:4558–4566

Guevara J (2010) Dimensionamiento y evaluación económica de la destilación extractiva del sistema Etanol-Agua usando glicerol como agente de separación. Universidad Nacional de Colombia

Kumar A, Gupta R (2003) Fundamentals of polymers, 2nd edn. McGraw Hill, New York, Chapter 4

Odian G (2004) Principles of polymerization, 4th edn. Wiley, New York

Seavey K, Liu YA (2008) Step-growth polymerization process modeling and product design. Wiley, New York, Chapter 15

Van Krevelen D (2009) Properties of polymers, 4th edn. Elsevier, Amsterdam

Wakabayashi C (2008) Fuzzy control of a nylon polymerization semi-batch reactor. Fuzzy Sets Syst 160(4):537–553

Index

A

Abrams, 59
Absorption, 242, 272–275, 302, 482
Acid(s)
    acetic, 68, 201, 203, 204, 211, 219, 225, 287
    adipic, 447, 448
    propionic, 68
    strong, 64
    weak, 64
Analysis
    chemical processes, 27
    conceptual, 172, 242
    convergence, 10–20
    operability and optimization, 2, 363
    operation, 295, 320
    stability, 278
    ternary mixture distillation, 74
    thermal and hydraulic, 250, 325–337
    topological, 18, 19, 37, 477
    vibration, 172
Antoine, 74
Aspen HTFS+®, 140, 155, 169, 507
Aspen Properties®, 76, 78–81, 83, 84, 88
Aspen Tech®, 62, 63, 447
Aspen Technology Inc, 55–57, 60, 62, 63
Azeotrope
    binary, 465
    miscible region, 288
    separation, 289
Aziz, K., 106–108

B

Baffles
    spacing, 148
    type(s), 148

Balance
    energy, 7, 8, 29, 33, 45, 65, 104,
        105, 110, 141, 150, 174,
        177, 200, 224, 241, 244, 246,
        303, 305
    material, 156, 199, 200
Bañares-Alcántara, R., 282, 284, 286
Base data, 91, 93
Baxendell, P.B., 107
Beggs, H.D., 107
Bernoulli, 104
Bieker, T., 290
Binary parameters, 56
Black, C., 281
Blank, C., 346
Born, 65–67
"Branch and Bounds" (B&B) method, 348
Brill, J.P., 107–108

C

Calculations
    columns, 338
    columns(short), 242
    exchangers(detailed), 250
    order, 3
Carlson, E., 53, 68, 70
Castillo, F., 288
Centrifuge, 426–427, 433
Chemical engineering problems, 343,
    346, 347
Chen, D.H.T., 65
Chien, I.L., 418, 419
Chiller, 141, 380, 419
Classifiers, spiral, 428
Colburn, 243

© Springer International Publishing Switzerland 2016
I.D.G. Chaves et al., *Process Analysis and Simulation in Chemical Engineering*,
DOI 10.1007/978-3-319-14812-0

Column(s)
  absorption, 272, 273, 482, 484, 485
  azeotropic, 463
  liquid–liquid extraction, 242
  simulation, 242, 251, 260, 301–303,
        310, 485
  thermal and hydraulic analysis, 250,
        325–337
Comminution, 429–431
Compressor, 111–114, 136, 480, 485
Conceptual phase design
  control, 7
  process monitoring, 7, 9
Condenser, horizontal, 155–156
Construction material, 169, 174, 356
Controller(s)
  feedback, 375–377, 379, 393, 407
  tuning methods, 419
Conversion, 195–197, 199, 201, 205, 212,
        219–221, 224, 225, 237, 415,
        459, 461, 478, 481
Cooler, 139, 140, 151–155, 251, 326, 469,
        480, 485

Cost(s)
  equipment, 352, 362, 487, 510
  fixed, 350, 351, 461
  operation, 344, 345
  operational, 176, 350, 352, 488
  pumping system, 353
  purchase, 346, 353, 354, 356
  raw materials, 351, 353, 362
  service, 325
  transport, 512
Crusher, 430, 434–440
Crystallization, 432
Cyclones, 426, 443, 445

D
Darcy equation, 106
Dennis, Y., 282, 288, 290
Degrees of freedom, 20, 257, 265, 266,
        343–345, 348, 373, 406
Design
  and evaluation mode, 250
  process, 7–9, 111, 146, 343, 366, 371,
        461, 487
  specification, 3, 7, 10, 11, 38, 46, 47, 296,
        298–300, 318, 320, 325, 329, 349,
        380, 473, 474, 476
Development
  born, 65

Pitzer-Debye-Hückel, 65, 66
Diamond, 277, 281, 285, 286
Diagram(s)
  Moody, 105
  process flow, 1, 27, 28, 32, 130, 221,
        291, 292, 294, 351, 380, 408, 415,
        456, 463
  typical process, 455
Digital computers, 241, 245
Distillation(s)
  Azeotropic, ethanol–water system, 291
  enhanced, 275
  region(s), 285, 287, 288, 290, 291
Distribution of services, 130
Doherty, M., 243, 302
Duns, H., 108

E
Economic indicators, 487, 510, 511
Edgar, T.F., 344, 348
Ekerdt, J.G., 227
Electricity, steam, 352, 353
Energy
  balance, 7, 8, 30, 33, 45, 65, 104, 105, 110,
        141, 150, 174, 177, 200, 224, 244,
        246, 303, 305
  Gibbs free energy, 200
Engineer, process, 2, 47, 139, 142, 171, 337,
        346, 347, 366, 377, 414
Equation oriented, 5–7, 10
Equation(s)
  Antoine, 74
  Bernoulli, 104
  energy, 104
  Ergun, 147
  exponential, 75
  linear, 346
  material balance, 156, 199, 200
  MESH, 244, 247, 248
  non-linear, 12
  Pitzer, 65
  rate, 305
  reactor design, 199–200
  Van der Waals, 54
Equilibrium
  liquid-liquid, 290
  vapor-liquid, 54, 64, 88, 135
  Vapor-liquid, 62, 69, 92, 277, 278, 280, 284
Esteban, H., 347, 348
Exchanger(s)
  detailed, 151, 154
  floating head, 144, 355

floating heat, 146
head, 9, 291, 293, 345, 347, 351, 352,
      354–355, 358
heat, 139, 140, 142–145, 148–153,
      155, 156, 160–165, 167, 170,
      172, 174, 176, 177, 180, 181,
      184, 186, 188, 380, 469,
      506, 507
shell and tube, 139, 142, 190–192
Shell and tube, 153, 156, 164, 171
thermodynamics, 154, 188
Expander(s), 111–112

F
Filtration, 425, 431–432, 445
Flow
   azeotropic ethanol, 353, 463
   cash, 489, 510
   constant, 243, 417
   continuous, 104
   distributed, 107
   intermittent, 107
   inviscid, 104
   laminar, 104
   molar, 121, 244, 272, 299, 362
   n-propanol, 172, 174
   physical, 4
   random, 104
   ratio, 43, 304
   stage, 246
   turbulent, 104
   two-phase, 106, 108
   vertical, 106–110
   water, 117, 155, 172–174, 224, 225, 274,
      275, 374, 406, 409, 482
Fluid handling equipment, 103–137
Fluid mechanics, 103, 105
Fogler, H.S., 196
Fraction(s)
   average segment, 61
   average surface area, 61
   binary interaction, 61
Fruehauf, P.S., 418

G
Gasoline blending, 385, 387, 388, 391–393,
      395, 398–400
Geddes, R.L., 245, 288
Geometry
   baffles, 143, 148–149, 159, 162
   exchanger, 149–151, 156, 171

head, 143
   methods to classify, 143
   nozzle, 159
   shell, 143, 147–148, 158, 162
   tubes, 143–148, 162–163, 171
Gilliland correlation, 243
Giudici, R., 447
Glasser, D., 243, 302
Gomis, 82, 88
Govier, 106
Grassi, V.G., 417
Gregory, G.A., 108
Guggenheim, 67
Gundersen, T., 175, 176
Gupta, R., 426, 447, 457

H
Hagedorn, A.R., 108
Handling
   gases, 114–129
   liquid, 130–136
Heat transfer area, 355, 508
Heater, 140, 151–153, 251, 293, 326
Henke, G., 246
Henley, E., 241, 242, 245–246, 248,
      282, 286
Holman, J., 141–149
Hückel, 65–67
Hydraulic
   analysis, 250, 325–337
   classifiers, 428, 429
Hydrocyclones, 426, 427, 433
Hypothetical compounds, 93

I
Income tax, 487, 488, 501
Industry
   chemical, 64, 143, 344, 345, 415,
      461, 477
   chemical process, 195
   crude oil, 81, 388
   fuel alcohol, 242, 488
   gas, 81, 109, 415
   hierarchy, 490
   petroleum, 220
   polymer, 448, 477
   polymeric, 62
Inflation, 487, 488, 501
Initialization streams, 474
   Interface Aspen Icarus®, 489
Ionic substances, 64

J
Jacobian, 11–14

K
Kemp, I., 179, 180
Kern, D., 155, 156
Kinematic viscosity, 104
Kinetics
    chemical, 1
    heterogeneous reactions, 198
    polymerization, 449
    process, 487
    Kister, H., 302
    Kiva, V., 302, 307
Kreul , L., 307
Krishna, R., 256
    Krishnamurth, R., 247, 248, 290, 307
Kumar, A., 447, 457

L
Language
    chemical process, 1
    programming, 284
Law(s)
    Newton, 104
    power, 197, 198, 202, 207, 226–229, 232,
        233, 236
    Raoult, 277, 278
Leiva, F., 291
Lewis, W.K., 245, 290
Libraries of subroutines
    models or blocks, 2
    procedures, 2
Linnhoff, B., 327
Liquid and vapor, 245
Liquid–liquid equilibrium (LLE), 75
Liu, Y.A., 449, 451
Local composition, 58, 59
Ludwig, E.E., 302
Luyben, W.L., 373, 417, 419

M
Map(s)
    residue curves, 7, 277–291, 464
    Malone, M., 302
    Manan, Z., 282, 284, 286
Marquardt, 13
Matheson, G.L., 245
Megan, J., 282, 288, 290

Method(s), 5, 11, 244, 376, 378,
        379, 412, 413
    2N Newton, 246–247
    active set, 347
    black boxes, 349, 350
    Boston, 248
    Branch and Bounds, 348
    bubble point, 242, 246, 247
    convergence
        Newton-Raphson and Broyden
            Quasi-Newton, 5, 11
        successive substitution or direct
            iteration, 5
        Wegstein, 5, 11
    depreciation, 501
    direct substitution, 14
    Direct substitution, 11, 20
    double iteration, 247–248
    factorial, 140
    Fenske, 243
    Homotopy–Continuation, 248
    iterative, 10, 11, 16, 25
    K. Venkateswara Rao-A.
            Ravlprasad, 243
    Lewis-Matheson, 245
    Naphtali and Sandholm, 247
    Newton, 13, 14, 308, 309
    numeric, 6, 348, 406
    properties, 14, 20, 33, 479
    relaxation, 15, 248, 348
    rigorous, minimal specification, 244
    Secant, 11
    sequential quadratic programming, 349
    short, 152
    simultaneous correction, 247
    SRK, 32
    stage by stage, 245–246
    Theta-q (Please insert symbol), 242
    Theta-θ, 246
    Thiele-Geddes, 245
    Tomich, 247
    tuning
        closed-loop, 378
        controller, 376, 378, 379
        open loop, 378, 412, 413
    Underwood, 243
    UNIQUAC, 59–61
    Wang and Henke, 246
Minimum wage, 489, 490, 500–502
Ministry
    Mines and energy, 70, 487, 488
    Social Protection, 487

Model(s), 69, 248
   activity coefficient available in Aspen
      Polymer Plus®, 60
   basic, 150
   Bromley-Pitzer, 65, 68
   calculation, 106, 149–151, 188, 248
   calculationModel(s), 248
   coefficient, 57–63, 65, 88, 91, 309
   Duns and Ros, 108, 109
   dynamics, 7, 371, 379, 407, 420
   equilibrium, 284, 301, 302, 307–310, 322
   Guggenheim, 67
   Hayden-O'Connell, 68
   heterogeneous, 236
   HTFS calculation, 109
   ideal, 53, 54, 57
   integrated, 1
   integration, 6, 68
   kinetic(s), 226, 227, 229–231, 233,
      236, 448
   LHHW, 198, 232
   linear, 6
   mathematics, 1–3, 5, 6, 139, 343–346, 372
   non equilibrium, 251, 301–325, 328,
      337, 338
   NRTL, 59, 60, 65–67, 69, 84, 90, 92,
      93, 292
   operation, 4
   Pitzer, 67, 68
   properties, availables in Aspen Plus®, 69
   selection, 33, 53, 68–79, 98, 157, 451
   semi-empirical, 65
   simulation package, 55, 56
   special, 62–68
   steady state, 398, 414, 415
   thermodynamic, 21, 22, 28, 33, 53, 68–76,
      88, 89, 91, 156, 211, 221, 232, 255,
      290, 391, 448, 451, 487
   Tulsa, 110
   Van Laar, 57
   Weighted and End Point, 149
   Wilson, 58, 59, 61
Modeling of the operation
   absorption, 241, 242, 249
   distillation, 241, 242, 249
   extraction, 241, 242, 363, 366
   stripper, 241
Module
   available in Aspen HYSYS®, 113, 114,
      117, 151–155, 201, 203,
      251–252, 432
   available in Aspen Plus®, 62, 112, 113,
      151–153, 200–202, 432

   duplicator, 211, 263
   heat exchange, 140, 151, 152,
      154, 156, 188
   intercoolers, 139
   mapping, 504
   polymer, 448
   pump(s), 356
   RadFrac, 246, 249–250, 256, 272,
      295, 310, 325, 328
   reactors, 201, 206, 210, 217,
      218, 407, 459
   short calculation, 190
   turbo expanders, 139
Molecular structure, 28, 97, 98
Mukherjee, R., 143, 144

N
Naphtali, 247
Nascimiento, C., 447
Navigation tree, 38, 46, 86, 132, 207, 227, 299,
      356, 474, 477, 493, 498
Newton, 5, 11–14, 246, 349, 350
   Newton-Raphson solution algorithm, 246
   Nieuwoudt, I., 282, 290
Non-polar mixtures, 71

O
Operation(s)
   closed loop, 375
   mixing, 29
   process, 1, 2, 338, 457

P
Package
   Aspen Engineering Suite, 2, 181
   simulation, 27, 55, 56, 68, 155, 167, 195,
      200, 268
   thermodynamic, 116, 203, 216, 226
Parameter
   adjustment, 79, 88–93
   binary, 33, 34, 56, 75, 90, 92, 391
Peng-Robinson equation, 56, 91, 92
Perry, R., 100, 289, 373
Pham, H., 243
Phase(s)
   liquid, 53–58, 61, 64, 68, 69, 74, 79, 108,
      199, 219–221, 244, 247, 250, 277,
      278, 281, 282, 284, 290, 301,
      303–307, 309, 321, 482
   solid, 53, 250

Phase(s) (*cont.*)
   vapor, 53, 55, 68, 69, 73, 74, 78, 220, 226,
      238, 244, 277, 278, 284, 290, 301,
      303–307, 309, 321, 478
Physical properties
   calorific, 309
   density, 73, 199, 309, 388
   enthalpy, 73
Pipe(s)
   calculation, 135
   horizontal, 107
   line, 109, 110, 130, 131, 133,
      135, 136, 478
   module, 130, 133
   network, 103, 114, 115, 119–121, 135
Plant(s)
   ethanol, 511
   operation of, 10, 347, 371
   pilot, 9
   process, 103
Poettman, F.H., 109
Point(s), stationary, 278
Polymeric chain growth, 448, 452, 456,
      457, 460, 461
Problem(s)
   cavett, 18, 19
   classification, 346–348
   description, 20, 114, 130, 155, 156,
      201–211, 225–226, 253–255, 265,
      272, 291, 447–449
   linear programming, 346–348
   non linear, 347, 348, 350
   operation, 330
   proposed, 320
Process
   chemical, 9, 47, 53, 275, 501
   in chemical engineering, 343
   commercialization, 9
   control, 371–375, 381, 414, 461
   controllability, 7, 8, 223
   distillation, 415, 463, 475
   dynamic, 372, 379, 385
   dynamic analysis, 7, 379, 408, 420
   dynamic behavior, 2, 7
   iterative, 4, 5, 10, 34, 47
   optimization, 2, 7
   unit, 3–5, 10, 366
Production(s)
   batch, 447, 448, 455, 457–461
   continuous, 447, 455–457
   ethylene glycol, 477
   ethylene oxide, 477
Product price, 488

Program(s)
   HTFS+®, 140
   independent, 139
   Mixed Integer Non Linear, 347–348
   sequential quadratic, 347, 350
   simulation, 140, 352
Pseudocomponents, 71
Pump(s), 25, 26, 29, 34, 36, 103, 110 111,
      130–132, 135, 136, 224, 253, 293,
      294, 297, 298, 352–354, 374, 391,
      392, 408, 415, 470, 508

R
Raphson, 5, 11, 246
Raw material
   increase, 489
   preparation process of, 195
Rawlings, J.B., 227
Reaction(s), 197, 461
   chemical, 20, 74, 196, 201, 211, 272, 434,
      448, 479, 481–482, 485
   conversion type, 214, 216
   multiple
      complex, 197
      independent, 197
      parallel, 197, 461
      series, 197
Reactor(s)
   in Aspen HYSYS®, 235
   batch, 200, 202, 457–462
   chemical, 461
   configuration, 205, 209, 216, 224
   continuous, 461
   conversion, 201, 215–217
   exothermic, 406
   fermentation, 373
   Gibbs, 201, 205–206, 211, 220
   kinetics, 216
   methanol reforming, 225–236, 481
   packed bed, 200
   propylene glycol, 220–225, 406, 410
   rigorous, 206
   stirred tank, 209–211
   stoichiometric, 204–205
   tubular(s), 209, 217, 237, 456
Reboiler, Kettle type, 148, 151, 339, 354
Recycle, 43
   nested, 18, 19
Redlich-Kwong (RK) equation, 54, 55
Refinery, 7, 247
Regime(s), 107, 109
   characteristics, 107

distributed, 107
intermittent, 107
models
    annular, 107, 109
    dispersed flow, 109
    stratified, 107, 109
segregated, 107
transition, 104
Renon, 59
Residual functions, 11
Resins, 447, 448, 456
Rooks, R., 277
Ros, N.C., 108
Ross, R., 415
Rules
    empiric, 419
    heuristic, 139, 143, 172
    tuning, 378, 410

S
Salt, 64, 70, 71, 238, 448, 456
Seader, J., 241, 242, 245–248, 282, 286
Seavey, K., 449, 451
Segura, H., 289, 290
Seider, W.D., 10–15, 21, 302, 303, 343,
    353–355, 367
Selection, properties model, 73, 76–79
Sensitivity analysis, 38–43
    in Aspen HYSYS®, 38, 43–45
    in Aspen Plus®, 38–43, 238
Sensor(s), 374, 375, 379
Services
    cooling, 176, 493
    electricit, 356
    electricity, 353
    industrial, 490
Shell
    diameter, 148, 149, 171, 190
    heat transfer, 149, 192
    operation, 148
    type, 147
Simmrock, K., 290
Simplifications, 18, 104, 140, 146,
    490–491
Simulation application(s)
    control, 290
    cost analysis, 290
    evaluation data, 290
    process adjustment, 290
    process modeling, 290
    process synthesis, 290
    system identification, 290

Simulation(s)
    applications, 2
    Aspen HYSYS®, 21–25, 82–85
    Aspen Plus®, 38–40, 86, 156–160
    azeotropic distillation, 462, 463, 468
    chemical process, 47
    chemical Process, 1
    dynamic, 9, 371, 383, 398, 403, 408, 433
    dynamic state, 379–381, 383, 385, 393,
        395, 402–404, 406–410
    nylon resin reactor, 447–461
    procedure, 291
    process, 463–468
    sequential, 3, 6, 48
    steady state, 181, 380, 385, 388, 390, 391,
        393, 407, 408
Simulator(s), 2, 5, 6
    Aspen HYSYS®, 3, 7
    Aspen Plus®, 3, 7
    chemical process, 2–3
    commercial and academics, 3
    flexible, 8
    hybrids, 6
    sequential modular, 4–5
    simultaneous, 2, 5–6
    steady state, 2, 7, 8, 20
    types, 3–7
Smith, J., 290
Smith, R., 179, 180
Soave-Redlich-Kwong (SRK), 55
Solids, 425–445
    operations, 425–445
Soujanya, J., 69, 76
Stage
    compression, 429
    equilibrium, 241, 242, 244, 245, 251, 275,
        301, 340
    feed, 256, 269, 327, 333, 347
    process design, 7
    process distillation, 251
    reaction, 482
    separation, 301
    top, 307
Standard, 12, 18, 75, 139, 140, 143, 165, 169,
    310, 401, 499
    TEMA, 143, 145, 165, 167, 171
Steady state, 2, 7, 20, 149, 150, 157, 181, 220,
    224, 225, 236, 237, 244, 245, 248,
    278, 287, 350, 372, 373, 377, 379,
    380, 383, 393, 399, 400, 406, 407,
    410, 416, 474, 475, 491
Stoichiometry, 196, 202, 203, 211, 214,
    215, 233

Straight lines balance, steps, 290
Strategy
    hybrid, 6
    sequential, 6, 7
    simultaneous, 3
    SQP, 349
Streams
    arrangements / combinations, 20, 148, 181
    in Aspen HTFS+®, 191
    ethanol and gasoline, 391
    feed, 3, 21, 23, 117, 156, 157, 211, 220,
        245, 264, 278, 310, 313, 323, 331,
        334, 339, 380, 381, 391, 392, 438,
        439, 441, 442, 444, 466, 468
    hot gas, 38
    initialization, 16
    liquid and vapor, 282, 301
    product, 3, 25, 43, 117, 197, 202, 217, 263,
        295, 296, 301, 319, 503
    recirculation, 4
    recycle, 5, 14, 16, 17, 463, 466, 470,
        473–475
    reflux, 466, 470, 473, 475
    side, 244, 304, 308
    tear, 1, 2, 5, 10, 11, 14, 16–20, 37, 38, 47,
        475, 477
    value, 1, 18, 33, 473
Study
    cases, 43, 44, 459, 478
    revamping, 10
    simulation, 8
Subroutine
    design mode, 3, 140
    rating mode, 3
Substances, 64
Sullivan, S.L., 247, 302
Surface
    geometric configuration, 210
    area, valves set point, 1
System, 64, 65
    algebraic equation, 5
    binary, 57–60
    chemical, 373
    complete, 27
    control, 2, 290, 379, 414
    distillation, 281
    electrolytic, 64–68
        thermodynamic properties, 64, 65
    equations, 1, 6, 406
    gas compression, 103
    heterogeneous, 277
    mechanic, 372
    methanol, 69

    nodes, 130
    pentane–hexane–heptane, 285
    pipes, 130
    polymeric, 62–63
    in the simulation, 191
    ternary, 288
    thermal/electric, 226
    transport, 114

T
Tao, L., 277
Tarquin, A.J., 346
Task(s)
    describe the components adequately, 82
    estimating any parameter, 141
    obtain and use the data, 188
    select polar components, 60
    validate the physical properties, 72, 73
Taylor, R., 12, 14, 247, 248, 256, 290, 303,
        306, 307
Temperature(s)
    adjustment, 337
    bubble, 246
    changes, 114
    feed, 42, 227, 334, 337, 339, 363
    interface, 307
    liquid, 308
    profile, 246, 260, 262, 268, 270, 271, 274,
        295, 322, 331, 416–419, 482
    stream, 190, 275
    vapor, 309
    variation, 334, 419
Theory
    atomic, 196
    Debye-Hückel, 67
    transfer, 139
Thermodynamic, 21, 37, 45, 139, 141,
        151–154, 167, 200, 202, 204, 205,
        209, 213, 220, 244, 247, 301, 302,
        306, 326, 327
    electrolytic, 65
Thiele, E.W., 245
Thomas, M., 243, 302
Timonen, J.A.M., 175
Tolsma, J., 282, 289
Tool(s)
    informatics, 2
    primary design, 241
    simulator, 383
Topological Analysis, 17–18
Towler, G., 10, 288
Treybal, R., 246, 301

Tube, 143, 145
  calculation, 105
  shape
      fixed, 143, 145
      U, 143, 145
  specifications, 130, 146
Type
  available programs, 139–140
  exchangers, 146, 151, 154, 355
  packaging, 306, 307
Tyreus, B.D., 419

U
Umbach, J.S., 177, 180
UNIFAC, 60–62, 95, 96, 391
Urdaneta, R., 272
Uses
  applications, 7–10, 62, 70, 98, 144,
      146–148, 152, 195, 290–291,
      434, 487
  in Aspen HYSYS®, 93–96
  in Aspen Plus®, 97–98
  process simulation, 8–10
Utilities
  calculator, 352, 359–361
  envelope utility, 82, 85, 87
Uyazan, A.M., 365

V
Valle del Cauca, 487, 490, 511
Values
  binary, 347
  optimal, 243
Valves
  control, 222, 223, 374, 376, 379, 380, 385,
      391, 392, 399, 402, 403, 406, 408,
      409, 415
  fast opening, 376
  relief, 114, 400, 401, 403, 406
  safety, 400
Vane, k, T., 243, 301, 302
Van Krevelen, D., 60, 457
Vapor phase, 53, 55, 68, 69, 73, 74, 78, 220,
      226, 238, 244, 277, 278, 284, 290,
      301, 303–307, 309, 321, 478

Vapor–liquid equilibrium (VLE), 75–76, 88
Vapor-liquid- liquid equilibrium (VLLE), 74
Variable(s)
  binary/integer, 347, 348
  composition, 1
  continuous, 347, 348
  controlled, 374, 375, 377, 378, 381, 385,
      395, 418
  dependent, 9, 43, 44
  design, 345, 359
  discrete, 347, 348
  disturbance, 374
  flows, 1, 132
  independent, 9, 33, 43, 44, 354, 363
  input, 3, 373–375
  iteration, 5, 245, 246
  manipulated, 40, 46, 320, 374, 375,
      381, 418, 419
  measured, 375
  MESH/column state, 245, 247
  operation, 2, 343, 345, 346, 356
  optimization, 187, 346, 350, 364, 365
  output, 3, 6, 373, 375
  pressure, 1
  problem, 344
  process, 6, 9, 120, 125, 325, 345, 347,
      363, 364, 371, 373, 378, 383
  relaxed, 348
  temperature, 1, 43
  type, 18, 40, 43, 347
Vectors, 3, 10, 12, 14, 284
Velocities
  fluids, 104, 105, 107, 148
  gas, 107, 108
Volatility, relative, 243, 244, 287, 290, 462

W
Wahnschafft , O., 285, 288
Wang, J., 246
Wilson, 58, 59, 61, 100, 101

Z
Zone(s)
  feasibility, 346
  instability, 225, 376

Printed in the United States
By Bookmasters